THE SCIENCE OF DESCRIBING

Title page of John Gerard, *The Herball* (1636). © Bibliothèque centrale MNHN Paris 2005.

THE SCIENCE OF DESCRIBING

Natural History in Renaissance Europe

BRIAN W. OGILVIE

THE UNIVERSITY OF CHICAGO PRESS
CHICAGO AND LONDON

The University of Chicago Press, Chicago 60637
The University of Chicago Press, Ltd., London

Paperback edition 2008
Printed in the United States of America

17 16 15 14 13 12 11 10 09 08   2 3 4 5 6

ISBN-13: 978-0-226-62087-9 (cloth)
ISBN-13: 978-0-226-62088-6 (paper)
ISBN-10: 0-226-62087-5 (cloth)
ISBN-10: 0-226-62088-3 (paper)

Library of Congress Cataloging-in-Publication Data

Ogilvie, Brian W.
The science of describing : natural history in Renaissance Europe / Brian W. Ogilvie.
p. cm.
Includes bibliographical references and index.
ISBN 0-226-62087-5 (cloth : alk. paper)
1. Science—Europe—History—16th century. 2. Natural history—Europe—History—16th century. 3. Science, Renaissance. I. Title.
Q127.E8045 2006
508′.094′09031–dc22

2005019254

⊗ The paper used in this publication meets the minimum requirements of the American National Standard for Information Sciences—Permanence of Paper for Printed Library Materials, ANSI Z39.48-1992.

TO MY PARENTS

Marvin Ogilvie and Beverly Eby,

and to Jennifer

(once more)

Ich will aber diss mals mein eigene gefahr/als Angst/ Sorg/ grosse Arbeit/ Hunger/ Durst/ Frost/ Hitz/ Schrecken/ lange sorgliche Reiß hien und wider/ durch vil ohnwege des Teutschenlands/ als inn Wälden/ Bergen/ Thälern/ und Ebenen feldern lassen bethüen/ dann was haben andere leut darmit zü thün? Der Lust und Gemeiner nutz haben mich solches alles zü dulden/ dahien verursacht.
At the moment, however, I will say nothing of my own dangers—fear, care, great exertion, hunger and thirst, frost and heat, terror, long involved travel here and there, through the German wilderness, its woods, mountains, valleys, and plains. For what do other people have to do with that? Pleasure and the common good have led me to bear them all.

—Hieronymus Bock, *New Herbal* (1539)

It's certain there is no fine thing
Since Adam's fall but needs much laboring.

—William Butler Yeats, "Adam's Curse"

CONTENTS

FIGURES

PREFACE AND ACKNOWLEDGMENTS

For the historian, Renaissance natural history poses a puzzle. Why did a significant number of sixteenth-century Europeans, far more than in the millennia before, devote much of their time and resources to discovering and cataloguing nature's productions? Natural history involves hard, tedious work. To the uninitiated, its rewards seem meager, its disputes trivial. Hieronymus Bock and many of his contemporaries appealed to a humanist commonplace to explain why they labored to discover and describe all kinds of plants: such work was both pleasant and useful. While not false, this commonplace hardly satisfies our desire to know. This book attempts to explore the two sides of Bock's explanation: Why did Renaissance naturalists think that natural history was pleasant and useful? What did they do, and why did they do it?

My interest in motives for the study of nature regards their cultural origins, not their psychological roots. I do not claim to know what secret impulses drove any particular sixteenth-century naturalist to observe, catalogue, and describe the world around him. Rather, I hope in this book to elucidate the origins of a broad interest in natural history—the conditions that made possible the existence not of one or a handful of observers of nature, each isolated in his or her passion, but of a community comprising hundreds of naturalists scattered across a continent.

For my focus on the Latin-writing, humanistically educated elite of the sixteenth century, I have no apologies. Renaissance natural history smelled equally of lampblack and the rich, black earth; naturalists were equally at home reading Aristotle on fishes and stalking through fish markets looking for novelties in the day's catch. Natural history as a discipline privileged certain forms of knowledge, usually the knowledge of the privileged. It excluded many people and their knowledge. This is the story of a discipline and its community, not of all knowledge of nature in the Renaissance. As moral individuals we may deplore how Renaissance naturalists restricted their community as much as possible to people like themselves; as historians, our task is to understand why they did so.

It is easy for intellectual historians to forget that their subjects were human beings, with individual virtues and flaws as well as those of their age. In this book, I have tried to keep that principle in mind. The individuals who play a role in this story were deeply committed to natural history: in important ways, it defined their lives. I have tried to be as charitable as possible, to take them at their word unless I have good reason not to. Historians are by nature suspicious, and rightly so, of public faces, of rationalizations and cover-ups; the hermeneutics of suspicion loom large in our mental apparatus. What people say and what they do are often two different things. My focus on practice, as well as texts, has, I hope, helped me avoid simply accepting Renaissance naturalists' self-portrayals. Yet good history, like good anthropology, should show how the past differs from the present, and the historian should be wary of rejecting his or her subjects' claims because they don't square with what we now think. As in any hermeneutic enterprise, these two principles pull the writer in different directions. What is important, though, is that I have tried not to forget them. Whether I have succeeded is for you to judge.

I have incurred many debts while writing this book. I take great pleasure in acknowledging them here.

Lorraine Daston provided criticism and advice throughout the research and writing of the dissertation on which this book is based. I owe a special debt to her and to Robert J. Richards and Constantin Fasolt, the other members of my dissertation committee. Paula Findlen read the entire manuscript and provided invaluable guidance in turning a dissertation into a book. Alix Cooper provided extensive comments on a conference paper based on chapter 4. My colleagues at the University of Massachusetts Amherst, especially Daniel Gordon and Larry Owens, have encouraged my research and commented on some of the manuscripts that coalesced into this book. I carried on a delightful correspondence with Claudia Swan regarding the *Libri Picturati* in Kraków. I would also like to thank the three anonymous readers for the University of Chicago Press, who saved me from several errors of commission and omission. Thomas Rushford and James Kelly provided last-minute transatlantic research assistance. If I have not adopted all of their suggestions, it is partly to avoid making a long book even longer.

I have presented portions of this book to audiences at the University of Chicago, the Max-Planck-Institut für Wissenschaftsgeschichte, the University of California at Davis, the University of Massachusetts Amherst, the Five-College History Seminar, and meetings of the History of Science Society and Frühe Neuzeit Interdiziplinär. For their probing questions and stimulating conversation, I would particularly like to thank Lorraine Daston, Michael Dietrich, Daniel Gordon, Wolfgang Lefèvre, Christoph Lüthy, Staffan Müller-Wille, Karen Reeds, Skúli Sigurdsson, Jessica Spector, Zeno Swijtink, Anke te Heesen, and Kathleen Whalen. I hope they will find that I have at least partially responded to their suggestions.

Others have contributed to the project in less direct but nonetheless important ways. Allen G. Debus encouraged my earliest research in Renaissance natural history. Julius Kirshner directed my early reading on humanism. Anthony Grafton suggested—rightly, though I had my doubts at the time—that I should focus my dissertation research on natural history. Charles Cohen directed my earliest studies of Renaissance art, and he taught me more than he probably realizes.

Susan Abrams, my first editor at the University of Chicago Press, took an active interest in this book from the days when it was a dissertation prospectus. Her encouragement, advice, and patience made it possible, and I sorely regret that she was not able to see it through to the end. Christie Henry has been the perfect successor; she has expertly balanced active intervention and benign neglect when each was appropriate. Mara Naselli's eagle eye caught many errors and inconsistencies that had escaped me. After talking with French friends, I am especially grateful for the care that the Press has put into this book.

From its beginnings as a dissertation, this project has benefited from generous financial support. My dissertation research was funded by a National Science Foundation Graduate Fellowship (1991–94); a University of Chicago Century Fellowship (1991–95); a University of Chicago–Mellon Summer Travel Grant (1994); a Max-Planck-Gesellschaft predoctoral research fellowship (through the Max-Planck-Institut für Wissenschaftsgeschichte, Berlin, 1995–96); and a University of Chicago–Mellon Dissertation Writing Fellowship (1996–97). A further research trip in the summer of 1999 was funded by a General Research Grant from the American Philosophical Society and a Faculty Research Grant from the University of Massachusetts Office of Research; the latter also paid plane fare for a final trip to Paris in the summer of 2001. In the summer of 1999 the Max-Planck-Institut für

Wissenschaftsgeschichte, Berlin, generously appointed me as a visiting scholar for a week during my travels. My colleagues on the Personnel Committee of the Department of History, University of Massachusetts Amherst, arranged for teaching release in the spring of 2000, allowing me to make substantial progress on the revisions. The final revisions were completed at the Columbia University Institute for Scholars at Reid Hall in Paris.

The Department of Special Collections at the University of Chicago Libraries, the Newberry Library, the Handschriftenabteilung of the Universitätsbibliothek Erlangen, the Handschriftenabteilung of the Universitätsbibliothek Basel, the Herzog-August-Bibliothek in Wolfenbüttel, the Bibliotheek Dousa (Western manuscripts) of the Universiteitsbibliotheek Leiden, the Bibliothèque centrale du Muséum National d'Histoire Naturelle and the Bibliothèque Nationale de France in Paris, and the graphic collection of the Biblioteka Jagiellonská in Kraków have all made my work even more pleasurable with their courteous and prompt assistance in finding the many books and manuscripts I have used and in hauling enormous folio volumes from the shelves to my reading table. The Bibliothèque Nationale de France has also provided invaluable service through Project Gallica, its digital library, which I have consulted in the final stages of preparing the book. The Interlibrary Loan office of the University of Chicago Libraries supplied microfilms, books, and articles. The Bibliothek of the Max-Planck-Institut für Wissenschaftsgeschichte was equally helpful in procuring the most obscure books and articles from all corners of Germany. After my move to Amherst I was pleasantly surprised to find many of the rare books I had used in Chicago, Paris, and Berlin could also be consulted at the Special Collections department at the Mt. Holyoke College Library in South Hadley, Massachusetts. The Interlibrary Loan Department of the W. E. B. DuBois Library at the University of Massachusetts Amherst has provided books and articles from near and far.

Portions of chapter 4 have appeared in different form in *The Power of Images in Early Modern Science* (Basel: Birkhäuser, 2003), *The Journal of the History of Ideas* vol. 64, no. 1 (2003), and *Pre-modern Encyclopaedic Texts* (Leiden: Brill, 1997). I would like to thank them for introducing some of my ideas to the wider world and, in the second case, for permission to repeat them here. I would also like to thank the Bibliothèque centrale du Muséum Nationale d'Histoire Naturelle (Paris), the British Museum, the Naturkundemuseum Kassel, the Universitätsbibliothek Erlangen-Nürnberg, the Universiteitsbibliotheek Leiden, and the Utah Museum of Natural History for supplying and permitting me to reproduce several photographs; individual credits appear in the captions.

History is a weird enterprise. We historians travel long distances to immerse ourselves in dusty libraries and archives, attempting to resurrect vibrant lives from the scraps they left behind. I am fortunate to share my life and my writing with someone who understands that: a fellow historian. Jennifer Heuer has read the entire book in its many versions; her suggestions have improved it immensely. She has also tolerated my moods and my interruptions of her own work during its writing. I dedicated the dissertation out of which this book grew to her, and I repeat that dedication with profound thanks. I also dedicate it to my parents in recognition of their unconditional support and love.

My friends, colleagues, and supporters bear no responsibility for any errors that may remain in this book. I have tried to be as accurate as possible, but like the naturalists I study, I have trusted in the claims of other scholars, both my contemporaries and my predecessors. History, like natural history, is a collective enterprise, and without a certain fundamental trust in my predecessors' accuracy and good will, I could not have written this book. When I could not check a claim, I have relied on my own sense of what was possible, and what was probable, in the Renaissance. I hope only that the gentle reader who discovers an error will not dwell on it, but rather keep in mind the good, and that those who correct my mistakes will do so not out of spite or malice but to benefit the Republic of Letters.

CONVENTIONS

Abbreviations

In the notes I use these abbreviations:

Basel UB = Basel (Switzerland), Universitätsbibliothek, Handschriftenabteilung
Erlangen UB = Erlangen (Germany), Universitätsbibliothek, Handschriftenabteilung
Kraków BJ = Kraków (Poland), Biblioteka Jagiellonská, Graphic Collection
Leiden UB = Leiden (Netherlands, Universiteitsbibliotheek, Bibliotheek Dousa [Western manuscripts]

References

In the footnotes, I have adopted an abbreviated reference style, generally citing the author's name, a brief form of the title, and, for the first appearance, the date of publication. Full references can be found in the bibliography, which is divided into manuscript sources, published primary sources, and secondary literature.

Nomenclature

Unless otherwise noted, Latin, non-English, and obsolete English vernacular names of plants and animals given in the text and notes are those used by Renaissance naturalists. Caspar Bauhin's *Pinax Theatri botanici* (1623) provides a comprehensive guide to the nomenclature; since later naturalists, including Linnaeus, included references to Bauhin's names, it is possible, if tedious, to determine the Linnaean equivalent of a given sixteenth-century name.[1] The plants described and illustrated in Rembert Dodoens's *Histoire des plantes* (1557) have been identified by the editor of the modern reprint, J.-E. Opsomer; the same has been done for Leonhart Fuchs by Frederick G. Meyer and for Hieronymus Bock by Brigitte Hoppe. For most readers, these will be more convenient references than Bauhin.[2] Nonobsolete English names are the current common names of the plants and animals.

When referring to Renaissance scholars, I have used the vernacular forms of their names in most instances: Leoniceno (not Leonicenus), Fuchs (not Fuchsius), Gessner (not Gesnerus), De l'Obel (not Lobelius), Dalechamps (not Dalechampius), Bauhin (not Bauhinus). When a scholar adopted a translation into Latin or Greek of his name, I have used the translated form if that scholar was known chiefly through Latin publications; hence I use Bock (not Tragus), since Bock wrote in German, but Camerarius (for Kammermeister). Throughout I use Clusius, not de l'Escluse, simply because the Latin form is more convenient in the case of this name, which appears frequently. The index contains cross-references.

CHAPTER ONE

Introduction

Setting the Stage: The Invention of Natural History in the Renaissance

Natural history was invented in the Renaissance. Of course, natural history had distinguished roots in classical antiquity and the Latin Middle Ages, and a namesake in the Roman encyclopedist Pliny the Elder's *Natural History* (*Naturalis historia*). But only in the middle of the sixteenth century did naturalists come to think of themselves as practitioners of a discipline that, though related to medicine and natural philosophy, was distinct from both. The naturalists who reached this self-consciousness did so in reflecting on the labors of their predecessors from the 1490s through the 1530s, who had undertaken to restore ancient Greek and Latin works on the history and medicinal uses of plants and animals and to identify the modern species described by the ancients. No single individual invented natural history; by its very nature, it could be the product only of a community. That community and the discipline it practiced are the subjects of this book.

Put another way, natural history meant something very different by the beginning of the seventeenth century than it had at the end of the fifteenth. In most of this book we will focus on what naturalists themselves did and said, but we can begin to grasp the significance of the Renaissance transformation of natural history by examining how scholars outside natural history wrote about it.[1] In the late fifteenth and early sixteenth centuries, scholars produced many surveys of the learned disciplines. For writers like Giorgio Valla, Polydore Vergil, and Juan Luis Vives, natural history had no place in the encyclopedia of the disciplines. It was subordinated to natural philosophy and medicine; at best, it represented a stage in the production of valid knowledge about nature. By the early seventeenth century, even for a philosopher like Francis Bacon, natural history was no longer simply an intermediate form of knowledge; it had become a distinct discipline.

Natural history as such was absent from the massive humanist encyclopedia compiled by Giorgio Valla (1447–1500), a Piacenzan humanist active in Venice, and published posthumously by his son in 1501.[2] In forty-nine books, Valla surveyed the whole of secular, liberal knowledge, from the quadrivial arts through "external things" such as glory and friendship. (He excluded divine learning and the mechanical arts.) The first eighteen books address the natural world; they are followed by seven books on medicine, a book of problems, and then nineteen books on the human world, from grammar to the benefits and problems of the body and the vagaries of reputation.[3]

Valla located animals and plants in several divisions of his encyclopedia. In his four books on "physiologia" and metaphysics, Valla discussed "nature," in the Aristotelian sense of the internal source of motion of a natural kind, and the natural world from the four elements to the cosmos as a whole. In these books of animals, he began with the soul (*anima*) and proceeded to its generation and growth; he did not consider individual species. In his books on medicine, on the other hand, he enumerated individual stones, animals and their parts, and plants, but he discussed only their properties as simple medicines—and, as with vipers, scorpions, and even bees, the harmful effects of their venom. Domesticated animals and cultivated plants found a place in the books on "Economy, or the administration of the home," which also included architecture.[4]

In our terms, then, Valla treated natural history under three distinct heads: natural philosophy, medicine, and "res rustica" (agriculture and husbandry). Though his encyclopedia contained the most up-to-date humanist learning, these categories had a long history behind them: as we will see in chapter 3, ancient and medieval students of plants and animals approached them in these three disciplines. Moreover, Valla was concerned with useful knowledge, in particular knowledge with moral value. He titled his work "Things to Seek and to Avoid," emphasizing the transformative power of learning on the student's character. Within this framework, natural history per se had no place. Natural philosophy could edify, medicine and agriculture were useful; mere facts and descriptions, on the other hand, had no conceivable place. Valla's son Gianpietro probably expressed his father's view when he compared the encyclopedia, food for the mind, with dinner: eat too much food and you need to be purged, implying that learning too much would result in mental indigestion.[5] Petrarch had earlier condemned knowledge for its own sake. Later humanists would disagree, but only after they had imbued detailed knowledge of nature with its own moral authority.[6]

In a widely read compendium *On Discovery* (*De inventoribus rerum*, 1499), the Italian scholar Polydore Vergil also treated plants and animals under several distinct heads analogous to Valla's. Vergil's most extended treatment comes in chapters on herbal medicines, agriculture, viticulture, and silviculture. He also discussed how the ancients used plants to write on and to start fires.[7] He discussed animals at length in chapters on horsemanship, hunting, and husbandry. Animals also taught human beings which plants heal. Their knuckle-bones were used as dice, their skins protect us from the cold, and they fought with condemned criminals to amuse Roman audiences.[8] For Vergil, as for Valla, plants and animals were used in several intellectual and practical disciplines, but they themselves were the object of none. Pliny the Elder's *Natural history* was one of his chief sources, but he considered "natural history" to be no more than the book's title. When he turned to the discipline of history, he included only human affairs. Vergil's utilitarian approach to the study of nature is summed up in a chapter on, of all things, marriage: man has dominion over all things of the earth, "for we till its soil, take its fruit, sail its waters, and have in our power fish, flying things, and four-footed beasts. So God made all things for man's sake since all are for man to use."[9] If Valla underscored the moral value of knowledge; Vergil emphasized—at least when discussing the living world—its practical benefits.

Unlike for Valla and Vergil, natural history was a distinct category for Juan Luis Vives, whose encyclopedia *De tradendis disciplinis* (1531) surveyed the whole of human knowledge from a schoolmaster's chair. But it was not a discipline; rather, natural history comprised the results of scientific investigation of any subject. Vives treated natural history as the product of a complex set of cognitive operations, beginning with "contemplation" of nature and understanding of metaphysical categories. Each of these corresponds to one of "the mind's two chief faculties," perception (*vis intuendi*) and judgment. Using these two faculties in tandem, the investigator of nature examines the exterior causes of things; afterwards, if he is talented also in "those things that completely escape the senses and can be investigated only by thinking," he can proceed to the investigation of "spiritual things." When this understanding is set down on paper, the result is "the history of nature," or "a copious narration of things, including not only effects but also causes, that aims to explain rather than investigate."[10]

For Vives, then, natural history was neither a discipline nor a method of inquiry. Instead, it was a specific stage in a broader investigation into nature; it bore the same relationship to that investigation as an oration does to the art of rhetoric. Behind Vives's definition of natural history looms

Pliny's great encyclopedia, which encompasses the cosmos; the earth; its animals, plants, and minerals; and its human inhabitants and their arts and crafts. A particular natural history might be much more limited. Aristotle's *History of Animals* was one such book. In *De causis corruptarum artium* (1531), a companion to *De tradendis disciplinis* that bitterly attacked the medieval scholars who had corrupted the pure doctrine of antiquity, Vives castigated the schoolmen for adopting Aristotle's most obscure works, from the *Physics* and *Metaphysics* to the *Meteorology*, while neglecting the more useful and reliable *History of Animals* and the *Problems.*[11] Even Aristotle was not perfect, however. While Aristotle was the best ancient authority in "the inspection of nature," others had made contributions that the schoolmen neglected, and Aristotle's errors were recognized and corrected by later "historians of nature" and even by common folk.[12] Vives recognized "history of nature" as a literary genre, and called its practitioners "historians"; he also called for a renewal of this kind of investigation of nature. But he did not identify it as a discipline with a distinct community of practitioners.

For these humanist encyclopedists of the late fifteenth and early sixteenth century, natural history did not qualify as a discipline. It might be a literary genre, as it was for Vives, but it entailed neither a clearly demarcated realm of phenomena nor a set of precepts and methods for study. By the early seventeenth century, natural history held a secure place in encyclopedias of the disciplines. In *The Advancement of Learning* (1605), Francis Bacon located natural history as a species of history, along with civil and ecclesiastical history. As such, natural history corresponded to the mental faculty of memory. It was concerned with facts rather than causal explanations, which Bacon—in good Aristotelian fashion, but unlike Vives—reserved to philosophy. Natural history, in turn, was subdivided into three kinds: history "of nature in course, of nature erring or varying, and of nature altered or wrought; that is, history of Creatures, history of Marvels, and history of Arts."[13] Bacon thought that his contemporaries had neglected the latter two categories, but that the first was satisfactory. He would change his mind in *De augmentis* (1623), where he concluded that existing natural histories were inadequate foundations for his new science.[14] And Bacon's own natural histories, in the *Sylva sylvarum* and other of his works, were quite different from those published by sixteenth-century naturalists. For the moment, though, it suffices to note that Bacon gave natural history a distinct place in the encyclopedia of the disciplines, a place that it had lacked a century earlier.

The contrast between Vives, for whom natural history was a form of writing about nature, and Bacon, for whom it was a kind of learning in its own right, shows in the coarsest form the transformation that the rest of this book will describe in detail. Of course, for Bacon, natural history was still a matter of books, just as Vives admitted that natural histories were kinds of books. But the books whose contents the English lord chancellor weighed, and eventually found wanting, were only the last stage in the production of knowledge. The encyclopedists' terminology itself reflects this sense: Vives's "disciplinae" and Bacon's "learning" implied that knowledge was not only a product but also a method to produce it. For Vives, natural history was not, itself, a discipline; for Bacon, it was.

By itself, Bacon's classification does not testify to the formation of a new discipline. Bacon was too subtle and forceful a thinker for that, too prone to create systems, as his correspondence between history and memory demonstrates. But it suggests a shift, as does Bacon's distinction between perfect and imperfect kinds of natural history. That is indeed the case. As a discipline, natural history was born in the sixteenth century. By 1600, Renaissance naturalists recognized themselves not only as producers of a particular kind of knowledge, natural history, but also as members of a community of naturalists and masters of its techniques. Natural history was not merely a collection of books that the encyclopedists could arrange tidily in their imaginary libraries: it was a cultural form, situated above all in a specific community, though like all cultural forms it was at least partially accessible to certain others.

What does it mean to call natural history a cultural form? One aspect of culture is to give meaning to experience. Natural history explained the world—or rather, a certain set of the world's phenomena. But that is only one sense of culture: the sense of belief. Culture goes deeper, because it includes practices that implicitly (rather than explicitly) grant meaning and pattern social reproduction.[15] Natural history was a cultural form in that sense as well. In the late Renaissance, becoming a naturalist meant mastering not only a set of concepts but also a specific set of techniques that granted meaning to interactions with the world. Naturalists also granted meaning to natural history as an activity; natural history was implicated in broader cultural forms and social formations, and participated in the reproduction of those forms. The term "discipline" seems especially appropriate for natural history: its Renaissance sense of a field of inquiry with accepted principles reflects the emergence of natural history as a recognized field, while its connotation in the twenty-first century suggests the processes of socialization and self-control that were required to make a serious naturalist.[16]

Calling natural history a discipline also helps us avoid the anachronistic distinction between professional and amateur science and scholarship.[17]

Like the humanist philology of the late fifteenth century, sixteenth-century natural history required training that was informal yet rigorous; it demanded, as we shall see, knowledge of a broad range of literature and mastery of precise skills of observation and description. Yet there were no professional qualifications for becoming a naturalist, nor was it possible for most naturalists to make a living from natural history. We could call naturalists amateurs, then, but only in a restricted sense: they practiced natural history out of devotion to the subject, but there was nothing amateurish about how the best of them pursued their studies. This combination of high standards and open boundaries often provoked anxiety about naturalists' status and occasionally led to polemics.[18]

Renaissance natural history was never static. The four generations of naturalists from 1490 to 1630 had changing concerns, relations, and practices; the way that Niccolò Leoniceno studied plants in the 1490s differed sharply from the way that Caspar Bauhin investigated the vegetable world a century later. Each had different concerns and participated in different communities, and they expressed their ideas in sharply contrasting forms. Nonetheless certain continuities characterize Renaissance natural history, and at the same time demarcate it from the natural history of antiquity and the Middle Ages on the one hand, and the later seventeenth century on the other.[19]

The most important of these continuities is the concern with description. It is no exaggeration to say that description, as both process and result, is the central concern of Renaissance natural history.[20] This is true in a negative and a positive sense. From the controversy over Pliny in the late fifteenth century through the botanical tables of the early seventeenth, description lay at the heart of Renaissance natural history. Unlike their medieval predecessors, Renaissance naturalists condemned the inaccurate or inadequate descriptions of the natural world that had been bequeathed from antiquity. The disparity between what they saw and what they read motivated careful investigation into the variety of the created world, and prompted the development of new descriptions modeled after the old. Initially, these descriptions were pictorial, but soon a technical descriptive language was elaborated that eventually took precedence, within the community, over pictures. From the 1530s through the 1630s, the task of natural history—as a discipline distinct from the mere art of gardening and the lofty science of nature—was describing nature, cataloguing its marvelous and mundane products.

Description, as a central problem, marks Renaissance natural history off from its medieval predecessors and its seventeenth-century successors. For Vives, whose notion of natural history was modeled after Pliny, natural history was a narration of both causes and effects. For Bacon, proper natural history would be "inductive," the basis for a new science of forms or causes. Bacon dismissed the natural history of the Renaissance as "more suitable for philosophers' table-talk than for establishing philosophy"; here, as elsewhere, he undervalued a discipline he did not fully understand. But he did recognize that contemporary natural history, done "for its own sake," consisted primarily in "the variety of natural species": that is, in description.[21] As we will see in chapter 3, medieval natural history cannot be considered a coherent discipline. In the Middle Ages, plants and animals were studied as a propaedeutic to either medicine or natural philosophy. In both cases, given the close bonds between natural philosophy and medical theory, medieval naturalists were concerned to discover the nature or essence of plants and animals—either as a basis for universal judgments or as a means to employ the essence in cures. Seventeenth-century naturalists, on the other hand, were increasingly concerned with classification and the causal explanation that Bacon advocated. Description remained a part of natural history, but it was increasingly routinized and systematized. In other words, description *concerned* Renaissance naturalists in a way that it did not concern their predecessors or successors. For that reason, this book is a history of the Renaissance science of describing.

In calling Renaissance natural history a science of describing, I have in mind a felicitous parallel between my account of natural history and Svetlana Alpers's interpretation of seventeenth-century Dutch art in *The Art of Describing.*[22] For Alpers, Dutch art of the Golden Age was chiefly descriptive: not only did painters strive to depict accurately the surface appearances of the natural world, they also struggled with the theoretical problems posed by the attempt to make canvas correspond with what they saw. To this historical claim about Dutch art, Alpers adds a provocative historiographical thesis: the discipline of art history, elaborated on the basis of Italian painting, is (or was) ill equipped to understand Dutch art. From the fifteenth century, Italian painting emphasized narrative and allegory; beginning with Giorgio Vasari in the sixteenth century, art historians focused on these elements as well as on technique and style, leading to a disciplinary preoccupation with iconography and iconology. Insofar as Dutch artists strove to depict the appearances of a moment, to describe the world pictorially, the Italian model of interpretation breaks down. Their works demand a new interpretive scheme. In the place of history and allegory, Alpers approaches Dutch art in terms of observation, optics, and mapping.

Like Dutch artists a century after them, Renaissance naturalists grappled with the practical and theoretical problems of description—especially the practical—as their predecessors had not and their successors would not need to. The existing historiography on natural history has not adequately grasped this point. There have been many fine studies published recently on aspects of Renaissance natural history and its points of intersection with more general Renaissance cultural trends: museums and collecting, gardening, commerce, and curiosity.[23] But general histories have treated Renaissance natural history as part of a broader history of biology, even though the word "biology" was coined only in the nineteenth century.[24] In so doing, they have overemphasized the theoretical and philosophical elements in natural history, particularly taxonomy and classification, while neglecting or treating as self-evidently worthwhile the specific achievements of Renaissance natural history: the recognition of the vast diversity of the animal and vegetable world (particularly the latter), the establishment of a community of scholars engaged in a common enterprise, and the elaboration within that community of methods for discovering new natural kinds and describing them precisely. This was the novel achievement of Renaissance naturalists. This book aims to demonstrate that their thoroughgoing, fine-grained empiricism took intense effort and intellectual resources; it was far from self-evident, and the problems of classification that characterized later natural history were a consequence of the attempt to describe the world. Empiricism, in short, is neither easy nor self-evident, and the novel empiricism of Renaissance natural history deserves to be at the center of its history.

Good humanists, Renaissance naturalists did not claim the dubious honor of having invented something new. Rather, they portrayed themselves as restorers of an ancient tradition, a tradition founded by ancient near eastern potentates like Solomon and Mithridates and brought to perfection by the Greeks. This tradition had fallen into desuetude during the barbarous middle age only to be revivified in Italy in the fifteenth century. Beginning in the last decades of the fifteenth century, the nearly forgotten learning of the ancients had been recovered and improved, according to Leonhart Fuchs, writing in 1542, and Thomas Johnson, writing ninety-one years later.[25] The engraved title page of Johnson's edition of John Gerard's *Herball* reflects this assertion of continuity (frontispiece).[26] God—represented by the Tetragrammaton—presides over the Garden of Eden, with a quotation from Genesis: "Behold, I have given you every herb bearing seed, which is."[27] A further inscription admonishes the reader, "Lest you forget the Author of the divine gift, every plant shows the presence of God." Ceres and Pomona, Roman goddesses of grain and fruit, flank Eden.[28] Below this rank are the human founders of the "historie of plantes": Theophrastus, the Peripatetic philosopher who wrote the *History of Plants* and *Causes of Plants*, and Pedanius Dioscorides of Anazarbos, physician in the Roman legions and author of the most important surviving ancient work on pharmacy. They represent the scholarly and practical skills united in the lowermost figure, Gerard himself, presented on the first page of his book as both a true Christian and a worthy successor to the ancient tradition.

Fuchs and Johnson portrayed the history of their discipline as an inverted arc: ancient greatness, medieval decline, and modern restoration. They did not invent the narrative: fifteenth-century scholars had applied it to Latin learning, while in the sixteenth century it was a commonplace in histories of the arts, crafts, and disciplines. Andreas Vesalius sketched the history of anatomy following the same pattern.[29] Giorgio Vasari applied it to art, while suggesting that the cycle of decline and rebirth applied not only to painting, sculpture, and architecture, but also to literature and other aspects of culture, and that the same causes were at work.[30] In the cases of Latin scholarship, painting, sculpture, and architecture, the narrative of decline and rebirth may be sound—if one accepts the humanists' criteria for judging decline and improvement. At the very least, these cultural traditions were pursued continuously throughout the Latin Middle Ages, even if their history shows often sudden stylistic discontinuities.[31] Anatomy and natural history are another matter.

Human anatomy was practiced, albeit infrequently, in the thirteenth and fourteenth centuries, while fifteenth-century artists frequently anatomized corpses to learn how best to depict sinew, muscle, and bone beneath the flesh in their paintings and sculptures.[32] Vesalius's *De humani corporis fabrica* (1543) went far beyond its published predecessors, but Leonardo da Vinci's anatomical drawings were even more detailed, if less systematic and learned, than Vesalius's woodcuts. The study of human anatomy through dissection was already being carried out when humanist Greek scholars rediscovered the "original professors of anatomy," that is, the Greek physicians, in their original form.[33] There had been a hiatus in the Western anatomical tradition, but its revival was not, in the first instance, due to the humanist recovery of ancient texts; a significant motive was the voracious appetite of European churches and chapels for the relics of saints, whose bodies could be very efficiently chopped up, preserved, and distributed by their pious venerators.[34] The humanist narrative, self-evidently true to Vesalius, seriously distorts the actual history of anatomy.

The humanist narrative distorts the history of natural history, too, but for another reason: it conflates textual tradition with cultural tradition—or, we might say, text with context. Before the age of print, a textual tradition involves multiplying manuscripts: copying them from predecessors and correcting them in light of other exemplars or one's native wit and knowledge of their subject.[35] Natural history, as we shall see in chapter 3, had several textual traditions, as medieval monks and scholars copied and recopied works that appeared useful to them in their studies in natural philosophy, medicine, and agriculture. The sixteenth-century scholar Conrad Gessner, in the prefaces to his *Historia animalium* (*History of Animals*) and the Latin version of Hieronymus Bock's *History of Plants*, drew together these textual traditions in his careful bibliographies of "those who have written about animals" and plants. An indefatigable bibliographer, Gessner could hardly avoid mentioning books, but other naturalists also presented their discipline as constituted, and thus legitimated, by its literary predecessors.[36] But natural history lacked an active cultural tradition in antiquity and the Middle Ages. Following Marshall Hodgson, a cultural tradition involves three phases: "a creative action, group commitment thereto, and cumulative interaction within the group."[37] The essential element of a cultural tradition is continuity: continuity within a group that shares a common commitment to an initial creative act—or, in the case of a scholarly discipline, an evolving problematic and method.[38] Natural history lacked such continuity before the late fifteenth century. Before then, it was not a cultural tradition. Since then, it has been and is. Although natural history has undergone significant transformations in the past five centuries, the tradition has continued to respond to the challenges posed by Renaissance practitioners of the science of describing.

If I insist on the novelty of Renaissance natural history, it is to counter a pervasive tendency in intellectual history and the history of science. All too often, natural history has been seen as a nearly continuous tradition from the investigations of Aristotle and his pupils at the Lyceum, through the foundation of modern systematic botany and zoology by Linnaeus, to the synthesis of systematics and evolution in the work of Darwin and his successors.[39] In this narrative, as in so many other narratives of Western history, the early Middle Ages represent something of a hiatus, but with the recovery of Greek learning in the twelfth century, the story picks up anew. This tendency has been reinforced by the fact that most histories of science, until quite recently, were written by scientists who, in general, stressed the continuities and focused primarily on intellectual content (i.e., theories).[40]

In other words, historians have bought Renaissance naturalists' imagined history of their discipline and continue to peddle it today.

There is, to be sure, some truth in this story, but it neglects the crucial roles of contingency and choice. In the case of specific problems regarding the living world, such as theories of generation, the continuities are clear, because later writers were in fact responding to the claims and arguments of their predecessors. Matters get much murkier, though, when one moves beyond specific problems to general issues—where traditional histories have recognized shifts of emphasis within disciplines. They become murkier still when one goes beyond theories to practices and asks how natural history related to other cultural traditions and what role it played in the intellectual life of the time. In chapter 3 we will consider briefly the place of natural history in ancient and medieval culture. For the moment, I would like to turn to its place in the learned culture of Renaissance Europe.

Natural History and Renaissance Culture: Five Aspects

If natural history was a coherent cultural tradition in the Renaissance, it cannot be sharply demarcated from Renaissance society and culture more generally. Its emergence as a discipline is intimately connected with other movements and developments within the learned culture of the Renaissance. These connections will be drawn out at length in subsequent chapters. Here I would like to sketch them briefly, concentrating on five areas: the cultural movement of humanism; the intense contemporary interest in empiricism and the fact; collecting and the culture of curiosity; the emergence of new forms of scientific organization; and the vexed question of the Renaissance "world view." In effect, I would like to justify the lines I have drawn to limit my subject. I hope to have carved off natural history from its setting at the joints—while recognizing that even the best butcher does have to hack through ligaments, tendons, veins, and arteries while carving.

Humanism

Humanism, that protean concept, is the broader intellectual framework within which Renaissance natural history took root and flowered.[41] In the late fifteenth and early sixteenth century, humanist textual scholarship provided one of the principal spurs to the development of natural history.[42] But the influence of humanism on natural history went far beyond this. On a basic level, the triumph of the humanist secondary school curriculum, first in Italy and then, in the sixteenth century, north of the Alps, provided naturalists with a set of attitudes toward language and classical texts that would infuse natural history.[43] Renaissance naturalists idealized classical antiquity even when they surpassed its achievements. They valued eloquence in their publications and in their letters; Conrad Gessner may not have written the most eloquent Latin, but he recognized this as a failing. Their books were prefaced with elegant—or pedestrian—epigraphs by their friends or colleagues, and they maintained in their correspondence an ideal of sincere friendship that had its roots in the humanist Republic of Letters.

By the middle of the sixteenth century, humanism was less a conscious movement than a tacit attitude motivating scholarly inquiry in disciplines that would eventually become the humanities.[44] The belief that antiquity provided a model for scholarly activity and a standard for judging its products had become an assumption, not a position that needed to be defended. Like other fields of sixteenth-century scholarship, natural history often shows only a superficial similarity to what the ancients actually did—but the same could be said of fifteenth-century humanism. Just as fifteenth-century humanists thought they were imitating the ancients when they created an alphabet based on Carolingian minuscule, their sixteenth-century successors in natural history appealed to antiquity to justify what was really a new discipline.[45] These attitudes can be described as a form of humanism, but with the caveat that we learn more about late Renaissance humanism by studying its particular manifestations, such as natural history, than we learn about those manifestations by simply labeling them as humanistic.[46]

Empiricism and the History of the Fact

Humanists encouraged fact-finding in many ways. Philologists who hunted manuscripts in order to correct texts, historians who turned to archives in order to discover the material for their narratives, physicians who translated anew the classical texts of Galen and compared them with what their own eyes revealed—all such humanists paid attention to the individual, sometimes recalcitrant fact. But empiricism in early modern Europe was not limited to humanists. Iconoclasts like Paracelsus, who urged physicians to burn their books and read only the Book of Nature; lawyers like Andrea Alciati, who treated the Roman civil law not as universal truth but as the product of specific historical circumstances; craftsmen like the potter-philosopher Bernard Palissy; surgeons like the prototeratologist Ambroise Paré—these and many others reveal to the modern historian how widespread was the early modern European cult of the fact.[47]

Yet the cult of the fact had different denominations. From the late medieval books of secrets to Francis Bacon's insistence that the secrets of nature were best observed when she was "out of course" or even "vexed," one tradition of empiricism looked for the elusive, marvelous, or recondite fact. This tradition produced not only compendia of wonders but also the modern notion of the experiment, a specific intervention into nature that ran counter to the received scholastic notion of experience as that which happened always or for the most part.[48] Naturalists' taste in facts was more catholic. Renaissance naturalists sought out not only rare, unusual, and strange plants and animals, but also everyday, common, literally garden-variety creatures. Their biblical antitype was not Daniel, who proclaimed God's "signs and wonders," but Solomon, who "spake of trees, from the cedar tree that is in Lebanon even unto the hyssop that springeth out of the wall."[49] Within particular disciplines, the impulse to gather facts produced several different empiricisms; a close study of natural history reveals how its disciplinary dynamic produced a science not only of facts but of descriptions.

Collecting and Curiosity

Early modern Europeans did not collect only facts; they also collected natural objects, human artifacts, and delightful mixtures of the two.[50] The history of collecting has been the focus of cultural historians interested not only in the use of collections for scientific research but also how collectors used their collections as a means of self-fashioning and social advancement.[51] Most collectors of natural objects were not themselves naturalists, but they were increasingly drawn to naturalists' publications in the late Renaissance. The use of collections for power, in whatever sense, was by the mid-sixteenth century predicated on the existence of natural history as knowledge. Collectors had to know about their *naturalia* in order to truly possess them. In this case, knowledge was power in a sense very different from that intended by Francis Bacon.

Recent analyses of the social structure of collecting and exchange in early modern Europe have focused on patronage networks and a proto-capitalist exchange system with "brokers" of material, connections, and knowledge.[52] Ultimately, this approach is founded on the notions of symbolic capital and its accumulation and exchange propounded by Pierre Bourdieu, or involves the same assumptions about culture.[53] Although such an analysis illuminates aspects of the early modern culture of collecting, it uses economic terms that were themselves contested in the late Renaissance. Sixteenth-century naturalists conceived of their study, when it was not ancillary to the noble art of medicine, as a liberal art in its own right, and they explicitly rejected protocapitalist intrusions into its domain.[54] In this regard, the natural history community of the sixteenth century displays a sharp contrast with the capitalist organization of bulb-selling that catered to the noble and bourgeois pleasure-gardeners of the seventeenth century.[55]

Collectors occupied an uneasy middle ground between the advancing forces of market values and the opposing notion of liberal studies. Naturalists could be vicious in their condemnation of those who sold what should be given away freely.[56] "Freely" is a relative term; naturalists operated within an informal system of exchange in which those who received without reciprocating would gradually be excluded from the community. And most naturalists were willing to pay others—gardeners, travelers, apothecaries, and other outsiders—for the obscure objects of their desire. In this regard, naturalists delimited their community socially as well as intellectually. Full members of the community were those who gave and received liberally, without besmirching their science with gold.

Forms of Scientific Sociability and Organization

The relation between collecting and natural history underscores that as a cultural form it was developed within a particular community. This community was not open to everyone. It was not a club with entrance requirements and a membership list; in many ways it was more permeable than other communities in the late Renaissance. Naturalists included not only gentlemen and professionals but also apothecaries and printers. Nonetheless, the mechanisms that formed and maintained the community also excluded certain groups, including some who might have had a lively interest in studying plants and animals.[57] This was especially true in the early sixteenth century, as humanistically trained physicians struggled to claim natural history for themselves; their remarks on the decline of the subject since the days of antiquity usually drip with contempt for the herbalists and wise women who had cultivated knowledge of medical botany in the Middle Ages. Naturalists were often introduced to the subject at a university, and Latin was the lingua franca of natural history, as it was in other areas of Renaissance scholarship. The community maintained ties through correspondence, whose rules of decorum often barred unknown individuals. Those who needed to support themselves from their activities in natural history, like apothecaries or gardeners, found themselves at odds with the explicit anticommercial attitude of naturalists. And with a handful of exceptions, Renaissance naturalists were men; women seldom had the opportunity to acquire the training and connections required to fit in.[58]

In this regard, natural history was part of the creation of the expert, collective knowledge that characterizes modern science.[59] It excluded groups whose knowledge of plants, animals, and minerals was more personal and extended, but also more local and, usually, more tacit than that of the self-proclaimed experts. They interacted with naturalists, who were keen to learn not only from personal observation but also from the experience of "rustics." But rustics' experience had to be validated by those who saw themselves as guardians of the heritage of the ancients and practitioners of a science that went beyond mere individual experience.

The relation between experience and knowledge was at the core of sixteenth- and seventeenth-century conceptions of science, and it played a key role in the transformation of natural philosophy that we retrospectively dub the scientific revolution. In their work, Renaissance naturalists struggled with ways to communicate and reproduce experience of nature. Furthermore, the sheer amount of empirical information they gathered had the effect of transforming their experience of nature's individual productions. From our vantage point in the egalitarian twenty-first century, we may deplore the mechanisms that excluded women and artisans from the naturalists' enterprise. But we cannot deny that the communication of local knowledge that characterized Renaissance natural history led naturalists to a new appreciation of that local knowledge. The experience of a naturalist who recognized a type of wild rose as peculiar to a small region was qualitatively different from that of a local who had never known another kind. Natural history thus prompts us to reflect on the whole notion of experience—that of the individual and that of a community.

The Renaissance "Worldview"

Beyond its connections to specific movements like humanism, empiricism, collecting, and scientific sociability, natural history has contributed to various attempts to educe a set of governing ideas, "worldview," or "episteme" that characterized Renaissance thought in general. Such attempts have generally given a neo-Platonic cast to Renaissance thought. Arthur Lovejoy's magisterial *Great Chain of Being* (1936) traced the notions of hierarchy and plenitude from classical antiquity to the eighteenth century, while E. M. W. Tillyard, drawing on Lovejoy, considered them an integral part of the Renaissance world view in England.[60] But the most ambitious attempt to establish a Renaissance world view or "episteme" was the project of Michel

Foucault, and natural history played a key part in Foucault's account. In the mid-1960s, Foucault analyzed brilliantly how some Renaissance thinkers understood the world in terms of correspondences between the microcosm and macrocosm, metaphoric correspondences that they took literally. His analysis was based on a narrow sample of sometimes idiosyncratic thinkers, but some of his followers, dazzled by the master's brilliance, have not only applied his interpretation more broadly than the evidence warrants but even elided distinctions on which sixteenth-century naturalists insisted.[61]

As humanists, Renaissance naturalists noted the ancients' stories about plants and animals, their appearance in proverbs and emblems, their mystical or allegorical meanings in certain texts, and other references to them that we consider part of culture rather than natural history per se. This interest is not, however, evidence for an "emblematic world view" in which the distinction we draw between nature and culture was elided.[62] Many writers dismissed such claims as arrant nonsense.[63] Even Conrad Gessner, who collected mountains of literary references to animals in his massive *History of Animals*, recognized a distinction between natural history proper and what he called philology: the latter, though of interest to him, was a concern not for naturalists but for literary scholars.[64] Other writers distinguished nature and culture in their own fashion: Joachim Camerarius, for instance, collected plants in his garden and animal and plant emblems in his study, but he published them separately, his descriptions of plants as a botanical treatise and his emblems as four emblem books. He was interested in both, and he acknowledged the connections between them, but he did not consider them to be the same.

This is not to deny that naturalists understand the world as a forest of symbols. Rather, it is to insist on the distinctions they made between the empirical study of nature and the symbolic interpretation of it—and to suggest that Foucault's particular interpretation excludes more widespread symbolic systems that Renaissance thinkers applied to nature. Natural theology was one: all naturalists accepted that God had created plants, animals, and minerals in the six days' work, and the notion that "every plant shows the hand of God" is almost ubiquitous on the engraved title pages of late Renaissance botanical books.[65] Some writers argued, against the astrological botany of writers like Leonhard Thurneisser von Thurn and Nicholas Culpeper, that according to Genesis plants were created before the stars, so that the latter could have no influence on the former.[66] The Book of Nature was a powerful metaphor in the sixteenth century, as before and after, but naturalists, as naturalists, were concerned with establishing its text, not deciphering its language.

Natural history does offer insights into broader culture, however, as many careful scholars have argued.[67] The most convincing of these interpretations have been based on what naturalists did or wrote, or how their works were used by others. In using natural history as a prism through which to view Renaissance culture, one should remember that as a group, naturalists shared certain interests and practices that they developed to serve specific ends, and that those practices and interests were not necessarily shared by the other communities with whom naturalists interacted. The rest of this book will examine this community and its practices. Readers who have little interest in the methodological reflections that have shaped my history may skip ahead to chapter two.

Methodological Problems: Experience and Practice

The notion of experience appears deceptively simple. Yet few concepts have received more, and more mutually contradictory, examinations in the philosophical tradition. From Aristotle to phenomenologists, by way of empiricists, Kantians, idealists, positivists, and several other philosophical troupes, experience has been defined, refined, and redefined to the point where for even the most experienced observer of the philosophical stage it may be impossible to define. Yet experience is the central concept for my study, and if I am to avoid confusing my readers as much as the philosophers have confused me, I cannot avoid attempting to pin it down. The easiest way to do so is to take it by surprise from the rear, by examining its use in distinct polarities and delimiting it negatively through its polar opposites. Four key axes seem to get at the most important uses of the word from Aristotle's day to the present: experience/theory, experience/perception, experience/inexperience, and experience/reality.

Aristotle's analysis of experience in the *Metaphysics* distinguishes experience from both theory and perception, as a middle term that is necessary but not sufficient for scientific knowledge. Experience is not theory, because an experienced man does not know the causes of the phenomena in which he is experienced. He may be able to cure patients but he doesn't know why, only how. This notion corresponds to the "ordinary experience" of twentieth-century philosophers: the sense in which we distinguish on-the-job experience from formal education. One should keep in mind that Aristotle denied that experience sufficed for scientific knowledge, but he held that experienced individuals without theory were better at dealing with specific cases than inexperienced men who knew causes, because action involves dealing with particulars, not universals.

Experience is not perception either, in Aristotle's account; he held that a single experience was formed from several perceptions of the same phenomenon, mediated by memory.[68] Here is where an Aristotelian account differs from modern accounts of experience, from the empiricists to the phenomenologists. Neither the former nor the latter wish to deny that experience includes the abstract schemes that repeated perception involves—that you have to see several trees before you can experience the "treeness" of any particular example—but they also include perception of objects qua objects in the realm of experience. The empiricists did so to underscore the notion that the mind is a *tabula rasa* before perception, while both Kantians and phenomenologists insist that perception is a form of experience in order to underscore the constitutive action of consciousness, the mental work that is required to retain a welter of sense perceptions and forge out of them the awareness of perceiving a single object that remains unchanging as perceptions shift. This difference is fundamental for understanding experience in Renaissance natural history. When Renaissance naturalists and philosophers discussed experience, they distinguished it from both theory and perception.

A further, ordinary-language use of experience is in contrast to inexperience, and distinguishes the experienced individual from the inexperienced one. Evidently this sense is related to the experience/perception axis, for one who has not perceived a phenomenon, either once or several times, is inexperienced with regard to it. But it is far from identical. On the one hand, "experienced" in this sense may refer either to personal experience or to vicarious experience mediated through oral or written reports. On the other hand, though, it also implies practical knowledge—the ability to do. Aristotle's example of the empirical physician includes this notion of experience as well as the epistemological sense. This is no accident. Only the individual who is experienced in a practical sense will be able to make useful experience out of his or her own perceptions and those of others mediated through reports. This sense of experience will also be critical for understanding Renaissance natural history, and especially, its limits when dealing with exotic objects.

A final axis contrasts experience with reality. In this sense, experience denotes the way things seem to an observer, while reality refers to the nature of things in themselves. Experience is subjective, reality is objective. This is the axis whose poles have concerned philosophers since Descartes—the empiricists postulated rules to get from reality to experience; Kant denied that reality per se was knowable; and some phenomenologists collapsed the axis by simply denying the distinction. It is also the sense in which historians use "experience" in the recent turn to studying the experiences of groups or individuals in the past: the African-American experience, women's experience, the immigrant experience, and the like. In this sense, experience is closely related to *mentalité*, *Weltanschauung*, and worldview, and its use reflects a growing concern with subjectivity in history, a concern whose philosophical roots lie in existentialism—itself a development from phenomenology.[69] Historians who write about various experiences wish to emphasize the difference between "objective" conditions—often identified with the subjectivity of the white, male historian—and how different groups in the past understood and responded to those conditions. They also wish to draw attention to the felt texture of life—what it was like to be a spinster in colonial New England, or a sailor on the Spanish Main, echoing in this regard R. G. Collingwood's claim that all history involves recreating the thoughts of the past.[70] Though applied to a far wider range of subjective experience than aesthetics, this usage has clear affinities with the notion of aesthetic experience.

These four senses of experience cannot easily be sharply demarcated; each bears echoes of the other. Nonetheless, we can try to use the word carefully. A philosopher would suggest using subscripts to distinguish them, or perhaps even coining three new words. Without going that far, I have tried to be precise in my own use of the term. In this book, I will draw on all the senses as necessary.

Natural history did not go beyond experience to theory, so (despite naturalists' protests) it was not a science in an Aristotelian sense. But naturalists needed to observe plants and animals several times in order to have experience of them. Only repeated observation could make them experienced naturalists, capable of assimilating new experiences and assessing the reports of others' experiences in letters, books, and oral testimony. And such conditioning inevitably shaped how they experienced the natural world. The very fact that I can write the preceding sentences points to the range of meanings of "experience"; the fact that they are intelligible points to how we distinguish in everyday use the different senses.

Historians of science have recently argued for a fifth polarity between experience and experiment. In this polarity, the latter was a new concept defining the scientific revolution of the seventeenth century. Peter Dear has argued that sixteenth-century natural philosophers conceived of experience as a general category, referring to commonsense notions of what usually happened in nature. For example, heavy objects fall when released, fire rises, and an arrow shot straight up returns to the same place whence it was shot.[71] This notion was consonant with Aristotle's oft-repeated dictum that the study of nature concerned itself with what happened "always or for the most part," not with nature's singularities.[72] Even in the case of controversial claims—for example, that two objects of unequal weight dropped from the same height fall at the same speed—the experience that justified them was general, based on the claim to have done something repeatedly and to be aware that it was true, rather than to have observed it in a particular instance.[73] The experiments of the early Royal Society, on the other hand, posited a notion of the *particular* experience that was localized in time and space. Such experiences might be corroborated by others, but they were always linked to the circumstances of their production in a way in which "fire goes up" was not.[74] In this view, the scientific revolution was predicated on a type of empiricism radically different from that of late medieval scholastic philosophy, characterized by Francis Bacon's dictum that "the secrets of nature reveal themselves more readily under the vexations of art than when they go their own way."[75]

In this account, experience can be analyzed in terms of two different polarities.[76] On the one hand, the Aristotelians' notion of experience as a general sense of what happens "always or for the most part" is opposed to the Royal Society's notion of the experiment as a singular event that establishes a matter of fact about nature. Dear is concerned above all with this opposition. But there is another polarity implied by this account of experience: the polarity between firsthand and reported experience. The two polarities appear to be independent: one can experience the sweet taste of an exotic fruit once or a thousand times, and one can also read that redwood trees grow to over two hundred feet in height or that a particular tree is three hundred and fifteen feet tall. In practice, the two are often connected: thus, on Dear's account, Giambattista Riccioli undermined the authority of the claim that bodies of unequal weights fall at the same speed by arguing that this general claim was, in fact, based only on one man's experience, which others had repeated as hearsay.[77]

If we turn to naturalists' own accounts of experience, we will not be enlightened. Sixteenth-century naturalists did not define what they meant by experience. A few instances in which it is compared and contrasted with other notions allow us to form a preliminary sense of its meaning. Olaus Magnus, writing in 1555, contrasted knowledge gained from "the assertions of others" with that acquired by "personal inspection, or experience."[78] Olaus's formulation echoes that used half a century earlier by Leoniceno. Leoniceno glossed "experientia" as "sense itself" and distinguished it from knowledge (and misinformation) gained by reading authorities.[79] In both cases, experience was personal, gained from the senses, rather than something written down or conveyed orally by another.

It is difficult to be more precise, however, for naturalists referred to "experience" in passing or treated it as an unproblematic resource. Carolus Clusius, for example, observed that "through long experience we have learned that alpine plants can be preserved better in gardens if they are placed in a shady but open location."[80] Dominicus Baudius compared Clusius to "Ulysses, who experienced much," emphasizing once more the personal element denoted by the word.[81] However, this leaves the question of the relationship between naturalists' experience and natural philosophers' Aristotelian category of experience unsettled.

By looking at naturalists' practices, however, we can get a clearer notion of what they meant by experience. In writing natural histories, they condensed their observations and experience into discrete, concise descriptions. In their descriptions, the notion of "experience" takes on many forms, depending on the information available, from singular events to summaries of numerous observations made over a long period of time. In both cases, however, there is little that is naive or commonplace about such experience: the good observer needed an educated eye and mastery of the discipline's practices. The naturalist's notion of experience thus assumed an intermediate position between commonplace statements about the world and claims to have observed singular occurrences; both of them grounded, as far as possible, in firsthand observation.

If naturalists were imprecise when they talked about experience, they developed many specific techniques aimed at producing it reliably and communicating it to others. In some cases, they even crossed plants and bred sports in ways that look surprisingly like experiments. Such cases were rare, but most of the time naturalists' notion of experience was much more particular than the supposed everyday, common-sense version of experience attributed by recent scholars to Aristotelian natural philosophers. Since a feature of many early seventeenth-century calls to reform natural philosophy was the emphasis on natural history, looking at how naturalists experienced the world sheds light on this important aspect of the scientific revolution.

Experience requires doing; it is process rather than product. It cannot be studied solely on the basis of books. Yet the standard literature in the history of natural history has a strong bookish cast. It presumes that naturalists are those scholars who published books, and that the history of natural history can be written primarily on the basis of those books.[82] Both of these assumptions are questionable. Books were only part of the apparatus of early modern natural history, and they were not only products of the investigation of nature but also tools used in the process of investigation. An adequate history must consider that process and the other tools employed in it, especially since examining the practices of knowledge production is essential for understanding its content.[83] Furthermore, by emphasizing books, earlier historians presupposed an ideal intellectual community populated by authors who shared the same concerns and goals. This notion of history as a great reading room is misleading on two counts. First, there were many naturalists who published little or nothing but whom contemporaries considered important colleagues. Second, there were very real intellectual communities in early modern natural history, both local and international, joined by particular bonds of conversation and correspondence. These communities, not a hypothetical ideal community of texts, are the proper *loci* of historical analysis.[84]

Another question that this study touches upon is the question of truth and trust.[85] Proponents of seventeenth-century new philosophies, from the most thoroughgoing Baconian empiricist to the purest Cartesian rationalist, were unanimous in rejecting authority and calling for a natural philosophy based on personal experience. Yet, as Steven Shapin has pointed out for the late seventeenth century, and as the briefest introspection will confirm, most of what we believe is taken on trust. Much of what other people tell us we count as true, not because we have verified it, but because we trust their accounts. The antiauthoritarianism of the new philosophies was an ideal—an important ideal, calling for the systematic reexamination of received truths[86]—but it could never be carried out in practice.

Based as it was largely on local observation and communication of those observations to a wider community, natural history offers prime ground for examining the mechanisms through which trust was generated in early modern intellectual communities. Travelers' tales had always been eagerly but suspiciously read by intellectuals.[87] Both the avidity with which they were read and the suspicion with which their claims were regarded arose from the same source: the emphasis on what was strange, wondrous, or marvelous. Sixteenth-century naturalists grappled with the problem of determining, in the case of second-hand accounts of often marvelous plants and animals, how to separate the true from the false. Their solutions, as I will show, were both social and scholarly. They were more likely to trust those whom they knew, especially those they knew to be careful observers themselves. In the case of unknown sources, separated from them and their social world by space or time, they applied humanist methods of collation and comparison in order to tease, if possible, the truth from vague or contradictory accounts.

The role of naturalists' practices in settling this question is of prime importance. Methods of finding and studying *naturalia*, and the "literary technologies" used to reproduce experience, played an important role in producing knowledge that others could trust. This history thus contributes to the general literature in the history and sociology of knowledge that is concerned with the practices through which knowledge is produced and legitimated. But it is also a history of a particular community, formed in a series of specific historical settings.

The remaining chapters of this book trace that community's history, the intellectual traditions on which it drew, its methods for investigating and describing nature, and the consequences of its labor, consequences that often led to unintended results. Chapter 2 delineates the Renaissance community of naturalists from the 1490s through the 1620s. Chapter 3 examines the ancient and medieval traditions that converged in Renaissance natural history, sketching how Renaissance naturalists created their own fictive past out of ancient and medieval natural philosophy, agricultural writing, and pharmacy. The chapter situates the immediate background to Renaissance natural history in humanists' approach to particulars and their positive valuation of nature for its own sake; it then traces the early history of Renaissance natural history, from the debate over Pliny in the 1490s to the humanistic natural history of the 1520s. Chapter 4 examines the techniques that Renaissance naturalists, especially from the 1530s through the 1610s, used to investigate and describe nature. It treats these approaches to nature not simply as tools but as elements in the formation of the naturalist's sensibility. Finally, chapter 5 notes the transformations of natural history in the early seventeenth century from a science of describing based on firsthand experience in nature to a descriptive enterprise in which classification was an increasingly important concern, and then examines how naturalists from the late sixteenth century to the late seventeenth century approached nature beyond Europe. Though natural history underwent many dramatic transformations after the Renaissance, it remained and still is, in many ways, a science of describing.

The techniques of the science of describing were worked out primarily in the realm of botany, which for that reason comprises the bulk of this study. Renaissance naturalists studied animals and minerals, but the core of Renaissance natural history was *res herbaria*, the study of plants. For every publishing undertaking like Georg Agricola's *De re metallica* or Conrad Gessner's *History of Animals*, there were half a dozen histories of plants. Moreover, as I show in chapter 2, important developments occurred first in botany and then spread to zoology and, to a lesser extent, mineralogy.[88] The reasons are manifold, including the medical background of many naturalists (chapters 2) and of early humanist natural history (chapter 3) but also, by the end of the sixteenth century, the cognitive challenges posed by studying plants (chapter 5). Nonetheless, the same people who studied plants also had an interest in animals, though they were more difficult to study and there were many fewer of them than plants (invertebrates, lumped into the categories of "insects" and "worms," interested very few naturalists before the late seventeenth century).[89] That interest was especially piqued in the case of exotic animals. The problems posed for naturalists in Europe by exotic plants and animals were strikingly similar; for that reason, Renaissance zoology receives more attention in the last chapter than in the rest of the book. Readers who are distressed by the relative absence of zoology from earlier chapters may be consoled by the thought that the book is, in consequence, significantly shorter than it might otherwise have been.[90]

For a similar combination of pragmatic and intellectual grounds, I have concentrated on Europe outside of Italy and Iberia. A number of important studies of Italian natural history have underscored its vitality and, in many respects, peculiarity when compared with northern Europe.[91] Hence Italian naturalists appear in these pages chiefly in the light of intellectual history. Iberia is somewhat different. One of the most important botanical excursions of the sixteenth century was Carolus Clusius's journey through the Iberian peninsula in the 1560s. But Iberia produced few naturalists who participated in the broader European community.[92] The explanation I learned in graduate school—that Spanish intellectual life suffered under the Inquisition and, in particular, its restrictions on foreign study—is being put to rest by a new generation of scholars who have, instead, emphasized the particular interests and institutional concerns of Iberian naturalists.[93] In many ways, Iberian natural history in the sixteenth century resembles the institutionalized natural history of eighteenth-century Europe, with its national botanical gardens and projects of acclimatization.[94] Be that as it may, Iberian authors receive little attention here because, except through heavily edited translations, they played little role in elaborating the Renaissance science of describing.

CHAPTER TWO

The World of Renaissance Natural History

In November 1609, Caspar Bauhin, professor of anatomy and botany at the University of Basel in the Swiss Republic, was the most famous naturalist in Europe. His chief rival, Carolus Clusius, the late Prefect of the Academic Garden at the University of Leiden in the Netherlands, had died the previous March. Clusius's *Rariorum plantarum historia* (1601) and *Exoticorum libri decem* (1605) had summed up his life's work investigating and describing natural objects—especially plants, above all, those previously unknown or undescribed.[1] Bauhin had, as yet, no such monument to his fame. He had displayed his scholarly acuity and breadth in the *Phytopinax* (1596), an incomplete yet magisterial guide to plants and their literature. Two years later he edited the *Opera omnia* (1598) of the Sienese physician Pietro Andrea Mattioli, the most prominent Italian naturalist of the later sixteenth century.[2] His 1601 pamphlet pointing out the many errors in the anonymous *Historia generalis plantarum* (1587–88) emphasized his critical sense—and his acerbic character.[3] These works led the community of naturalists to anticipate eagerly the general history of plants that Bauhin had promised in his *Phytopinax*.[4]

On November 20, according to the new Gregorian calendar in effect in Catholic lands, a young Venetian physician named Angelo Busti took pen in hand to write to Bauhin.

> You will not find it hard to believe that the rumors of your wisdom in the knowledge of plants and other *naturalia* have led me, though absent, to love and esteem you. If it is true, as many historians say, that some men have come to love women whom they have never seen because their character and beauty were praised by others, how much more can beauty of spirit, which unlike physical beauty is incorruptible

> and immortal . . . lead men to love those whom they know through their books and fame?

Busti's love of Bauhin—*amare* is the Latin verb he used—was due to Bauhin's knowledge of "naturalia," of "animals, plants, stones, and underground things." As a young man, Busti had been driven to observe and collect *naturalia* before he even knew the word; as he grew older and learned more, his pleasure in them grew apace. On beginning university studies at Padua, he augmented his pleasure by conversing about *naturalia* with his teachers Antonio Cortuso and Prosper Alpino. Contemplation, cognition, and conversation each had a particular delight; combined they were magnified, impelling Busti to love and honor those who were esteemed in this science. Hence his love letter to Bauhin: he hoped that the famous Basel professor would deign to acknowledge him with a small reply.[5] This kind of flattery could only please Bauhin, a prickly man whose relations with equals were strained but who basked in his students' adulation: Bauhin responded on March 3 (according to the Julian calendar [o.s.]: Basel had been reformed in the 1520s). Not surprisingly, he asked Busti to send him some botanical specimens for his work.[6]

Busti's breathless Latin and his flattery, shameless to modern readers and excessive even for its time, suggest a naïf. But his letter reveals a clear sense of the origins and intellectual development of Renaissance natural history. He began with Aristotle's dictum that all men desire naturally to know. This impulse to know comes from the pleasure of knowledge, Busti continued, not from any utility, and we take our greatest pleasure from the knowledge of particular things. For this kind of knowledge, vision is the most important sense, because it identifies the distinctions (*differentiae*) among objects. For many centuries, "heroes" like Aristotle and Dioscorides had devoted themselves to this knowledge, but somehow the natural impulse to study nature had become stunted after their time.

> God restored this knowledge through famous men of recent times, who did not spare effort and long hours, but as often happens with new things, it was diminished and still imperfect. For that reason other famous men, imitating their predecessors, attempted to restore it to its original splendor. They devoted the most work to the histories of plants, about which many have written many books, taking care to observe everything and to add what their predecessors observed badly or not at all, or what was omitted.

Bauhin, at long last, would bring this knowledge to perfection and put an end to future labors with his *Theatrum botanicum* (*Theater of Plants*), about which Busti had learned from the Veronese apothecary Giovanni Pona.[7]

For all its artful flattery, Busti's letter offers deep insight into the character of Renaissance natural history as its practitioners and aspiring practitioners understood it. Naturalists formed a community of curious observers. In appealing to Bauhin, Busti portrayed himself as an intimate of two famous sixteenth-century Italian naturalists, Antonio Cortuso and Prosper Alpino, and his contemporary Pona, author of a famous description of Mount Baldo, who was already one of Bauhin's correspondents.[8] Busti's intimacy with these men recommended him to Bauhin, as did his familiarity with Bauhin's work, published and projected. He was already a member of the community; only as such could he hope for a response from its star. The social bonds of this community, forged and maintained through correspondence and travel, were based on two factors, one affective and one cognitive: voluntary association based on "friendship" or even "love," and curious—that is, painstaking and inquisitive—observation of *naturalia,* above all plants, and careful description of those that were unknown to their predecessors. Observation and description was the community's *raison d'être,* but it was sustained, in the absence of more than a handful of professional positions in natural history, through its affective bonds, bonds expressed in the humanist idiom of friendship, of *amicitia* and *amor.*

This community had existed for little more than a century when Bauhin and Busti started their correspondence. It had taken shape in northern Italy in the last years of the fifteenth century, and had spread north with medical students from Germany as they returned home from Italian universities. Bauhin and Busti were part of its fourth generation. This chapter examines the contours of the community of Renaissance naturalists in shape and time. It begins by tracing four generations of naturalists from the controversy over Pliny in the 1490s to Bauhin's death in 1624. After examining their contours, it looks at how one became a member of the community, who was on the margins, and who was excluded. It then turns to the geography of Renaissance natural history: the division between Italy and the north, the importance of local communities, and the bonds forged by travel and correspondence that made Renaissance naturalists into citizens of the same province of the Republic of Letters, committed to a common enterprise, a common purpose, and common methods.

Four Generations of Renaissance Naturalists

If description, as a central concern, united Renaissance naturalists from the 1490s through the 1620s, the composition of the discipline and its emphases changed significantly. Earlier scholars have seen these changes as a shift from philology to independent observation. Charles Nauert traced them by examining the study of Pliny the Elder's *Natural History:* taken seriously as a source for natural history in the early sixteenth century, by the century's end it was studied by philologists but not by naturalists, who had turned to observation.[9] Carlo Maccagni too divided Renaissance natural history into two phases. The first, from the middle of the fifteenth to the middle of the sixteenth century, was dominated by the humanistic activity of "discovery, study, evaluation, and publication of classical texts, carried out exclusively with philological techniques." The second, from the middle of the sixteenth to the end of the seventeenth century, was characterized by a shift from erudition to pure research, embodied in the formation of collections and botanical gardens, public and private, and the publication of increasingly more books. In the earlier phase, naturalists produced new editions of Aristotle, Theophrastus, Pliny, and Dioscorides, and argued about whether Dioscorides was to be preferred to Pliny. The latter period is marked by "the botanical gardens of Pisa and Padua, the museums of Calzeolari in Verona and Imperato in Naples with their printed catalogues, the edition of Dioscorides completed by Mattioli using the *Placiti* of Luca Ghini, and Guilandino's *Papyrus,* where three chapters from Pliny (*Natural History,* XIII, 11–13) are weighed against the author's own experience as a traveler and naturalist."[10] In this view, natural history's transformation in the sixteenth century is a matter of emancipation: naturalists threw the crutches of ancient learning and found they could walk better without them.

In broad terms these scholars are correct. Around 1500, European naturalists took Aristotle, Theophrastus, and Dioscorides as the starting points for natural history. (Some had doubts about Pliny.) By 1600, these texts were studied only for specific facts, or in modern editions that corrected and supplemented them. But looking at the transformation from this bird's-eye view is misleading. It emphasizes the resemblance between natural history and other early modern scientific disciplines, like astronomy or natural philosophy, whose practitioners rejected the ancients' texts and turned to the Book of Nature instead. But Renaissance naturalists were reformers, not revolutionaries.[11] They moved away from the ancients not because they consciously rejected them but because, in the course of their engagement with nature, they came to pose questions and demand answers for which the ancients' texts no longer provided guidance. Even at the beginning, the ancients were not crutches; they were tools. When the tools proved no longer useful, naturalists found or invented others. To understand why natural history changed so dramatically in the sixteenth century, we need to look at it more closely.

To do so, I identify four scholarly generations. The first, active in Italy from the 1490s to the 1530s, comprised humanists and physicians, men whose primary concerns were identifying the medical plants described by ancient Greek and Roman writers and reforming medical education on the basis of those writers, above all Galen. They turned to the natural world to identify the medical materials described by the ancients and to reconcile contradictions between ancient texts. The next generation, from the 1530s to the 1560s, included northern Europeans who studied in Italy but quickly learned, on returning home, that the plants that grew around them were not always the same as those described by the ancients. While they were often physicians, and their books had a strong medical bent, they were increasingly interested in studying the natural world for its own sake, and much of their investigation and description was only tenuously connected with the healing arts. The third generation, from the 1560s to the 1590s, pursued their predecessors' descriptive program in increasing detail and forged connections with collectors and the patrons of magnificent gardens. Many, if not most, continued to learn the elements of natural history in medical school, where lecturers in *materia medica* continued to pay lip service to the notion that natural history was the handmaid of medicine. But those who acquired and kept a passion for natural history pursued it for reasons having little to do with medicine. The fourth generation, active from the 1590s through the 1620s, dealt with the legacy of the third. By 1590, naturalists had discovered and named so many different species (mostly plants) that beginners were overwhelmed and even experts found it hard to keep up with the variety of things—and the even greater variety of names—in natural history books. Naturalists of the fourth generation tried to make sense out of this confusion, beginning the emphasis on taxonomy and classification that would henceforth characterize natural history. At the same time, the close bonds between natural history, collecting, and gardening began to unravel. Collectors of curiosities and flower lovers found little to engage them in naturalists' forbidding tomes, and they developed their own literature. Natural history had finally become autonomous—at the cost of becoming obscure.

These divisions are necessarily rough. Nonetheless 1530, 1560, and 1590 cut the world of Renaissance natural history at the joints as closely as any historian is ever able to establish sharp divisions. Much of the justification for my choices will come in chapters 4 and 5. Here I wish to set out the social composition of these generations and begin to note how their socialization into the world of natural history differed. In broad terms, each generation quickly mastered the achievements of its predecessor, while developing new problems of its own; these common experiences, as well as common readings, gave each generation of naturalists its own character, within the broader, continuing concerns of the science of describing.[12]

Medical Humanists and Critics, 1490–1530

The first generation of Renaissance naturalists might not have thought of themselves as such. Its key members were northern Italian scholars, and if they had to identify themselves as part of a tradition, they would have chosen classical scholarship, medicine, or both. Their publications located them firmly within one or both of those traditions, as did their institutional locations and intellectual circles. Nonetheless they quickly came to agree that investigation of the natural world through firsthand observation and experience was the only way to understand natural history, and they taught those methods to their students. Those students, in turn, would form the core of the next generation of naturalists, the first to begin to think of their study of nature as something distinct from medicine and scholarship, if still intimately linked with the disciplines that gave it birth.

The most important member of this generation—old enough to be its literal as well as its spiritual father—was the Italian humanist and physician Niccolò Leoniceno (1428–1524).[13] Leoniceno taught moral philosophy and then medicine at the University of Ferrara, where he strove mightily to replace Arab medical texts with the purer medical learning of the ancient Greeks. Medical botany, he thought, had been especially corrupted by the barbarous Arabs and their ignorant medieval followers, but it had also been damaged by the mistranslations and mistakes that Pliny the Elder had introduced into Latin medical botany in his *Natural History.* Stung by this challenge to the authority of a classical Roman writer, the Florentine court humanist Angelo Poliziano set an acquaintance of his, the lawyer and historian Pandolfo Collenuccio, the task of defending Pliny from Leoniceno's attacks. In a series of pamphlets published between 1492 and 1509, the combatants debated the relative merits of ancient sources and the precise meanings of the Greek words that Pliny had translated or transliterated—accurately or not—into Latin.[14]

Leoniceno and Collenuccio approached the debate from subtly different perspectives. Both agreed that ancient texts were key to settling the question, and after the publication of Ermolao Barbaro's massive *Castigationes plinianae* in 1492, Collenuccio could not deny that the received text of Pliny contained many scribal errors. But the stakes of the debate differed for each side. As humanists and critics, Collenuccio and Poliziano wished to defend the honor of the ancient Roman; they found it distasteful that Leoniceno lumped Pliny together with Avicenna, Mesue, and other medieval Arab physicians. For his part, Leoniceno wanted to purify medicine and make sure that apothecaries were using the right plants when they prepared drugs according to ancient recipes. If Pliny had erred, his errors were pernicious, and his text should be excluded from medical education in favor of the Greeks.[15]

While Barbaro had corrected Pliny's text philologically, neither Leoniceno nor Collenuccio was satisfied with this appeal to texts.[16] They agreed that the only way to determine whether Pliny was right was observation: to collect the plants that he described and compare them with his and others' descriptions. As we will see in chapter 3, this process of observation and comparison of plants with texts was the key element in the formation of Renaissance natural history. Observation and comparison would be pursued by Leoniceno's students, like the German Euricius Cordus, whose *Botanologicon* (1534) sums up Leoniceno's method and applies it to the flora around Marburg. Cordus introduced Leoniceno's method to Leonhart Fuchs, a leading light of the next generation of naturalists. Leoniceno's students and correspondents, and their own intellectual circles, spread the methods that the Ferrarese professor advocated and, in turn, formulated the problems and trained the community that Fuchs's generation would adopt.

The medical and philological interests of the first generation of naturalists were reflected in the books they read and wrote. Born in 1428, Leoniceno began his career in the age of manuscript. His interest in Greek medicine and science ensured that manuscripts remained an essential part of his library until his death in 1524, after more than a quarter-century of Greek printing in Italy.[17] Leoniceno amassed one of the largest private libraries of his day: a posthumous inventory, probably incomplete, lists 345 items, 117 of them in Greek, totaling 482 works (many shorter books were bound together).[18] His books of natural history and *materia medica* consisted chiefly of ancient Latin and Greek works in manuscript and printed editions, translations, and commentaries. He owned several Greek manuscripts of Aristotle's shorter works on animals, though he possessed the *History of Animals* only in Theodore Gaza's Latin translation. He had a printed edition of Pliny, along with Barbaro's *Castigationes* and a separate printed index. But the weight of his collection was in manuscripts and printed versions of Galen, Dioscorides, and Theophrastus. He had a manuscript and a printed translation of Theophrastus's works on plants.[19] His collection of Dioscorides was more extensive. At least three manuscript versions in Latin and Greek rubbed shoulders with three printed Latin versions: the medieval Latin translation and the contemporary versions by Barbaro and the Florentine chancellor Marcello Virgilio. The set was completed by Barbaro's corollaries to the text and an index to Barbaro's translation. The works by Galen in Leoniceno's collection are too numerous to enumerate: in his last decades Leoniceno was engaged in a project to translate all of Galen afresh into Latin.

Other classical works and postclassical texts occupied much less space on Leoniceno's shelves. He had manuscripts of Columella, Aelian's *On Animals,* and a Latin translation of Oppian's work on fishes, along with the *Physiologus,* a late antique bestiary.[20] He possessed Serapion's work on medicine (in Latin) but no other Arab medical texts. Finally, he had two medieval Western works: a book by "the herbalist Jacopo de Bosco with the figures of plants, very old and tattered, bound in brass and leather," and a book, probably a manuscript, on falconry.[21] Latin and German herbals had been published in the last decades of the fifteenth century, including an illustrated Latin herbal published in Venice in 1499, but Leoniceno did not own any.[22] In keeping with his desire to reform medicine and medical botany on the basis of the ancient Greek physicians, he apparently considered modern authors unworthy of his attention.

The only moderns who played a significant role in his scholarship were translators and commentators. For Leoniceno and other medical humanists, most ancient medical and natural history books had to be approached through state-of-the-art textual scholarship and the new translations that resulted.[23] Ignorant or careless medieval copyists had "corrupted" classical texts; the critic doctored them, curing them of their wounds. Ermolao Barbaro claimed to have corrected more than five thousand errors in Pliny's text.[24] In his commentaries on Dioscorides, Barbaro continued this work, explaining obscure references and correcting scribal errors. Marcello Virgilio's Dioscorides also contained much philological commentary.[25] Their translations of Dioscorides and Theodore Gaza's Latin versions of Aristotle on animals and Theophrastus on plants provided the best Greek natural history to contemporaries who could not read the original. Another, better version of Dioscorides, translated by the French physician Jean Ruel, would follow in 1516. The popularity of these translations and commentaries is underscored by the northern European reprints of the 1520s and 1530s.[26] Through the 1520s, the main task of the nascent community of naturalists was reforming medicine on the basis of these newly restored ancient texts; though the next generation would abandon this goal, its members would continue to apply the methods of their teachers and to draw on ancient texts as models.

The next generation could not have been formed, however, without the work of other medical reformers, intent not on reforming *materia medica* on a classical basis but on establishing medical faculties' rights to oversee the profession of apothecaries.[27] If physicians were to judge how competent apothecaries were in their trades, they had to know not only the formulas for compound medicines but also how to identify the simple medicines that went into compounds. Most of those simples came from plants. Hence leading medical faculties came to require courses in *materia medica,* above all medical botany, of their students—courses taught in the traditional method of commenting on an authoritative text, with demonstrations. The first such courses began at Padua in 1533. At Bologna, Luca Ghini began lecturing on *De simplicibus* in the 1534–35 academic year. Montpellier instituted lectures on Galen's and Dioscorides's *materia medica* in the 1530s and 1540s.[28] These courses and requirements institutionalized the informal demonstrations of Leoniceno, Euricius Cordus, and others in the first generation. Such courses encountered some resistance from conservative medical faculties, but by the 1550s they were commonplace in European medical faculties, and new universities—such as Leiden, founded in 1575, and Helmstedt, founded in 1576—included *materia medica* in their medical curricula.[29] These courses were followed by university medical gardens, beginning with Padua and Pisa in the 1540s.[30] Lectures on botanical medicine and demonstrations in university gardens ensured that Leoniceno's method would be taught to a new generation of students—at least some of whom would turn that method to the new end of cataloguing the variety of nature.

Medicine lay at the center of the nascent community of naturalists—a community as yet without a conscious identity, but one that was nonetheless forming. But medicine was not the only interest that bound this generation together. Leoniceno's opponent in the debate over Pliny, Pandolfo Collenuccio, was a lawyer, not a physician. Collenuccio was encouraged in his challenge by Angelo Poliziano, the poet and philologist, and another philologist, Ermolao Barbaro, made vital contributions to the study of nature. The ancients' books, everyone agreed, could unlock the secrets of nature. The issue was which books to trust. If we can trace a diminution in the explicit connections between medicine and natural history over the next century, it is important to keep in mind that even in its beginnings, natural history engaged not only physicians but also professionals in other fields whose interest in the natural world was an avocation—an avocation often pursued with great seriousness.

The First Phytographers, 1530–60

The next generation established natural history firmly as a discipline with a recognized set of authoritative texts and techniques, practiced by a vibrant, diverse community of scholars.[31] Its membership was international: its leading lights included the German "fathers of botany," Otto Brunfels, Hieronymus Bock, and Leonhart Fuchs; Pietro Andrea Mattioli, the Sienese physician whose commentary on Dioscorides became the standard herbal of sixteenth-century Italy; Jean Ruel, author of three books on the history of plants as well as a translation of Dioscorides; and Euricius Cordus's son Valerius (though his important publications were chiefly posthumous). Rembert Dodoens (1517–85), the oldest of the great trio of south Netherlandish naturalists, also belongs with this generation. Their principal concern was comprehensively studying the natural and imported plants and animals of the areas where they lived, and identifying them with creatures described by the ancients. While their publications situated this activity in the context of medicine and pharmacy, they also indicate that natural history for its own sake was increasingly important.

The contours of the community of naturalists in this generation can be seen in the correspondence and works of Conrad Gessner. Gessner's extensive correspondence and polymathic interests kept him in contact with people from all walks of life. Like most other sixteenth-century scholars, he kept an *album amicorum,* a small notebook in which his new acquaintances wrote their names and a motto; the surviving volume includes 227 autographs.[32] Gessner often added a note describing the profession of the writer and the circumstances under which they met. Those with an interest in natural history were mostly physicians, medical students, and apothecaries, but Gessner's publications indicate that much of his information on plants and animals came from those without a medical background. In his *Historia animalium* (1551–58), Gessner listed the correspondents who contributed information for the book.[33] Physicians were prominent: they comprised thirty-three of the sixty-three whose professions are known. However, only one was an apothecary, Adrianus Marsilius of Ulm. Eleven were theologians, ministers, or professors of sacred literature; six professors or schoolteachers in profane letters; four lawyers, and three printers. The remaining four were a consul (Georg Agricola), a town councilor, a merchant, and a Swiss politician and historian (Aegidius Tschudi). Gessner's 1560 list of the gardens of Germany conveys a similar picture: the prominent Germans who cultivated "philosophical" gardens with rare plants included not only physicians and apothecaries, but also merchants (including the Fugger family), patricians, pastors, and professors.[34]

It is no surprise that physicians comprised the largest group. As we have seen, *materia medica* had recently become a required subject for medical students, and the very possibility that the discipline could expand to the scale it did was predicated on having a body of men who were forced to study it. Those who published books of natural history were usually practicing physicians or medical professors—men like Conrad Gessner, Leonhart Fuchs, Rembert Dodoens, and Guillaume Rondelet. But many doctors and medical teachers contributed to the activity of natural history without publishing themselves. The most famous is Luca Ghini. Ghini, who never published, was apparently the great popularizer of the technique of drying plants for herbaria, first described in print by Adriaan van de Spiegel at the beginning of the seventeenth century.[35]

Apothecaries also took an active interest in natural history, for obvious reasons. Valerius Cordus's brother-in-law Johannes Ralla was an apothecary, as were several other naturalists in Saxony.[36] Dirk Cluyt of Delft, later prefect of the Leiden botanical garden, was an apothecary in contact with Clusius, as were Giovanni Pona from Verona and Christianus Porretus.[37] Though apothecaries occasionally published on natural history—Ferrante Imperato, for example—they were more important for their work in collecting material. As merchants in *materia medica* they had many distant connections, as well as growing their own products in gardens.[38] Pieter Coudenbergh, an Antwerp apothecary and "most diligent investigator of all medicinal simples," possessed a garden "filled with many and rare plants through his effort."[39] Sixteenth-century physicians could be suspicious of apothecaries as a group of tradesmen—Euricius Cordus was fond of pointing out how much of what apothecaries knew was false, while Paracelsus called them liars and cheats[40]—and as businessmen, they were not always interested in spreading knowledge freely. But several of them, actively in communication with other naturalists, were members of the community.[41]

Also involved were the intellectual elite of the late Renaissance: theologians, professors of philosophy and letters, and lawyers. Humanistically educated and university trained, they had the skills, resources, and leisure to pursue interests in natural history.[42] In some cases, their work was directly related to the subject: the philologist Julius Caesar Scaliger wrote commentaries on Aristotle's and Theophrastus's natural histories.[43] Other philologists occasionally addressed natural history in their work, like Joachim Camerarius the Elder, who discussed a plant mentioned by Dioscorides in his correspondence with Leonhart Fuchs.[44] But many pursued natural history as a leisure activity—like most physicians, except those who taught courses in *materia medica*.[45] Conrad Gessner mentioned several lawyers, theologians, and pastors as contributors to his *Historia animalium*.[46] Sebastian Münster, professor of Hebrew at Basel, and Olaus Magnus, titular archbishop of Uppsala, made important, if problematic, contributions.[47]

Natural history in the period through the 1550s was thus actively practiced by a wide range of people. But the breadth of naturalists' origins and interests was not necessarily matched by the depth of their interest or the strength of their numbers. Printed works from this period show that publishers were quite conservative. The earliest printed works in natural history fell into two categories: editions of ancient texts and herbals.[48] Early Renaissance printed herbals were editions or adaptations of their medieval predecessors. Eminently practical, they listed plants—and occasionally, animals and minerals—and noted their medicinal virtues.[49] Except for Pliny's natural history and Aristotle's history of animals, the ancient texts in natural history were also medically-oriented. The early herbals attracted a popular audience, whereas medical education in pharmacy remained firmly based in Galen, Avicenna, and to an increasing degree Dioscorides.

The new herbals of the 1530s and 1540s, written by humanist physicians like Otto Brunfels, Hieronymus Bock, and Leonhart Fuchs, were to a great degree a fusion of these elements. Brunfels collated ancient texts with modern plant illustrations by Dürer's disciple Hans Weiditz.[50] Fuchs wrote his own descriptions or plagiarized them from the ancients, presenting them in conjunction with carefully-drawn illustrations of plants from life.[51] Bock did not illustrate the first edition of his book, but his descriptions, which he composed himself, were much more precise than those of his contemporaries.[52] Bock's book was especially popular; it was translated into Latin and published, with illustrations, in 1552. The illustrations were also used in later vernacular editions, which served as home medical manuals.[53]

These herbals were avowedly practical. They described plants, noted where and when they grew, and listed their medicinal virtues and, occasionally, their harmful side-effects. The first botanical work of Rembert Dodoens, a transitional figure between this and the next generation, was a herbal in the same style; originally published in Flemish in 1554, it was translated into French (1557) and reissued numerous times in Flemish, with additions and corrections.[54] Like Bock's German herbals, the success of Dodoens's work points to the continued existence of a vernacular market for practical manuals. The same market existed in England, where John Gerard's plagiarized translation of Dodoens was immensely popular.[55]

It is easy to understand why herbals were popular. In an age when learned physicians' fees were unaffordable to most people, and most medicines were compounded of plants, vernacular herbals provided a relatively inexpensive source of up-to-date medical information. By the 1560s, most herbals included an index to maladies, so that the proper cure could easily be found.[56] Latin herbals, like the translation of Bock and the immensely popular commentaries on Dioscorides by the Sienese Pietro Andrea Mattioli, were used in medical faculties for teaching *materia medica.*[57] They continued to be reprinted, often with only minor revisions, into the eighteenth century.

At the same time, the second generation of Renaissance naturalists emphasized the contributions of their teachers by reissuing their works. The Basel house of Henricpetri brought out Leoniceno's collected works in 1529. Barbaro's commentary on Dioscorides, along with a translation by the Florentine Marcello Virgilio, was reissued the same year by the Cologne printer Johann Soter; Barbaro's Plinian castigations, meanwhile, were reprinted in Basel in 1534.[58] Brunfels excerpted their works—often at great length—in the second volume of his *Herbarum vivae eicones* (1532).[59] In these editions, the phytographers could learn about the method of investigating nature that they and their contemporaries applied in compiling their herbals. By recognizing their predecessors, naturalists of the second generation showed that they were participants in a common cultural tradition founded at the end of the previous century.

Nonetheless, there were signs by 1560 of new directions in natural history publications. The herbals of the 1530s and 1540s towered over the field like mighty trees in the forest, but new shoots were emerging that would soon strangle and choke the works of Brunfels, Bock, Fuchs, and the other herbal writers. An example is Conrad Gessner's pamphlet *Little Commentary on Rare and Admirable Plants Called Lunariae, Either Because They Glow at Night or for Some Other Reason.*[60] Published in 1556 along with his description of the ascent of Mons Fractus near Lucerne, this work said little about the medical uses of plants, instead employing a range of personal observations and textual citations to sort out the distinctions between plants with similar names.

Cataloguers and Collectors, 1560–90

By 1560, natural history had crystallized as a discipline. The state of the field was summed up in the herbals of the 1530s and 1540s, and in Conrad Gessner's massive *Historia animalium,* published in four parts from 1551 to 1558. These works were the product of an international correspondence network whose members continued to expand the boundaries of knowledge—especially in botany. Gessner's *Historia animalium* remained a useful reference work for over a century, but botanical books had to be updated or fade away. The generation active from 1560 to 1590 would continue to intensively investigate nature, in the process often abandoning even the pretense that their studies were useful for medicine.

To be sure, most naturalists in the later sixteenth century had at least one foot firmly planted in the world of medicine. By the second half of the sixteenth century, as noted above, courses in *materia medica* were *de rigueur* in medical schools: as a consequence, the number of physicians trained in natural history expanded enormously. Furthermore, the small number of top-notch medical faculties, and the growing importance of the *peregrinatio academica* for northern medical students, led to the formation of intense social bonds among physicians, bonds that were maintained for decades through correspondence. As university-trained naturalists became professors in their own turn, they introduced the new curriculum in *materia medica* in their own universities. Felix Platter, who had studied in Montpellier, and Theodor Zwinger, a former Padua student, introduced formal botanical excursions for Basel medical students in 1575.[61] These field demonstrations were supposed to deepen students' understanding of the healing powers of medicinal plants. Beyond introductory courses, though, natural history was largely isolated from the practice of medicine.

Despite the preponderance of physicians and apothecaries among those involved in the *res herbaria,* medicine now defined natural history neither professionally nor intellectually. There were no formal qualifications for the subject—unlike, for instance, natural philosophy, taught in the universities as part of the arts course. In the 1580s Dirk Cluyt, an apothecary, could be considered as a candidate to be prefect of the Leiden botanical garden, a post equal in honor and remuneration to that of a medical professor.[62] In the end, the position was offered to Carolus Clusius, who had studied law at Wittenberg and medicine at Montpellier but had neither taken a medical degree nor practiced. By the 1550s, when Clusius was engaged in his university studies, medicine had taken a back seat to the discovery of new plants.[63] Conrad Gessner asserted in 1561 that physicians preferred to stick with well-known remedies; they had little professional interest in new discoveries.[64] Half a century later, Adriaan van de Spiegel, in his *Isagoge in rem herbariam* (1606), complained bitterly about the direction taken by botany since Gessner's day:

> If the herbalists of our age had devoted as much effort to examining the medicinal virtues of those plants whose pictures they offer to us, as they have to describing their forms and generation, it would have been better for human health.[65]

In the classroom, even in Spiegel's day, lecturers like Caspar Bauhin continued to concentrate on those simples that were in fact commonly used in medicine, but for *studiosi,* medicine was of secondary interest.[66]

Clusius's circle in the southern Netherlands is a case in point. Clusius had studied medicine, and his acquaintances in the Low Countries after his return from Montpellier included Dodoens and the Antwerp apothecary Pieter Coudenbergh. But Clusius was also in contact with the printer Christoffel Plantin, who later published Clusius's most important books.[67] And he spent many years as the houseguest of Jean de Brancion, a noble residing in Mechelen, who shared his passion for plants and gardening.[68] Even Coudenbergh's garden was not merely a professional tool; it contained many rare plants that its owner had laboriously gathered together. When Spanish troops destroyed it in the 1584–85 siege of Antwerp, Coudenbergh must have been heartbroken.[69]

What united Dodoens, Clusius, Coudenbergh, Plantin, and Brancion was not medicine but collecting. For it was in the second half of the sixteenth century that natural history was caught up in the growing European passion for collecting.[70] This was the first great age of exotic horticulture and collections of curiosities. Of course, *naturalia* had been collected on a small scale for centuries: princes and prelates had long assembled wonders and curiosities, both secular and profane. Some fifteenth-century churches hung whale bones alongside their relics as a means for attracting and entertaining parishioners.[71] In the 1510s, the bishop of Trondheim in northern Norway sent a walrus head to Pope Leo X in Rome: this natural wonder was intended not for natural history research but as a token of patronage and a representation of the reach of papal authority.[72] By the early sixteenth century, rich burghers also possessed such objects. But they were often of uncertain provenance. In 1550s Zürich, it was still possible for a collector to have an animal horn in his collection without knowing from which animal it came: its mere size qualified it as a natural curiosity worthy of possession.[73]

As collecting became more and more popular, collectors could no longer get away with merely owning curiosities. By the 1560s, a collector also had to possess his collection intellectually: he had to be able to explain his wonders to visitors.[74] According to contemporaries, the collector Bernardus Paludanus could find and explain in his collection almost any object described in books.[75] Rudolf II's collection in Prague included not only a wide range of *naturalia,* but also a large number of books of natural history, and the inventory of the collection from the early seventeenth century includes learned annotations. For instance, the emperor possessed "a great, long foreign bean-plant in a strong, black, woody, rough shell, seventeen inches [*zöll*] long, which is called *lobus fabae arborescentis* by Clusius in book 3, chapter 6."[76] Visitors to collections expected to hear such explanations. In the early sixteenth century, knowledge of nature was prized by many humanists, but it was not in great demand in court and bourgeois society. Though Gargantua urged his son Pantagruel to study nature as part of his humanistic education, Castiglione's courtier did not need to know about nature.[77] But his seventeenth-century successor did: for instance, Etienne Binet's *Essay of Nature's Marvels and of the Most Noble Artifices* (1621) listed such knowledge as a prerequisite for courtly conversation.[78]

Courtiers were required to learn how to talk politely about nature because natural history collecting had become a passion among princes and great nobles. The Habsburg emperors are a case in point. Maximilian II was interested enough in natural history to hire Clusius to establish a medical garden in Vienna. Though Maximilian's successor Rudolf did not retain Clusius, he maintained a lively interest in natural history, particularly exotic animals and plants.[79] Lesser princes and nobles, too, collected *naturalia,* sometimes drawing on their collections to ingratiate themselves with their superiors. The cassowary brought back to Amsterdam by a Dutch captain in 1597 underwent a veritable *cursus honorum:* after being shown to paying customers by the captain, it was acquired by George Everard, Count of Solms. When the count had tired of his pet, he gave it to the elector of Cologne, who in turn gave it to the emperor.[80] Rudolf's agents also snatched up the first birds of paradise brought back to Europe with the feet attached.[81] Rudolf's collection in Prague shows the extent of his interests: the emperor possessed hundreds of *naturalia* and natural history books, in addition to other marvels and wonders.[82] Knowledge and power were intimately entwined for the saturnine Emperor: he saw his collection as nothing less than a microcosm, a symbol of the world over which he exercised his power. It also served his court artists as the basis for their own elaborate mingling of nature and art.[83] Princely and noble collectors created a demand for *naturalia* and books to explain them, and they provided one of the few opportunities for a naturalist to support himself from his studies.

Many naturalists in the late sixteenth and early seventeenth centuries were themselves passionate collectors, if necessarily on a smaller scale than princes and emperors. Felix Platter is emblematic of them. The Basel town physician had a collection in sixty drawers, containing minerals, plants and plant parts, animal parts, insects, coins, *objets d'art,* antiquities, and seven volumes of drawings.[84] Platter's cabinet was visited by travelers passing through Basel, such as Jacques-Auguste de Thou and Michel de Montaigne, and at least some visitors paid him an honorarium.[85] But the income from these visitors was probably offset by Platter's expenses maintaining the cabinet and acquiring objects for it: for example, he bought a great deal of Conrad Gessner's material from the Zürich scholar's heirs, including the original illustrations for the *Historia animalium* and Gessner's fossil collection.[86]

Platter's collection was in turn dwarfed by that of Ulisse Aldrovandi in Bologna.[87] Another great collection was that of Bernardus Paludanus in Enkhuizen. One contemporary described his visit in 1594 to this cabinet of wonders:

> The other day I visited Paludanus. . . . He showed me his collection, which had such varied and numerous items that I scarcely believed they existed in nature. Nature herself seems to have moved into his house, entire and unmutilated, and there is nothing written down in books that he cannot present to your eyes. That is why the great man Joseph Scaliger gave all his rarities (which were both numerous and spectacular) to Paludanus, saying "Here are your things, which I have possessed unjustly."[88]

Many other sixteenth- and seventeenth-century naturalists were known throughout the learned world for their collections, like Ferrante Imperato in Naples, the brothers Contant in Poitiers, and Ole Worm in Copenhagen. By the late seventeenth century, collections of *naturalia* were commonplace throughout Europe.[89] Many local naturalists made a point of collecting and organizing as much material from their locality as possible.[90] In these cases, natural history overlapped significantly with the culture of collecting.

Not every naturalist, though, was a collector. If Platter, Paludanus, and Aldrovandi were enthusiastic collectors of *naturalia* and *artificialia,* the other end of the pole is represented by Carolus Clusius. Clusius was an enthusiastic gardener. But he possessed few *artificialia* and he had little interest in dried plants. In his negotiations with the curators of Leiden University, he begged off teaching on minerals with the excuse that he knew nothing about them.[91] Clusius was driven to intellectual, rather than physical, possession of nature. His goal was to explore nature, reproduce it in his garden, and describe it, and he delighted not in cabinets full of dry, dusty remains but gardens with bright tulips, irises, and lilies.[92] It is worth remembering that Clusius wrote several books and translated many others; Aldrovandi, on the other hand, spent his life organizing and increasing his collection, and began to publish his natural history only at the end of his life. Most of the volumes bearing his name were, in fact, edited posthumously by his friends and students.

The relations between natural history and the culture of curiosity are complex. As we will see in chapter 4, the naturalists of the third generation could not have continued their intellectual labors without collecting. Gardens, herbaria (collections of dried plants), cabinets with animal bones—and collections of notes and drawings—provided much of the empirical substrate of late sixteenth century natural history. But naturalists like Clusius approached these collections with a different sensibility than collectors. For the latter, the collection was a treasure to preserve and display; for the former, it was a tool to dismember and describe. Collectors fashioned themselves as *virtuosi* by gathering and showing rare and precious items; naturalists fashioned knowledge by studying the productions of nature, both exotic and humble, rare and common.

Conrad Gessner recognized the difference when he classified and catalogued gardens. Some were cultivated for fruits and vegetables, others for medicinal plants. Some were planted "only for ornament and elegance," with contrasting colors, labyrinths, fountains, and topiary. Some served to show off the owner's wealth. And some were planted with a wide range of plants, including rarities, "for the admiration and contemplation of nature."[93] As Gessner recognized, some Renaissance gardeners (or garden owners) delighted in the world of appearances, in the bright colors and varied forms of plants or in the projection of their own wealth and standing that the garden embodied. But others approached the garden intellectually. They did so with pleasure—natural history was not a profession, after all—but their pleasure had a strong whiff of the lamp.

Herbaria are another case in point. The herbarium has clear similarities to cabinets of curiosity, and indeed a collection of dried plants was de rigueur for Baroque virtuosi.[94] But there were important differences, at least for naturalists. Most importantly, the herbarium was a means to knowledge, not an end in itself. And its proponents saw it as the second-best solution to a problem. The name "winter garden" points to one common attitude: plants were best seen at first hand and alive, but since that was impossible in winter, dried plants were better than nothing.[95] Clusius, who collected and dried plants on his travels through Spain, was increasingly pessimistic about their usefulness in descriptive botany.[96] Though some herbarium collectors could be quite possessive, others were less so. Leonhard Rauwolf, after describing the plants he had collected on his travels in the Near East, was willing to offer his herbarium to a generous prince.[97] Felix Platter and Jacob Zwinger regularly lent pages of their collection to Caspar Bauhin,[98] and Platter even arranged his according to Bauhin's *Phytopinax*, for which it probably served as an important reference.[99] For dedicated naturalists, the herbarium served to extend experience and was not an end in itself.

The same distinction between collecting as a means and as an end can be seen in other aspects of early modern museum culture. Antiquarians, for example, often relied upon the collections of their patrons or friends.[100] Jacob Spon, one of the most famous seventeenth-century antiquarians, shocked his contemporaries by selling off the ancient coins he had gathered in his travels. In justification, he claimed, "It's by the large number of medals that pass through your hands that you become skilled" in antiquities: what is important for the scholar is knowing, not possessing.[101] Knowledge itself was, of course, a sort of possession, more enduring than mere physical ownership. And the scholar's collection, fixed in the mind, could extend far beyond the drawers of a cabinet or the walls of a study.

Of course, few naturalists were not bitten by the collecting bug, and most collectors wanted to know something about their curiosities. Usually naturalists were located somewhere between the two poles. Caspar Bauhin used his herbarium above all a research tool, but in his *Pinax* he advertised that it was open to visitors.[102] University botanical gardens were also planted with decorative flowers. They served the communities in which they were located by providing room for promenades, conversations, and courting, as well as for study.[103] Nonetheless, collectors and gardeners on the one hand, and scholarly naturalists on the other, were driven by different impulses. A small group of scholars produced knowledge; this group included not only those who wrote or edited books but also the correspondents and conversation partners who provided them with descriptions and material. A much larger group of collectors consumed this knowledge so that they could reproduce it on demand when showing off their collections.

This dialectic between producing and consuming knowledge created a market for books on natural history in the late sixteenth and early seventeenth century. Cabinets of curiosity invariably included bookcases for descriptive literature (and a few herbarium volumes).[104] And beginning in the 1550s, publishers issued natural history books whose practical value was far less clear than that of the herbals. The works of Conrad Gessner and Carolus Clusius are exemplary. Gessner's own *Historia animalium,* already mentioned several times, was an enormous compendium of lore on every animal he knew, firsthand or otherwise. For most animals, Gessner included a picture, a learned account of its name in several languages, a physical description, an account of its habits, the medical use of its parts, its use in food, and a large section he called philology that summarized references to the animal in history, literature, and art.[105] Though the medical sections were a clear bow to physicians, the work was far too bulky and erudite to serve as a practical reference. Moreover, it was quickly translated into German, without much of the philological material that appealed primarily to a humanist audience capable of reading the original Latin. The bulky work was reprinted several times and remained a zoological reference for the next two hundred years.

Clusius's histories of "uncommon plants" do not impose upon the reader like Gessner's folios. They were thick octavos, capable of being held before the reader or slipped in a travel pouch.[106] But they are equally distant from the herbals. Clusius made no pretensions to comprehensiveness; he specifically excluded descriptions of plants that were already widely known.[107] He also made few bows to medicine; though trained in medicine at Montpellier, Clusius never practiced and may not have taken a degree. Even as an old man he scrupulously corrected correspondents who addressed him as a physician.[108] In his *Rariorum aliquot stirpium per Hispanias observatarum Historia* (*History of some uncommon plants observed in Spain*), published in 1576, Clusius referred occasionally to plants' medicinal properties, but he was often content to note that a plant had no known medical use. Even these brief references vanished by 1583, when he published his *Rariorum aliquot stirpium, per Pannoniam, Austriam, & vivinas quasdam provincias observatarum historia* (*History of some uncommon plants observed in Austria, Pannonia, and their environs*). In the place of detailed notes on qualities and virtues, Clusius included detailed accounts of plants' growth and form, their native locales, the circumstances under which he had observed them, and often his success or failure at cultivating them. Clusius was not alone in publishing such works; after his more traditional herbal, Rembert Dodoens turned to a treatise on decorative flowers.[109] Mathieu de l'Obel, another Netherlandish botanist, published similar works describing plants he had seen but making no pretence to comprehensiveness or medical usefulness.[110]

Properly marketed, these books sold to a wide audience. Christoffel Plantin, the Antwerp printer, was Clusius's publisher from 1567 to Plantin's death; he also published the later works of Dodoens and De l'Obel.[111]

Clusius, who was especially attentive to the quality of print and illustration in his books, chose Plantin for his care in printing natural history books.[112] Plantin, a shrewd businessman, accepted Clusius's books because he thought they would sell.[113] And by building up a list in natural history, Plantin could simplify his marketing and economize on production by sharing woodcuts among books by different naturalists.[114]

The importance of marketing natural history books is underscored by a publishing disaster of the 1570s. In 1571, Mathieu de l'Obel and Pierre Pena published their *Stirpium adversaria nova* in London, probably on their own account.[115] For whatever reason, De l'Obel ended up with most of the press run of 3,000 in his own hands. In 1576, he managed to sell 800 copies to Plantin, his new publisher; Plantin, in turn, bound them with De l'Obel's new *Plantarum seu stirpium historia* (1576), apparently hoping that demand for the new work would help sell the old. But De l'Obel still had most of the remainder. As late as 1603, he complained to Clusius that he still had 2,050 unsold copies in his possession: if the initial press run was 3,000 (an enormous number), that meant that De l'Obel had managed to sell only 150 copies of the work, apart from those sold after 1576 by Plantin.[116] And De l'Obel must have given away many copies, though presumably he anticipated receiving gifts or favors in return: Clusius, for example, owned a copy of the first edition, probably a gift from the author.[117]

There was no professional natural history in the sixteenth century—though the profession of "simples," or medicinal plants, had emerged in Italy in connection with medical gardens and lectureships.[118] But the number of serious naturalists, those actively involved in the pursuit of knowledge, was probably a few hundred—somewhere between the one hundred fifty who bought or received copies of De l'Obel's first book and the eight hundred or so purchasers needed to exhaust a typical early modern book edition. Few natural history books published between 1560 and 1590 went into multiple editions, unlike the earlier herbals of Bock, Lonizer, and Mattioli, which were frequently reprinted. Mattioli claimed that thirty thousand copies of his herbal had already been sold by 1568.[119] His and other herbals clearly appealed to a broad audience in need of medical references and textbooks. Mattioli's book was frequently required reading for students in *materia medica* courses; a condensed version for students was published by Joachim Camerarius in 1586, while Caspar Bauhin produced a critically annotated version in 1598, testimony to its popularity and Bauhin's desire to correct its errors.[120] The more specialized works by Gessner, Clusius, De l'Obel, and their contemporaries appealed to a different audience: those eager to read descriptions of new plants, regardless of their medical use.

A large library might contain dozens of these books. Joannes Sambucus (1531–84), a Hungarian physician and humanist resident in Vienna, owned a hundred printed books on natural history (in a library of over three thousand volumes).[121] In Sambucus's library, Dioscorides, Aristotle, Theophrastus, Pliny, and a handful of other ancients rubbed shoulders with Fuchs, both Corduses, Gessner, Clusius, Dodoens, De l'Obel, and other moderns such as Apollonius Menabeni, author of a study of the "great animal" (the elk), and Paolo Giovio, who wrote a book on fishes as well as the histories and biographies for which he is best known.

The second half of the sixteenth century was the heyday of Renaissance natural history. No longer limited to a small group of physicians, apothecaries, medical students, and humanists, the community of naturalists included collectors great and small, served by gardeners, cabinetmakers, illustrators, and other artisans who did much of the hard work that went into building and maintaining a collection. But with the very growth of natural history came differentiation and divergence within the community. A small group of scholars wanted to discover and describe novelties. Collectors, growing in number, sought to amass and display rarities. Brought together after 1550 when *naturalia* had first come into vogue, when scholars were the key to understanding what went into a collection, the two groups would grow increasingly separate by the end of the century, when the detailed scholarly work of naturalists no longer seemed relevant to a new generation of collectors and gardeners.

Systematizers, 1590–1620

By the last decade of the 1590s, scholarly naturalists' circles of correspondence were progressively limited to their peers. Of the forty-four correspondents whom Caspar Bauhin thanked for contributions to his *Phytopinax* (1596), four in five were physicians, medical students, or apothecaries. Two of the remaining eight were professional gardeners, one was a doctor of philosophy, three were magistri, one was a noble, and one was a patrician.[122] As in the previous generation, this community included many scholars who never published a word themselves: of the physicians listed most never published a book, certainly not one in natural history, while of the six apothecaries, only one (Ferrante Imperato) did so. But compared to Gessner's contributors half a century earlier, Bauhin's circle of correspondents appears narrow. Naturalists from the older generation, such as Carolus Clusius, continued to cultivate a wide variety of correspondents; in his last years, spent at the University of Leiden, Clusius corresponded with Matteo

Caccini, a Florentine horticulturist, and the young Nicolas-Claude Fabri de Peiresc.[123] But increasingly, specialized naturalists like Bauhin carried on their work in a community of like-minded scholars, while those interested in collecting and gardening relied on the market in books and *naturalia.*

This process continued into the seventeenth century. By that time, the market for natural history books was large enough, and differentiated enough, for the emergence of a new genre: the florilegium. Florilegia included detailed copperplate engravings or etchings of flowers, often from particular gardens, such as Basilius Besler's *Hortus Eystattensis* (1613), which depicted the episcopal gardens in Eichstätt, or Crispijn van de Passe's *Hortus floridus* (1614).[124] The latter was the first illustrated bulb catalogue: the flower fancier could order plants from it for his or her garden.[125] These were only the first of dozens of picture-books catering to the *amateurs de fleurs* in Baroque Europe.[126] The engraved pictures in florilegia could show much more detail than the woodcuts used in earlier botanical publications, but there was a cost: woodcuts could be printed on the same page as text in one run through the press, but engravings had to be printed separately. Considering that the audience for the books comprised flower-lovers, not scholars, it is not surprising that the text often atrophied.[127] In his *Paradisus terrestris* (1629), John Parkinson drew on the Latin and English scholarly botanical works for his garden book, but he explicitly eliminated scholarly controversies or careful accounts of precise places and times. His readers wanted practical information about plants that could and should be grown in English gardens.[128] Florilegia were written neither by nor for scholars, while scholars no longer wrote for collectors.

Instead, the leading naturalists of the early seventeenth century produced more specialized, detailed studies of plants and their geography. The local floras of the early seventeenth century make a sober contrast to the richly illustrated florilegia: they were lists of plants, sometimes with brief descriptions, aimed at university students. The earliest included Ludwig Jungerman's catalogue of plants growing around Altdorf, and Caspar Bauhin's catalogue of plants found within a mile of Basel.[129] Florilegia were large-format books, quartos or folios: these duodecimo catalogues, on the other hand, could be slipped into a student's pocket, and they were interleaved with blank paper for note taking. These two genres instantiate the increasing differentiation between natural history as a means of investigating the world and collecting as a type of self-fashioning in the early seventeenth century.[130]

At the same time, scholarly naturalists emphasized the demands of their studies. By 1600, natural history was hard work. As Clusius admonished his friend Camerarius, "The study of botany demands almost total devotion."[131] Camerarius was a physician, but that profession could take valuable time away from the study of plants. When natural history demanded its practitioners' full attention, the *studium* had lost much of its ludic quality. Collectors might turn to their herbaria, antlers, fossils and stones, shells, and other *naturalia* as a pastime—albeit often a serious one—but the scholar was becoming a drudge.

Drudgery may characterize all scholarship, but the drudgery of early seventeenth-century natural history emerged from specific developments in the previous generation. The Lyon herbal of 1587–88 had described more than two thousand plants, and though its publisher claimed that it was comprehensive, critics pointed to many errors and omissions.[132] Caspar Bauhin, one of those critics, devoted more than thirty years of his life, from the preparatory work for his *Phytopinax* (1596) through his *Pinax Theatri botanici* (1623), to the immense task of combing through the botanical publications of the sixteenth and early seventeenth century, comparing them with his herbarium, and pronouncing definitively on the synonyms and descriptions of every species of plant known to European naturalists.[133] Clusius's publications, a generation earlier, sparkled with their author's joy at tramping through the woods and fields, watching plants, plucking some to take home and transplant and leaving others to observe the following season. Bauhin's descriptions, on the other hand, were precise, concise, and lifeless.[134] They were the work of a man for whom the individual plant was important primarily as an element in a greater system, for whom the description was a preliminary to classification.[135] Natural history would continue to be a descriptive science, but as taxonomy and systematics came to the center of naturalists' attention, it was no longer simply a science of describing.

Scholars have occasionally noted a decline in natural history in the middle of the seventeenth century.[136] The political and military turmoil of the times might have contributed to this: the Thirty Years War disrupted travel and communications even in those areas that were not devastated by fighting, while the Civil War and Interregnum affected English naturalists. But other parts of Europe remained relatively free of such disruption. It is perhaps more precise to say that the community of descriptive naturalists, by isolating itself from horticulturists and collectors, could no longer draw on their resources and vitality. Other subjects, such as animal and insect anatomy, were vigorously pursued by scholars in the middle of the seventeenth century; such subjects would be integrated with descriptive and taxonomic natural history by the century's end.[137] Horticulture, both amateur and commercial, went from strength to strength—even in places like the fledgling Dutch Republic, at war from 1568 to 1648 with Habsburg Spain, save a twelve-year truce from 1609 to 1621. Descriptive natural history itself continued, albeit without the vigor of the sixteenth and early seventeenth century; it would be revived by the last decades of the century by scholars like Tournefort and Ray, who would turn once more to the now-classic works of the Renaissance phytographers and systematists.

The Preeminence of Botany

My periodization has focused on botany, the chief focus of natural history in the Renaissance.[138] Renaissance natural history emerged from *materia medica,* though it quickly went beyond physicians' and apothecaries' concerns, and it swiftly made an alliance with horticulture. Plants cannot hide from the naturalist, and they can be grown in gardens and dried in herbaria. But Renaissance naturalists also studied the animal kingdom. In research on Renaissance zoology, Laurent Pinon has identified four periods that closely resemble my own generational analysis, while showing that developments in zoology lagged behind those in the more vigorous realm of botany.

Pinon's first period, "the period of the ancients" (1450–1520), saw printed editions of ancient authors, their format based on medieval manuscripts. These initial efforts were followed by "the period of correspondences" (c. 1520–50), in which publications attempted to establish correspondences between ancient and modern names, at the same time adopting formats that deviated from the medieval manuscript model. The third, brief period focused on "the registration of nature in images" (1550–60). This period, which saw the publication of Conrad Gessner's magisterial *Historia animalium* and other encyclopedias, marked the end of great zoological syntheses. "After 1560, after that extraordinary group of original publications, no further great synthesis of zoology appeared," though monographs on specific subjects continued to issue from the press. Finally, the "period of explorations" beginning around 1610 saw a turn away from late Renaissance encyclopedism toward anatomical and theoretical reflections.[139]

Pinon's stages of zoology resemble those I have found in Renaissance botany—not surprisingly, for many scholars were active in both areas, though far more attention was paid to botany than zoology. As Pinon notes, zoological publications were somewhat behind botanical publications in the drive to provide a pictorial record of nature.[140] In fact, zoology generally lagged behind botany in the Renaissance; developments that occurred in botany often took a decade or two to appear in zoology. Though important editions and translations of ancient botanical authors appeared as late as Ruel's 1516 Dioscorides, students of plants had been debating the correspondences between ancient texts and modern plants since the dispute over Pliny in the 1490s. By the 1530s, Valerius Cordus, Hieronymus Bock, and Leonhart Fuchs were instead attempting to catalogue nature as they saw it. The following generation intensified the effort, continuing to produce new encyclopedic works such as the Lyon herbal (1587–88). By the 1590s, so many new plants had been described that writers across Europe were calling for systems to organize them; Adam Zaluziansky's *Methodus herbariae* (1592) and Caspar Bauhin's *Phytopinax* (1595) were two responses. Zoological investigations "overtook" botany—from our perspective—only at the end of the period, when according to Pinon theoretical and physiological investigations came to the forefront.

In short, the parallels between Pinon's and my periodization suggest both that we have—independently—hit on important transformations in Renaissance natural history, and that botanical studies were the main focus of Renaissance naturalists. From the task of collating ancient descriptions with modern plants, a task firmly grounded in humanistic medicine, to the largely autonomous cataloguing, collecting, and systematizing of the late sixteenth and early seventeenth centuries, botany was at the forefront of Renaissance natural history. In the beginning, this interest in botany was driven by medical concerns; animal parts had their place in the pharmacopoeia, but it was mostly vegetable. Moreover, there were far more vascular plants to be identified, described, and catalogued than vertebrate animals.

And the invertebrates scarcely interested Renaissance zoologists until the late sixteenth century. In the middle of the century, Thomas Penny began to collect material for the history of "insects"—in modern terms, terrestrial arthropods and, at least sometimes, annelids and other phyla. But his work was not published until Thomas Moffett's posthumous *Theater of Insects* (1634).[141] Moffett had acquired Penny's notes, along with notes by Edward Wotton and Conrad Gessner, and arranged them for publication, but he himself died in 1604 without having found a publisher.[142] Jacob Zwinger had been collecting material for a history of insects, but he died of plague in 1610.[143] Ulisse Aldrovandi's *De animalibus insectis* (1602) and Moffett's posthumous work, meantime, were the only significant histories of "insects" before Jan Swammerdam's 1669 *Historia insectorum generalis.*[144] As these abortive developments show, the one group of easily available creatures that rivaled plants in diversity of number and morphology was of little interest to most Renaissance naturalists. The situation would change only with the renewed interest in the seventeenth century in revisiting ancient physiology, when the study of animals—if not their description and classification—gained strength within natural history.[145] By then, the science of describing was giving way to classification and explanation.

A Collective Enterprise

By the 1550s, naturalists agreed that their discipline was fundamentally a collective enterprise. Of course, earlier naturalists had conceded that no one could know everything. In his *Botanologicon* (1534), Euricius Cordus had proclaimed that he was "ready at any moment, and without a blush, to be taught by any of you who may know better than I."[146] His scholarly humility, whether honest or feigned, was hardly anything new. But in the preface to his *Cruydeboeck,* Rembert Dodoens moved beyond Cordus's willingness to have any particular belief of his corrected. Dodoens pointed out the difficulties of this science even for Dioscorides, who had erred, let alone for moderns who had to contend with the confusion of words as well as the variety of things.

> The magnitude or difficulty of this science is such that it cannot be comprehended without early and careful examination of all plants and exact reading of many ancient writers—that is, without great labor, long travel, and continuous devotion. Furthermore, it is scarcely possible that the life and diligence of one or a few men could be equal to the task. Hence it should not be surprising that, despite the number of moderns who have diligently investigated this subject, others come along who desire to increase this science, and that I too publish my history of plants.
>
> For no one has brought this science to perfection: everyone has omitted many things, thereby leaving to posterity the opportunity to add much to the discoveries and observations of their predecessors and to increase the knowledge of plants.[147]

Dodoens presented the *res herbaria* as not only incomplete but, more importantly, incapable of completion. It was too great for one or a few people to master: there had been and there always would be an opportunity to contribute to it. Dodoens's preface must be read in light of the tradition of the *laudatio artis,* in which the difficulty of the subject is presented as an element of its dignity; it is far from a transparent expression of his beliefs.[148] But his particular formulation implies more than the difficulty of the subject: it presupposes its inexhaustibility. The world of plants had burst the borders imposed on it by ancient and medieval *materia medica*.

This was not an idiosyncratic view. Two decades later, Carolus Clusius wrote in the dedication to his work on Spanish plants,

> So many works of learned men who have contributed to botany exist today that they seem to cover the entire history of plants: nonetheless their variety is so immense (I could almost say infinite) that I believe that diligent observation and study can always add something new to it and provide more complete knowledge of them.[149]

Clusius's work bore out the claim: he described about two hundred previously unknown plants.[150] Considering that a naturalist such as Leoniceno or Euricius Cordus knew perhaps a total of 500 species, one can see the magnitude of the novelties Clusius described in this book. A few years later, Andrea Cesalpino remarked that "every day" some new plant was brought to the botanical garden in Pisa, "just like the proverb, 'Africa always bring something new'—not that nature produces new forms or brings forth new beauties, but that because of the immense number something new is manifested every day."[151] Ironically enough, Cesalpino published that remark the same year that Clusius brought out his second history of plants, describing yet more new species.[152] Naturalists in following generations discovered even more plant species, demonstrating *post hoc* the validity of Dodoens's and Clusius's claims.[153] In the early seventeenth century, Adriaan van de Spiegel repeated what was, by then, a commonplace:

> No human mind, however hard-working, will ever come to a wholly perfect knowledge of plants, for their variety is infinite in form, use, and other accidents. We see that rivers, swamps, the sea, mountains, fields valleys, sand, walls, stones, meadows, vineyards, woods, and uncultivated places—to sum up, Europe, Asia, Africa, and the Indies—always produce something new.[154]

Hence, by the mid-sixteenth century naturalists saw collaboration not as a social fact but as an intellectual necessity. But this necessity resulted itself from earlier collaborative efforts: for it was only by comparing the plants known to one author with those known to another that the range of variation in the plant world was firmly established. Botany, which Leoniceno and Cordus had seen as correctly identifying plants known to the ancients, had become an open-ended activity that depended on a community. And this community, in turn, provided the individual naturalist with the audience and the justification for his own naturalizing.

Collaboration can take many forms. Some collaboration is inescapable in intellectual life—discussing one another's works, translating and editing publications, and co-writing books. Some seems a natural consequence of a particular discipline—for example, naturalists' joint excursions into the field. But other forms of collaboration have deeper intellectual grounds. Nineteenth- and twentieth-century scientists, historians, and philosophers promoted collaboration as a guarantor of dispassionate objectivity: an individual's passions and errors would be compensated by those of fellow practitioners, resulting in knowledge that was objective because consensual.[155] Still other nineteenth-century scholars—particularly in observational sciences like astronomy—promoted an "mechanical" objectivity, in which an individual observer's reports would be standardized so that they could be employed in collaborative research, such as star mapping, that required coordinating many hands and eyes.[156]

Though Renaissance naturalists would not have put it this way, their collaboration rested on a tacit sense of such instrumental objectivity. They collaborated because they were aware that they had to identify and describe an enormous, swiftly-growing number of plant and animal species; no one naturalist, no matter how indefatigable, could suffice. The catalogue of nature would be the work of many hands. And for those hands to produce a useful product, their work had to be standardized. To be useful, descriptions had to use a common language and follow a common format. Otherwise it would be difficult, if not impossible, to forge ahead in the collective enterprise.

For descriptions to be standardized, in turn, naturalists had to be disciplined. This disciplining had three parts. Naturalists had to imagine themselves as members of an ideal community, characterized by a sense of devotion to common studies. Second, they had to be socialized into a particular local community, for they could master the methods and assumptions of the discipline only through example and precept. Third, at least some members of the local community needed to participate in the international community of naturalists, that province of the Republic of Letters that was devoted to the science of describing. In what follows, we will examine these three moments of collective activity: what made a naturalist, how local communities were sustained, and how the international community formed and retained its coherence.

What Made a Naturalist?

To become a naturalist, a scholar had to adopt an ethos comprising certain norms that were accepted by other members of the community.[157] The naturalist's ethos had several components. Naturalists were devoted to common studies. They were familiar with the established classics of the discipline, both ancient and contemporary. They participated in networks of correspondence and exchange. They could read and write Latin. And most of all, they had mastered an evolving set of dispositions and techniques.

The term "naturalist" itself is not contemporary, but it maps closely onto contemporary categories.[158] A common Renaissance term is "studiosus rei botanicae" or "rerum naturae": literally, someone devoted to botany or to nature.[159] Carolus Clusius addressed his history of Austrian and Pannonian plants to the "studiosis rei botanicae," while in his *Exoticorum libri decem,* he hoped to provide an example to youth interested in the "studium" of exotic things.[160] The collector Bernardus Paludanus described his interests as the "studium Rerum naturalium," while Jacob Cargill, who was interested more particularly in plants, reported to his teacher Caspar Bauhin on the "studia botanica" in London.[161] Perhaps striving for variety, one writer referred to "phytologicum studium."[162] Writers occasionally referred to natural history as a species of natural philosophy—Aldrovandi called it "the study of the natural philosophy of sensible things," and Adriaan van de Spiegel also insisted that it was a kind of philosophy—but this was uncommon.[163] Such claims to intellectual status would have been rejected by more careful thinkers like Conrad Gessner, himself trained in natural and moral philosophy as well as natural history and medicine, while truly philosophical treatments of natural history, like Andrea Cesalpino's *De plantis libri XVI,* were rare.

The noun "studium" and the adjective "studiosus" do not correspond exactly to the English "study" and "student." For humanists, schooled in Ciceronian Latin, the word "studium" still carried its original connotation of devotion. Writers signed letters "tui studiosissimus"—a Renaissance equivalent of "yours devotedly."[164] Ole Worm located the "studium botanicum" in his heart, not his head.[165] We have already seen how Angelo Busti used the language of love in his letter of 1609 to Caspar Bauhin. Renaissance naturalists, then, were passionate for natural things and devoted to learning about them. This is especially important because natural history was neither a profession, like medicine, nor a trade, like gardening.[166] Investigating, collecting, and describing nature's productions required a great deal of painstaking, tedious work. There was little hope of material gain in this study, though it offered some possibilities for social advancement. Hence passion was a primary motivating force—passion for nature and for the social relations of the Republic of Letters.

Passion and knowledge alone, however, did not make a naturalist. Many who had intimate knowledge of the natural world were not members of the same social network to which Gessner, Clusius, Bauhin, and other naturalists belonged. Apothecaries like Giovanni Pona participated in this community, but many others did not. Neither did midwives, country wise women and men, foresters, and others who possessed detailed local knowledge of local plants and animals. Or, to be more precise, they did not participate directly in it. The pages of Renaissance natural histories include numerous references, both direct and indirect, to such sources of local knowledge. But these men and women rarely participated as equals in the communications of the *studiosi rerum naturalium,* and their contributions were increasingly effaced over the course of the sixteenth century.

This was not a process of conscious exclusion. Rather, naturalists saw themselves as engaged in a tradition of inquiry that had flourished in antiquity and had dwindled to almost nothing in the Middle Ages, as Busti claimed in his letter to Bauhin. This history, I will argue in chapter 3, is largely imaginary; Renaissance naturalists constructed it out of disparate traditions in natural philosophy, medicine, and agriculture. But imagined traditions evoke the strongest allegiance, and knowledge of the canon of natural history was an important marker of the naturalist.[167] Apothecaries like Pona, Imperato, Coudenbergh, and Cluyt, with at least a smattering of Latin and social ties with other naturalists, fit in. Other apothecaries, and traditional healers, did not. Renaissance natural history owed a great deal to local informants, but they did not take part in learned naturalists' give and take.[168]

Indeed, naturalists defined themselves—or, more precisely, their compeers—by participation in networks of correspondence and exchange. Hence they excluded not only root-cutters, wise women, and many apothecaries, but also many contemporary writers of natural history books. Spanish and Portuguese writers in particular did not often participate in the dense networks of correspondence and exchange that characterized natural history in Italy, the Holy Roman Empire, France, and England.[169] The works of Oviedo, Acosta, Garcia da Orta, Monardes, and others were read, translated, annotated, and excerpted by naturalists outside of Iberia and her colonial possessions, but their authors were not full members of the natural history community.[170] Though their status as physicians or scholars accorded them more credibility to the *studiosi rerum naturalium* than the claims of peasant healers, their claims were subject to closer analysis than those of naturalists who were known personally to their contemporaries.

At the periphery of this hard core of naturalists—those whose words counted—were the yeoman laborers, whose contributions could be used but who were not competent to pronounce definitively on them. This group included sailors, farmers, fishermen, and even merchants. Adriaan van de Spiegel was willing to learn from peasants, and Conrad Gessner carefully noted sailors' tales of sea monsters.[171] Clusius noted carefully the information that sailors to the Indies provided on the provenance of the strange nuts and plants they carried back with them. Closer to home, he acknowledged that experienced woodsmen easily distinguished similar species of oak.[172] But such data were subject to careful scrutiny, both for plausibility and for the general reliability of the source. Spiegel, for instance, refused to believe folk reports on the medicinal virtues of plants until confirmed by several sources.[173] Experience alone did not guarantee knowledge; it had to be situated in a tradition.

Full access to that tradition, in turn, was denied to those who could not read Latin, the *lingua franca* of sixteenth-century scholarship.[174] Latin was far from the only language used by naturalists. In his *Phytopinax*, Bauhin cited works not only in Latin (though these comprised the vast majority) but also Greek, French, German, Italian, and Spanish.[175] Clusius corresponded in Latin, Italian, French, German, Dutch, and Spanish.[176] Nonetheless, Latin possessed advantages for an international audience: it was far easier to learn one foreign language than several, and it ensured that scholars from different areas possessed at least one language in common.[177] Felix Platter spoke Latin with some of his traveling companions on his way from Basel to Montpellier to study medicine, as well as with his apothecary host once he had arrived (until he learned French).[178] Even though Clusius could write Italian, he corresponded with Giovanni Vincenti Pinelli in Latin.[179] Furthermore, from the late fifteenth century, reading knowledge of Latin was expected of apothecaries.[180] Humanist educators' emphasis on Latin grammar and composition ensured that bourgeois and noble children had a fair command of the language; German humanists, for instance, switched easily between German and Latin in their correspondence, preferring the latter when writing to other humanists.[181] The brisk business in translations of natural history books indicates the existence of a market for vernacular texts, but many books were also translated from the vernacular into Latin in order to reach an international audience.[182]

Fluent in Latin, familiar with the classics of the field, and familiar too with fellow students of nature, Renaissance naturalists usually came from one of two backgrounds: aristocrats or members of the urban elite with the leisure to study nature, or physicians and apothecaries for whom such study was related, if sometimes tenuously, to their profession. Consequently, the community of naturalists was more homogeneous than Renaissance society. Clusius, who was particularly attentive to social distinctions, chose honorific epithets that suited the status of his correspondents. Apothecaries were at the bottom of the social hierarchy of natural history: Clusius usually used "honestissimus vir" to refer to apothecaries (or, less rarely, the positive degree "honestus"), though he frequently used "ornatissimus."[183] He usually identified physicians and other academics with the more honorable "Clarissimus Vir," or "Cl. V." And he clearly set off nobles, "viri magnificissimi," from the rest. Clusius was horrified to learn, late in life, that he had been addressing the Count of Arembergh only as a count, when he was in fact also a prince in his own right.[184] And Clusius's former landlady, the widow of his friend Dr. Aicholz, addressed him circumspectly as "der herr," presumably out of respect for his noble title, though he had lived in her house for years.[185] In the late sixteenth century, when social distinctions were becoming more marked, personal interactions must have been even more tinged by status considerations.[186]

Nonetheless, the *studium commune* of natural history created an ideal social space in which social distinctions could be suppressed, at least temporarily.[187] Clusius, for example, was born to the lesser nobility and became, on his father's death, Seigneur de Watènes. He insisted on behaving according to this status; when appointed by the emperor Maximilian II to establish a medical garden in Vienna, he insisted that he be named a courtier rather than a gardener, which carried unpleasant connotations of paid physical labor.[188] But Clusius was on familiar terms with apothecaries and printers, who were definitely below him, going so far as to dedicate one of his publications to the Antwerp apothecary Pieter Coudenbergh.[189] Late in life he counted as intimate friends both Justus Lipsius, like Clusius a passionate gardener but distinguished by learning not birth, and Joseph Justus Scaliger, who dressed and acted nobly so as to maintain his father's fictitious connection to the La Scala family of Verona—a connection that his friends accepted only out of politeness.[190] In the world of learning and natural history, achievement counted for as much as birth.

The young Basel medical student Jacob Zwinger (1569–1610) displayed a similar indifference to social distinctions between naturalists. In 1592, Zwinger studied medicine in Padua and traveled through Italy to meet other naturalists. He visited both the Bolognese patrician and physician Ulisse Aldrovandi and the Neapolitan apothecary Ferrante Imperato, mentioning them, in the same breath and without any notion of different social status, in a letter to his teacher (and later colleague) in Basel, Caspar Bauhin. Zwinger praised the *humanitas* of both scholars; the only difference he mentioned was that Imperato was easier to get along with and more generous with gifts.[191] Certainly Imperato was no ordinary apothecary—but it was precisely due to his interest in natural history that he stood out, and it was that interest which brought the famous and learned of Europe to see his collection and discuss natural history with him.[192]

Naturalists did not simply ignore social distinctions for politeness' sake; as we will see, they imagined their community in opposition to both the hierarchical society of the court and the increasingly commercial society of the town and marketplace. Natural history, as its practitioners insisted, was a liberal art: within the bounds of their community, naturalists were free and equal, unsullied by either servitude or filthy lucre. Such presumptions formed an essential part of the affective bonds that sustained the international Republic of Letters in early modern Europe. Before examining the international community of naturalists as it worked in practice, however, we must linger on the local communities, for it was there that naturalists learned the ropes and became qualified, in their peers' eyes, to contribute to the public good.

The Geography of Natural History

The larger community of Renaissance natural history was dispersed throughout western and central Europe. Through the early seventeenth century, naturalists were concentrated in Italy, France, southern England, the Low Countries, the Holy Roman Empire and Bohemia, and, more tenuously, the Venetian possessions in Crete.[193] Naturalists occasionally traveled beyond these regions; Leonhard Rauwolf and Prospero Alpino described their travels in Egypt and the Levant, while Carolus Clusius made an important trip through the Iberian peninsula in the 1560s.[194] But in quantitative terms, the explosion of knowledge in sixteenth- and early seventeenth-century natural history was due above all to careful examination of the flora and fauna of Europe.

Italy occupied a special place.[195] Italy was the birthplace of Renaissance natural history, and it was home to the most important medical schools of sixteenth century Europe. Its only rivals in medicine were Paris, Montpellier, and, after 1575, the new university in Leiden. Italian universities were the first to establish public medical gardens for teaching *materia medica* and botany.[196] The central figure in sixteenth-century Italian natural history was the Sienese physician Pier Andrea Mattioli, whose translation of Dioscorides, supplemented by Mattioli's own investigations and those of his correspondents, was the bestselling natural history book of the century. Mattioli strove to make himself the arbiter of natural history, praising his friends and attacking—or, worse, neglecting—his enemies in successive editions of *his* Dioscorides. And the group around Mattioli made enemies of foreign naturalists, disdaining them and disparaging their works. From Mattioli's perspective—even after he accepted an appointment as imperial physician and relocated to Prague in the late 1550s—only Italians could be full members of the imagined community of natural history.[197]

Needless to say, northern naturalists disagreed. Leonhart Fuchs, one of Mattioli's favorite targets, gave as good as he got. In letters to the elder Joachim Camerarius, Fuchs insisted that in the revised, three-volume version of his herbal he would demonstrate Mattioli's mendacity; "unless he is stupid, he will have to repent his misrepresentations of my position." Indeed, Fuchs continued, "that trifling Italian will learn that there are men in Germany who recognize his trifles and paint him in his true colors."[198] Perhaps Fuchs feared that Mattioli's success would make printers hesitate to publish his new edition. Conrad Gessner maintained more cordial relations with Mattioli, exchanging both letters and specimens, though it appears that Mattioli became less friendly as his reputation grew.[199] Other northern naturalists, less personally involved with Mattioli, recognized the importance of his books while pointing out errors. As Joachim Camerarius prepared an epitome of Mattioli in the early 1580s, his correspondents warned him about inaccuracies in the woodcut illustrations.[200] Camerarius's epitome, published in 1586, was followed twelve years later by Caspar Bauhin's edition of Mattioli's works, with corrections to the Greek text and "innumerable additions" to Mattioli's commentaries.[201] When Mattioli was safely dead, no longer able to skewer his enemies, his contributions could be weighed and assessed more equitably.

Even during his lifetime, Mattioli was extreme in his disdain for northerners. But his attitude was not unique; it expressed a strain of Italian cultural arrogance in the face of increasing irrelevance. Some Italian scholars, content with the political dominance of the new Italian nobility, drifted into complacency, satisfied with exhausting the accumulated intellectual capital of the previous two centuries.[202] Italian intellectual capital was still immense, but Mattioli's vehemence might have been driven by a suspicion that Italy was no longer as important as it had been. Its period of decline lay in the future, but the trend was already underway.[203] Natural history, like other aspects of the Renaissance, was being domesticated in northern

Europe, and one consequence was a relative decline in the importance of Italy.[204]

But there were empirical reasons that led Italians and northerners to have different concerns. Italians dealt with the same Mediterranean flora that had occupied ancient writers like Pliny, Dioscorides, and Galen; only for that reason could Mattioli publish his natural history as a commentary on Dioscorides. Northern naturalists, on the other hand, had learned by the end of the 1530s that they had to describe a flora that was largely unknown to the ancients. By the second half of the sixteenth century, naturalists had gone beyond a simple opposition between Italian and northern floras, developing instead an increasingly fine-grained sense of species and varieties with, in some instances, very precise ranges. Exchanges between Italians and northerners contributed to this fine-grained sense, for not every Italian was as arrogant as Mattioli, nor every northerner as defensively patriotic as Fuchs. Carolus Clusius never visited Italy but he corresponded, in Italian or Latin, with many Italian naturalists and horticulturalists, from Ulisse Aldrovandi to Giovanni Vincenzo Pinelli. Northerners who had studied in Italy had even more contacts. Jacob Zwinger exchanged letters with famous scholars and collectors, such as Aldrovandi, Cesalpino, and Imperato, and younger naturalists like Edmund Bruzo. Religious differences did not seem to hinder such exchanges, though some scholars had to be careful: when writing to the younger Joachim Camerarius, son of a noted heresiarch of the same name, the keeper of the Vatican metal collection, Michele Mercati, was careful to address Camerarius as "Anastasius Quaestor," presumably to avoid any unpleasantness with inquisitors.[205] Even if Italian and northern naturalists had to deal with different native floras, they remained in contact, if sometimes in tension, throughout the century.

Whether Italians or northerners, naturalists were not evenly dispersed throughout western and central Europe. They were concentrated in particular places, usually for reasons that had little to do with natural history. The distribution of naturalists determined the regions whose flora and fauna became well known, and the location of a particular individual often had a great effect on his participation in the community. To take an especially clear example, the writers on the New World were, due to their geographical separation from Europe, not integrated into the community of naturalists in Europe. As a result, their work made its way into the community only when approved by someone with standing in that community—often Carolus Clusius, who made his reputation at first on his annotated epitome of Garcia da Orta's colloquies on the drugs and simples of the Indies.[206] But the same principle applied to naturalists in Europe: only those regions where members of the community were active or had traveled entered the Renaissance catalogue of nature.[207]

Conrad Gessner's correspondents, for example, were scattered unevenly. Two of his correspondents were from as far east as Kraków and Warsaw; there were several from France, and a number from England as well.[208] The largest number, however, were concentrated in Switzerland, southern Germany, and Italy. Eight were located in Basel, six in Strasbourg, four in Zürich (as well as Gessner himself), three in Meißen, three in Augsburg, and two in Lausanne. The Italian correspondents were found in Milan, Venice and Padua, Florence, Pisa, Bologna, Ferrara, and Rome: the rich cities of the north.[209] This map is enhanced by another of Gessner's works, *De hortis Germaniae* (1561). Despite its title, Gessner also mentioned the most important gardens of Italy and France in the "names of certain men who are devoted to gardening." He began by going down the Rhine. In Chur there were four "who are more skilled in botany than the common run." He listed two in and around Lindau, four in Zürich, four in Basel, three in Strasbourg, two in Frankfurt, and the garden of Pieter Coudenbergh in Antwerp—though the region between the last two was unknown to him. Then, moving to the areas further from the Rhine, he mentioned gardens in Bern, Augsburg, Munich, Schorndorf, Esslingen, Nuremberg, Bratislava, Meißen, Torgau, and Stolberg (near Dresden). "I don't know the gardens of France," he added, "apart from a few that are celebrated in the letters of erudite men": all three of these were in Paris.[210] In Italy, on the other hand, "there are, without a doubt, more and richer gardens than in any other region at this time. For the soil there is the most fertile, and the climate is temperate; furthermore, the men are rich and devoted to pleasures, and medicine is in a good state."[211] As a consequence, he named only a few famous ones, in Padua, Venice, Rome, and Florence.

Gessner's indications can be expanded by other sources, in particular those focusing on the Low Countries and France, areas he did not know well (except for Montpellier). Pieter Coudenbergh was not the only naturalist in the southern Netherlands in the 1550s; as mentioned above, Dodoens, Clusius, and Christoffel Plantin were also there. Coudenbergh and Plantin had urged Clusius to translate a Florentine *Ricettario* into Latin, following on his French translation of Dodoens's *Cruydtboek*.[212] Furthermore, Clusius's close friend and protector Jean de Brancion, of Mechelen, had a garden that Clusius characterized as "omnium instructissimus toto Belgico."[213]

In sum, the centers of natural history activity in the early to mid sixteenth century were in southern Germany and Switzerland, along the Rhine and upper Danube, the southern Netherlands, and then in the rich cities of northern Italy, finally in Paris and Montpellier within France. Moreover, in certain of these cities, in particular Padua, Venice, Florence, Basel, Zürich, Strasbourg, and Paris, there were many *studiosi* who corresponded with other naturalists and sent them objects. Not surprisingly, these parts of Europe were precisely the most urbanized and most rich in the mid sixteenth century.[214] The one exception is Spain; Clusius complained in his Spanish flora about the relatively unexamined plants in the area.[215] Natural history was pursued actively in Habsburg Spain, but above all in Spain's American and Asian colonies, in connection with its commercial and medical value; Spanish naturalists were not engaged in the same enterprise as their contemporaries to the east.[216]

Caspar Bauhin's list of those who contributed seeds for his *Phytopinax* provides a contrasting snapshot from the end of the century. Thirty-three of the forty-four who contributed seeds are identified by location. Of these, four were in Basel, another nine in Germany (two in the north, the others in the south), two in eastern Europe, two in the Low Countries, six in France, and eight in Italy. Only one was in England, and one from outside of Europe: the Venetian Onorio Bello, who lived and worked in Crete.[217] This sketch does not necessarily indicate where the seeds themselves came from, since they often passed through several hands before reaching their final destination.[218] It does, however, indicate where members of the late sixteenth-century community of naturalists were to be found. In comparison with the range of Gessner's correspondents, there is more focus on the Atlantic coast—Bauhin had correspondents in Leiden, Enkhuizen, Hamburg, and Lübeck—and on the south, where material came from Ferrante Imperato in Naples as well as Bello in Crete. The northern areas were growing in economic importance in the course of the sixteenth century, which made correspondence and exchange possible and also provided more opportunities for university-educated professionals to establish themselves.[219]

Clusius's correspondents in the last decade of the sixteenth century and the first decade of the seventeenth reveal a similar geographical range, one that included the core areas in which Gessner's correspondents were active but extended beyond it. Clusius and his landlord Dr. Aicholz had formed the nucleus of a natural history community in Vienna in the 1570s and 1580s, and Clusius continued to correspond with naturalists in Vienna, Prague, Breslau, and other eastern cities from Frankfurt and Leiden. He also wrote frequently to an interrelated group of naturalists in Bordeaux and Poitiers and to others who were based in London. Like Bauhin, Clusius corresponded with Bello in Crete. But the overall map that emerges of European natural history at the beginning of the seventeenth century is similar, if broader, than that of half a century earlier. The great age of world natural history still lay in the future. In the Renaissance, natural history was a product of Europe. To understand its intellectual geography, we need to examine it at two levels: the local community, within which the daily activities of naturalists were located, and the international Republic of Letters.[220]

Natural History in Local Communities

As with other aspects of early modern urban history, local natural history communities varied a great deal from place to place. Three examples, each associated with a different institutional center, convey a sense of this variation, as well as the precariousness of natural history at the local level: Basel, whose naturalists were associated above all with the university; Vienna and Prague, where naturalists were courtiers of the Habsburg emperors; and Bordeaux, where a small community was sustained only by mutual interest and correspondence with other centers of natural history.

Basel, c. 1550–1620

Basel, a rich city located on an important trade route between Italy and the Rhineland, was an important center of natural history in the later sixteenth century. From his vantage point in neighboring Zürich, Conrad Gessner described the situation in the 1550s. Caelio Secundo Curio and Conrad Lycosthenes, both humanist theologians, cultivated some foreign plants in their gardens. Lycosthenes's stepson, the philosopher and physician Theodor Zwinger, had recently returned from Italy and brought with him many seeds from that country. In St.-Johanns-Vorstadt, Johann Jacob Loss cultivated a "magnificent" garden with citrus trees.[221] Gessner also reported some zoological interest in Basel at the same time; the Basel residents Hieronymus Froben and Johann Oporinus, both printers, and the polymathic Hebraist Sebastian Münster all contributed images or descriptions to Gessner's *Historia animalium*.[222]

The botanical interests of Basel naturalists were continued by the next generations. Felix Platter, a contemporary of Basilius Amerbach who studied in Montpellier in the 1550s, returned to his native city after a brief tour of Europe and quickly established himself as a prosperous physician. Platter had begun to collect plants during his student days; by the time he died, in 1614, his herbarium comprised some nineteen volumes.[223] Platter also began to collect and display curiosities, especially *naturalia*. His collection was visited by Michel de Montaigne and Auguste de Thou; after his death, his younger half-brother Thomas inherited it, and the collection remained in the family—gradually decreasing in size and prestige—into the eighteenth century.[224] As rector of the University of Basel, Platter was instrumental in establishing a university medical garden. His younger colleague Caspar Bauhin, born in 1560, was by the mid-1580s teaching botany to medical students in Basel. Platter, Bauhin, Theodor Zwinger's son Jacob, and the elder Zwinger's son-in-law Martin Chmielecius formed a nucleus of naturalists who routinely exchanged books, objects, and information.[225] These Basel naturalists were in frequent contact with Bauhin's elder half-brother Johann, court physician to the duke of Württemberg at Montbéliard and an avid horticulturist and botanist, as well as Johann's Scottish assistant Jacob Cargill, a graduate of the Basel medical school. Though the Basel group's main focus was on plants, both Jacob Zwinger and his friend Johann Heinrich Cherler collected insects and intended to write their natural history until premature death put an end to their plans.[226]

The pole around which Basel natural history revolved was the university. Though the botanical garden and the course in *materia medica* were the only formal institutions for natural history at the university, they served to initiate new generations of naturalists. The university's garden was complemented by professors' private gardens. And the university brought together professors and students whose interests in natural history might otherwise have developed in isolation. Some of those naturalists remained in Basel for their careers, while others returned home, strengthening contacts within the naturalists' Republic of Letters while helping standardize methods and norms of observation and description.

University communities such as Basel, Montpellier, Padua, and Leiden, with their formal institutions to support basic natural history and their positions for one or two naturalists who might be called professionals—a lecturer in *materia medica* and a prefect of the botanical garden—were the most important sites for creating and disseminating Renaissance natural history. But they were not the only ones. Carolus Clusius's career as a naturalist began with his studies at the universities of Wittenberg and Montpellier, and it ended with a largely honorary professorship at the University of Leiden, which he occupied from 1593, when he was already sixty-seven years old, to his death in 1609. He made his name as a naturalist, though, for translations and descriptions that he composed as a tutor to young aristocrats, a courtier to Emperor Maximilian II, and as a private individual living off of rents and the gifts of his friends and patrons. Though the court and the private home had disadvantages, both could both be productive centers for natural history.

The Eastern Habsburg Lands, c. 1560–88

By the 1560s, the center of Viennese intellectual life was not the university but the court. Under the patronage of Maximilian I in the late fifteenth and early sixteenth century, the University of Vienna had shone with humanist luminaries such as Conrad Celtis.[227] But the plague of 1527 and the Turkish threat from 1529, combined with the negative effects of the Reformation on enrollment in most German universities, discouraged any illustrious successors to Celtis's generation. Though Maximilian II supported the university during his imperial reign, the imperial court was where the most important intellectuals gathered.[228] In 1533, Ferdinand I established his royal court in the city and sponsored a massive building program to repair damage done during the siege of 1529 and provide room for court officials and bureaucracy. After the abdication of Charles V, Ferdinand became Holy Roman Emperor, and the number and importance of the officials in Vienna increased.[229] By the reign of Maximilian II, the *Altstadt* within the protective walls was occupied largely by court officials, while the unprotected suburbs were the domain of trade.[230]

This was the milieu that Carolus Clusius entered in 1573, when he accepted the emperor's summons to establish an imperial medical garden in Vienna.[231] Under Maximilian, a significant number of scholars were associated with the court—among others, Johann Crato von Kraftheim, Wolfgang Lazius, Hugo Blotius, Rembert Dodoens, and Jacopo Strada.[232] Crato and Dodoens were avid naturalists. Strada, the Emperor's architect and antiquary, maintained an elegant residence that served as an informal meeting-place for the court's scholars.[233] Clusius took up residence with Dr. Johann Aicholz, a member of the university medical faculty and an avid horticulturist who accompanied Clusius on many herborizing expeditions in the neighboring regions.[234]

Clusius's main activities from 1574 to Maximilian's death in 1576 were centered around the imperial medical garden and his explorations of the native Austrian flora. The garden kept him in Vienna during the emperor's absences, so that his interactions with other courtly naturalists were interrupted from time to time.[235] Attendance at court was sometimes onerous, robbing him of time he would have preferred to spend in his studies or excursions, while the costs merely of dressing his part as a courtier consumed most of his income from his property of Watènes.[236] Nonetheless, Clusius was relatively secure during the emperor's life. But Maximilian's successor Rudolf had a different view of natural history than Clusius. In 1577, Rudolf had the imperial garden in Vienna razed, and Clusius lost his position at the court, which had relocated to Prague, through the machinations of the Hofmarschalk, Adam von Dietrichstein.[237] Clusius remained in Vienna through 1588, continuing his excursions with Aicholz, corresponding with and visiting his Hungarian friends, and redacting his history of Austrian and Hungarian plants. Though the imperial garden no longer existed, he maintained a small private garden for his own plants, which he exchanged with other Viennese horticulturists, or occasionally purchased from them.[238]

The new emperor, however, continued to gather naturalists at and around his court. Gardens were popular in imperial Prague, and both Rudolf and his brother, the archduke Ferdinand, laid out spectacular pleasure and display gardens, with rare plants and menageries for exotic animals. The botanist Adam Zaluziansky taught *materia medica* at the Charles University, which experienced a revival in the late sixteenth century.[239] As we have seen, Rudolf's interest in natural history extended to his extensive cabinet of curiosities, full of marvels acquired for the emperor by agents throughout Europe. But natural history at Rudolf's court was pursued in a different key than the descriptive work of Clusius. The emperor wished to symbolically possess the world.[240] His collection served as a microcosm, subject to his will, that represented the ideal sway of universal monarchy. Access to the imperial collection was limited: only highly-ranked visitors could see its wonders. Though Rudolf's intendants drew on the literature of natural history to describe his possessions, they made no contribution themselves to that literature. In short, the emperor consumed public knowledge for his private purposes. Artists at the imperial court did have access to the collection, and they reproduced its contents in naturalistic drawings. Giuseppe Arcimboldo is only the best-known of many of Rudolf's court painters. But Arcimboldo did not intend to communicate descriptions to the Republic of Letters. He prepared his drawings as studies for his contributions to imperial propaganda: they were the basis of the carefully portrayed vegetables, flowers, and animals in Arcimboldo's series of paintings of the four elements, the four seasons, and Rudolf II as Vertumnus.

The fortunes of natural history in Rudolfine Prague remind us that local scenes of inquiry might be only loosely integrated into the international community of naturalists. Common interests and methods were not given; they had to be identified, propagated, and maintained by local practitioners, sometimes across large expanses of time and space. A patron's interests—especially when the patron dominated his clients like Rudolf did—might overwhelm a commitment to the broader community. This was especially likely when the locality in question could support a large number of naturalists who could be engaged in the same activities.

Bordeaux, circa 1600

Some local scenes of natural history, on the other hand, could persist only through their connections with the broader Republic of Letters. Clusius's correspondence reveals a small group of naturalists and horticulturists in Bordeaux who were in constant contact with their counterparts elsewhere in Europe—Poitiers, Leiden, Basel, and elsewhere.[241] The principal members of this community were Jacques Levenier; M. de Chezac, a president of the parlement (sovereign court) of Bordeaux; and a certain Sieur Ducasse, whose relations with other Bordeaux naturalists were strained. The community swims into historical focus in December 1597, when Levenier first wrote to Clusius. The two men did not know one another, but Levenier had a lively desire to acquire "rare flowers and other plants that smell good," and he decided that the best way to fulfill this desire was to write to "skilled men in the subject."[242] Levenier was already familiar with Clusius's Hispanic and Austrian histories; moreover, he knew that Clusius was preparing a revised, expanded version of those works. Clusius received the letter on the tenth of February 1598 and replied three days later. From Levenier's next letter, it appears that Clusius both offered bulbs and tubers from his garden and refused the payment that Levenier had offered.[243] As their correspondence continued, other members of the Bordeaux group appeared in Levenier's letters. Writing in the fall of 1598, Levenier thanked Clusius for a gift of plants. He had nothing himself with which to reciprocate, but Chezac had recently acquired a mountain lily with white flowers; Chezac would be happy to send a specimen to Clusius. Levenier had already sent two bulbs to Jean Robin in Paris.

In these early letters, Levenier and Chezac appear to prefigure the *amateurs de fleurs* who would characterize seventeenth-century French horticulture, more interested in adorning their garden with colorful varieties than with expanding botanical knowledge.[244] But this impression is belied by Levenier's later letters. In the fall of 1599 Levenier sent Clusius some plants along with an apology for not sending the seeds that he had promised: he had been bedridden for three and a half months, which had prevented him from gathering seeds in the wild. In the summer of 1601 Levenier spent two weeks looking for new plants in the Pyrenees; he undertook expeditions to the same range in 1602 and 1603, and possibly in 1605. Already in 1601 he was confident enough of his botanical knowledge to be fairly certain that he had found a new species, a kind of anemone or pulsatilla that differed significantly in its leaves and roots from the published descriptions he had consulted. In 1602 he was visited in the Pyrenees by Pierre Richer de Belleval, professor of medicine at Montpellier, who (Levenier claimed) had brought some students to discuss plants with Levenier. (Before returning to Montpellier, Richer also helped himself to many of Levenier's specimens.) Some years later he listened with detached amusement to travelers' tales about plants that they claimed to have seen on a mountain in Aragon; when he insisted that some of the plants could not have been in flower at the time and others were exceedingly rare, his interlocutors backpedaled, admitting that they may have been mistaken and that Levenier was not a man who could be easily fooled. By 1606, Levenier was employing agents: he had sent "my man Bachelier" to Constantinople and the Levant in search of "beautiful and rare things that have not yet been seen in Europe." He had also received some plants from a friend of his in Mexico—perhaps a fellow plant-hunter whom he met in the Pyrenees earlier.[245] Whether Levenier's interest in new plants predated his correspondence with Clusius, or whether his contact with the eminent botanist converted him from a garden enthusiast to a dedicated plant-hunter and naturalist, he shows how a small local group of naturalists could be drawn into the concerns of the international community.

Perhaps Levenier turned to Clusius, Robin, and other correspondents because of tensions within the Bordeaux group. Clusius was apparently also corresponding with another Bordeaux horticulturist, the Sieur Ducasse, for in early 1600 Levenier assured Clusius that despite his promises Ducasse did not have what Clusius wanted. By the end of that year Levenier was more open about the tensions between himself and Chezac on the one hand, and Ducasse and another, unnamed man on the other. Ducasse had allegedly insulted Clusius on seeing the latter's gifts to Levenier. If Levenier was trying to turn Clusius against Ducasse, someone else was returning the favor: in early 1601 Levenier was at pains to assure Clusius that he had made all his discoveries himself, contrary to what Clusius had heard.[246] In April, Clusius sent Levenier a copy of the poison-pen letter he had received; Levenier responded indignantly in July that he was not a monk, but he had reached the age of forty-five "sans reproche" and that everything alleged against him was false. If Levenier had not revealed the source of his plants—apparently the principal accusation against him—it was not that he had bought or stolen them but that he wanted to continue to explore the mountain regions where he had found them without interference.[247] In his next letter to Clusius Levenier accused an unnamed man, possibly the poison-pen writer, of being a plant thief: this individual was suspected by Chezac of stealing plants and had been barred from Chezac's garden. He had also been barred from the garden of M. de Bouville, a councilor of the parlement of Paris, and he had even tried to impersonate Jean Robin's valet in order to steal a box of plants! Levenier was canny enough not to name this person in the letter, but he implied that Ducasse, if it was not him, was from the same mold: he too had been barred from Chezac's garden.[248]

Such tensions, even antipathies or hatreds, characterized other natural history communities. In Basel, Martin Chmielecius complained about Felix Platter's avarice and jealousy.[249] And after initially warm correspondence, the relationship between Carolus Clusius and Pieter Paaw after Clusius's move to Leiden was cool if not cold.[250] Tensions are inevitable in any closed community. Nonetheless, competing interests and clashing personalities reveal the shared devotion to natural history that characterized these groups. Naturalists could not ignore even what their enemies in the field were doing. Although my history is largely a history of consensus about the boundaries of natural history and cooperation within those boundaries, the fact that controversy arose within those boundaries underscores their contemporary reality.

The preceding brief account of natural history in Basel, the Habsburg lands, and Bordeaux underscores the complexity of the new discipline of natural history. The Renaissance community of naturalists might best be imagined as a series of intersecting circles. University teachers, the court of a drowsy emperor, and jealous plant-hunters did not approach nature with identical interests or concerns. Nonetheless, they shared not only their interest in plants and animals but also a developing language, methods of investigating nature, and practical and intellectual problems. Frequent exchanges between local communities, both personal and epistolary, maintained a sense of community, of participation in a province of the Republic of Letters. Jacques Levenier, for example, corresponded not only with Clusius but also the Poitevin naturalists François Vertunien le Vau and Vertunien's son-in-law Pascal Le Coq. Vertunien, meanwhile, was in contact with Clusius, while Le Coq, who had studied at Basel, maintained his connections with that community through Jacob Zwinger—who corresponded occasionally with Clusius. Connections like these were vital for the construction and maintenance of natural history as a discipline. Equally important, though, were the shared activities that characterized the discipline but that could be carried out only in a particular place. Before turning to the naturalists' Republic of Letters we will linger for a more systematic look at the most important of these activities.

Shared Activities: Herborizing

Local communities were bound not only by a shared interest in natural history but also, more intensely, by shared activities: above all, the study of nature in gardens and on excursions. The botanizing excursion was particularly important, marked by passing forbidding town gates, then leaving the main highways to wander through forests, along streams, and up narrow rocky trails. The sober prose of natural history texts disguised this labor, except in the occasional prefatory remark, but it was an important means for the community of naturalists to constitute itself.

Conrad Gessner was one of the few naturalists to write at length about the hardships and joys of natural history excursions. He began his "Description of Mount Pilatus" by noting that it was his custom, every year or two, to undertake a journey, especially in the mountains, "for the sake of my mind and my health." The first part of the ascent went easy, but as he approached the slope, it was more difficult. He and his companions refreshed themselves with cold spring water, which struck the tired, hot, and thirsty men as the greatest pleasure. On the ascent and descent, Gessner's party stayed with the herdsmen who pastured their flocks there in the summer months. They shared the locals' bread, cheese, and milk, and slept in the hay. The ascent and descent were grueling and dangerous, the food simple, and the lodging rustic, admitted Gessner. But to a good man, liberally educated and trained in natural history, travel in the mountains was an unalloyed joy.[251]

Gessner's defense of mountain expeditions against the objections of the "soft and effeminate" smacks of humanist classroom exercises, as does the "Stockhornias" of Joannes Rhellicanus, an account of the ascent of Mount Stockhorn written in hexameters.[252] But there is no reason to doubt that the sentiment was genuine. We have already seen how Clusius, though increasingly weak, undertook several expeditions into the Alps around Vienna, braving inclement weather and other hardships, in order to learn more about their flora. Thomas Platter returned from more than one collecting expedition in southern France soaked to the bone, his plants ruined. On one of those trips, suspicious villagers refused shelter to his party, obliging them to continue for another hour in pouring rain until they reached the next village.[253] Platter, unlike Gessner, did not wax eloquent about the joys of travel, but he must have experienced great pleasure in finding "all kinds of strange plants" in order to repeatedly endure such journeys.

Field expeditions almost always involved groups. Gessner took with him three young companions: a sculptor, an apothecary, and a painter, presumably all interested in natural history. Clusius traveled in the Vienna woods with Dr. Aicholz and in northwestern Hungary with his friend Stephanus Beythe, a minister knowledgeable about the local plants.[254] In addition to excursions with fellow medical students, Thomas Platter went on long trips with his fellow Basilean, Johann Heinrich Cherler. Another young native of Basel, Jacob Zwinger, traveled to Padua to study in the company of Pascal Le Coq, botanizing along the way.[255]

Group expeditions strengthened the sense of belonging to a community of like-minded scholars.[256] In many cases, they also allowed visiting naturalists to take advantage of local knowledge. Clusius learned a great deal from his Hungarian preacher.[257] In 1596, Thomas Platter and Johann Cherler undertook a six-day expedition from Montpellier into the Cevennes, to the so-called Hortus Dei ("God's Garden," an area especially rich in plant species). In the small town of Ganges they were shown plants by an old physician,

> who had for many years explored the plants in the area and had, many years before, hiked through God's Garden with De l'Obel and shown him the plants to be found there. Hence he is called Dioscorides. He showed us the plants in his house: most of them were plants that are sold to apothecaries, like Angelica, Gentian and the like. He was certainly over eighty years old, from his appearance and speech.[258]

Naturalists were willing to learn even from uneducated peasants. Adriaan van de Spiegel, dressed in rustic clothes so as not to scare his interlocutors or attract highwaymen, quizzed the rural folk of northern Italy on the local plants and the folk remedies in which they were used. One doubts that the Tuscan peasants were taken in by this Fleming's disguise; neither was the thief who stole his notebooks.[259] Furthermore, Spiegel did not treat his local informants as equals; he demanded corroboration before trusting their claims. Nonetheless, this sort of ethnobotany *ante litteram* brought invaluable local knowledge to the community of naturalists.

The first generations, those of Leoniceno, Bock, and Valerius Cordus, had to learn the practices involved for successful field observation on their own or informally from their friends and colleagues. The situation changed for their successors. At northern Italian universities, as well as Montpellier in southern France, teachers led students on botanizing expeditions to expose them to these sources of vegetable drugs in the wild.[260] As with other aspects of Renaissance learning, the theoretical treatment of such education lagged far behind practice. It was only in 1606 that the first textbook on the elements of botany was published in Padua by Spiegel.[261] But documentary evidence—including correspondence and, in particular, the travel diaries of Felix and Thomas Platter—allows us a glimpse at the activity of pedagogical botanizing as it was practiced in the latter half of the century.

Such expeditions were led by a medical professor. Guillaume Rondelet was one of the most famous, and in their works his students frequently mentioned plants they had seen in his company. In the 1590s, Rondelet's successor was Pierre Richer de Belleval, founder of the Montpellier botanical garden (1593). Like Rondelet before him, Richer also took students on botanizing expeditions—though less frequently and less willingly, for his first priority was to establish a botanical garden.[262] Montpellier was not the only university where regular botanical expeditions were part of medical education. At Leiden, leading such expeditions was not an official duty of the praefectus horti, but the demands of several medical students, after Dirk Cluyt's death in 1598, to appoint his son Outger as successor stressed the young man's ability to accompany them on expeditions in the fields and woods of Holland.[263] At Basel, on the other hand, botanical expeditions were formally part of the curriculum after 1575.[264]

Hence, in the latter half of the sixteenth century, increasing numbers of medical students were forced to accompany their professors not only to patients' bedsides and apothecaries' shops but also into the wild, to be shown plants in their natural state. Naturally, the students were not always as concerned with the subject as their teachers. Martinus Chmielecius, a professor at Basel, complained to his young friend Jacob Zwinger that "no one among us wants to even go out the door to botanize, not even Cherler. Recently I just succeeded in forcing them to take the waterfall road, and I noticed how they behaved. Scarcely one out of twenty-one brought back ten plants. The others went to walk, not to herborize."[265] But others were more enthusiastic: for example Thomas Platter, whose first herborization at Montpellier took place only six days after his arrival, even before he had matriculated.[266] The number of students on such expeditions could be large—Chmielecius mentioned twenty-one, while Pascal Le Coq led parties of sixty or more during his days in Montpellier.[267]

Unlike more advanced phytographers' expeditions, these excursions emphasized practical botany. Students did not herborize in order to collect rare plants—though those who were botanical hobbyists, such as Clusius in his student days at Montpellier, certainly did so. Rather, teachers demonstrated the common medicinal plants that formed part of the recognized pharmacopoeia. The "Dioscorides of the Cevennes" mentioned above demonstrated common medicinal plants to the Montpellier students who visited him in Ganges.[268]

While only students at Montpellier had access to this living Dioscorides, the ancient one was still used as a textbook. Rondelet's former students remembered that he had worn out his copy in teaching.[269] As mentioned above, Joachim Camerarius brought out an epitome of Mattioli's Dioscorides in 1586, while Caspar Bauhin reedited the complete text in 1598. The curators of the new University of Leiden prescribed Dioscorides, also in Mattioli's commentary, as botanical textbook in 1598 and ordered a few years later that a copy be placed, securely chained, in the porticus of the botanical garden.[270] However, the flaws of Dioscorides—even in Mattioli's heavily altered "commentaries"—were clear. Already in 1601, the Leiden professor Pieter Paaw was alternating between Dioscorides and his own botanical notes for his summer lectures.[271] And the book was heavy, difficult to carry around on a long expedition, as Euricius Cordus's companions had long before noted.

Since students were expected to learn botany from long and frequent exposure to the plants themselves, they needed books that would aid that very task. When they were not to hand, students and teachers adapted others to that purpose. Mathieu de l'Obel's *Plantarum icones* (1581), a collection of woodcuts published by the Plantin press along with the names of plants but no descriptions, was apparently a popular field manual.[272] In 1611, Caspar Bauhin wrote to his Leiden counterpart Everard Vorst in regard to De l'Obel's book, now long out of print. Since it was so practical for excursions, Bauhin urged Vorst to take up the matter with the current generation of Plantins, proposing that they publish a similar book with De l'Obel's images and those since published by Clusius.[273] Bauhin even offered to edit the book for free, if it would mean getting this useful tool for students published.

That project was never realized, due to the Plantin heirs' lack of interest, so Bauhin soon wrote his own textbook.[274] In some respects, it was even more austere than De l'Obel's *Icones:* where De l'Obel had given only pictures and names, Bauhin renounced the images. His slim *Catalogus plantarum circa Basileam sponte nascentium* (1622) listed plants growing around Basel, their synonyms, and where they were to be found.[275] The book was interleaved with blank pages for students' notes, presumably including notes from herborizing expedition, and at least some students took advantage of the opportunity.[276] In many respects this book was the first local flora. But its title page proclaimed that it was intended for medical students and included Galen's remark that "a physician should have knowledge of all plants: but if that is not possible, he should at least know those frequently used."[277] And, given the practically nonexistent commentary and copious space for notes, Bauhin clearly intended his book to be a companion in the lecture hall, in the botanical garden, and on field trips.

When students, accompanied by a professor and equipped with makeshift textbooks and notebooks, went on botanizing expeditions, they were not merely repeating what Euricius Cordus and his friends had done earlier. They had to pay attention to the same range of variation, and the same problems of geographical distribution, that concerned their teachers. But the goal was different. Not only were students first exposed to the most common medicinal plants, they were also supposed to learn the most representative types rather than the rarest. Spiegel's textbook instructed teachers to choose the "middle species" of each kind of plant, point out its characteristic features, and then compare it to a few others of its genus. In this way the idea of the genus will be learned as effectively as possible. The teacher's other task, Spiegel added, was to know when to stop: like Bauhin, he cited Galen's wise words and said that for many students, the basics would suffice.[278] These basics, however, needed to be repeated again and again, and the student had to take careful notes to avoid the pitfalls of faulty memory.

Pedagogical excursions taught the fundamentals of botany to a large audience. Though most medical students did not devote themselves to natural history (even Spiegel gave the subject up in later years), the standardized training dispensed by courses in *materia medica* ensured that those who pursued the subject further had a common language and shared techniques for discovering and communicating facts and material in natural history. One consequence was a process of differentiation and specialization within natural history. As we have seen, naturalists in the 1550s were as likely to be ministers or humanist scholars as physicians.[279] By the end of the century, serious naturalists such as Caspar Bauhin corresponded primarily with others who had received the same training.[280] While the category of the "professional naturalist" did not exist in 1600, a group of trained specialists was evident.

The Community of Naturalists and the Republic of Letters

Local communities were the sites of intensive, often daily engagement with natural history, but they were fragile, often dependent on the efforts of a few, or even one, organizer. As Clusius's Vienna shows, such communities could change dramatically in a brief period of time, especially when they depended on a prince whose heir did not share his interests. Fortunately for Clusius, and for natural history as a whole, local communities were not the only arenas in which their interests found an audience. Travel, correspondence, and exchange allowed many naturalists—not only humanistically-educated scholars—to participate in an international community.

Correspondence and travel bound local communities into a larger whole. They established contacts between otherwise far-flung naturalists and permitted them to maintain those contacts for years. Despite the religious and political divisions of the sixteenth century, Europe was, in many ways, growing smaller: despite its material hassles, travel was becoming more common. And travelers were taking to the road for reasons beyond profit and pilgrimage.[281] Natural history participated in this broader movement.

In the sixteenth century, travelers usually went from town to town. In the more densely populated areas of western and central Europe, cities were located, at farthest, approximately a day's travel on foot or water from one another. The normal trade and travel routes went directly between towns: only these population centers were willing and able to provide for the needs of travelers.[282] Erasmus's route from Basel to Louvain in 1518 is typical: the prince of humanists took a boat down the Rhine to Strasbourg, thence overland to Mainz, further down the Rhine to Cologne, and then by horse and carriage to Louvain. He spent the night in the cities of Breisach, Speyer, Mainz, Bonn, Bedburg, Aachen, Maastricht, and Tongres.[283] Erasmus had nothing but bad words for most of the inns he encountered on the way. Justus Lipsius, traveling through Westphalia in the late sixteenth century, provoked a literary storm with his unkind remarks about that country's public houses.[284]

Travelers had to put up with more than bad accommodations. Aside from inclement weather and bad roads, there was always the chance that highwaymen would rob them, or worse. Erasmus was distracted from the dangers of robbery only by illness.[285] Felix Platter noted with morbid fascination the gallows trees he passed on his journey from Basel to Montpellier in 1552, erected along the roads as a warning to potential cutthroats.[286] Even off the highways, travelers had to be careful; Jacques Levenier came back nearly empty-handed from a plant-hunting expedition to the Pyrenees in 1602 because the bandits were so thick he did not dare visit his favorite collecting areas.[287] Given these conditions, it was natural for travelers to band together. Platter, like so many of his contemporaries, delayed his journey in several cities in order to wait for others going in the same direction.[288]

Medieval travelers endured such hardships for their purse or their soul: they were usually merchants or pilgrims, and the pilgrimage was the intellectual model for non-commercial travel in the Middle Ages.[289] In the sixteenth century the notion of travel for the soul's sake was succeeded, in humanist travel guides, with a new ideal of *peregrinatio animi causa:* travel for the mind's sake. As Erasmus's follower Joachim Fortius Ringelbergius expressed this ideal, only through travel could one avoid becoming small-minded and commonplace.[290] Montaigne thought that children should travel as part of their education: the world was to be their book.[291]

Most naturalists who were physicians had traveled in their student days. The medical schools of Montpellier, Padua, and, somewhat later, Basel and Leiden attracted students from as far north as Scotland and as far east as Poland. Felix and Thomas Platter traveled from Basel to Montpellier to study; Clusius, born in Arras, studied theology at Wittenberg and medicine at Montpellier, where he befriended Platter. Jacob Zwinger studied in Basel, Heidelberg, and Padua in the 1580s and 1590s before taking up a professorship back in Basel.[292] In their years of study naturalists often formed lasting friendships. And scholars who never met often shared the bonds of having studied in the same place with the same teacher.[293] The students who gathered around Guillaume Rondelet at Montpellier in the 1540s and 1550s forged bonds of friendship that lasted, in many cases, the rest of their lives. In 1593, Felix Platter wrote to Joachim Camerarius that he had renewed his acquaintanceship with Carolus Clusius, which had begun forty years earlier in Montpellier.[294] Jacob Zwinger formed close contacts with the Poitevin Pascal Le Coq and with several Italian scholars during his studies in Basel and Padua.[295] Felix Platter's younger half-brother Thomas also had close contact with his fellow medical students in Montpellier in the 1590s.[296]

Such intense relations were liable to fade with time, unless maintained by visits or correspondence. Clusius wrote to Camerarius in 1583, "I hadn't heard of Laurent Joubert's death. I knew him well when we were in Montpellier, and we ate our meals together at Dr. Rondelet's house."[297] Leonhard Dold wrote to Zwinger with fond memories of their student days in Basel, and stressed that he would not let their friendship die due to lack of correspondence—hence he was writing even though he had little news to communicate.[298] Johann Papius, writing from Königsberg in Prussia to Jacob Zwinger in Basel, apologized for the infrequency of his letters, due to the long distance involved. This letter was carried by his son, who was going to Basel to study.[299]

This marked another form of contact between local communities: the exchange of students. Since Nuremberg did not have a local university until 1623, promising young men were often sent to Basel to study. The coming and going of these students provided opportunities for friends to receive word-of-mouth news of one another's activities. Caspar Bauhin sent Joachim Camerarius's son back to Nuremberg in 1593 with praise for his abilities.[300] In 1587, Bauhin had sent another student to G. Weyer with a letter, mentioning that the student could report on Bauhin's activities, "since we are very close."[301] Other scholars recognized the importance of such word-of-mouth contact: Leonhard Dold in Nuremberg asked Jacob Zwinger in Basel whether any students returning to the latter city from Italy could report on the progress of Andrea Cesalpino's commentaries on Dioscorides.[302] Such communication was frequent at a time when the *peregrinatio academica* was becoming important.[303] Many of the earliest studies of Provençal plants were carried out by students on their way to and from the famed medical school at Montpellier.[304]

Shorter distances allowed for more frequent contact. In the 1530s, Otto Brunfels walked from Strasbourg to Hornbach in order to visit Hieronymus Bock and discuss the *res herbaria*—a distance of more than forty miles as the crow flies, much longer as the minister walks.[305] Some decades later, Felix Platter visited Rennward Cysat in Lucerne, although his occupations as city physician and university professor in Basel did not leave him as much time for such journeys as he wished.[306] Platter also maintained relations with Alsatian nobles, and occasionally traveled to their country seats to converse with and advise them.[307]

Rarer but important were long-distance journeys to visit colleagues. Often combined with herborizing, such trips could take weeks or months. Clusius, ever the restless soul, made four trips to England in the course of his career—two in 1579 and 1580, when he was residing in Vienna. On the second, he had originally planned to go only as far as the Netherlands, but on learning that Francis Drake's expedition had returned to Plymouth after circumnavigating the world, he took ship across the Channel to meet the explorer and his crew.[308] In his *Exoticorum libri* (1605), Clusius described many of the objects he had acquired on that trip, including a root that he named after Drake.[309] En route, he visited friends, colleagues, and patrons, including Wilhelm, Landgrave of Hessen, noted for his interest in the observational sciences.[310] While a professor at Leiden, barely able to walk, he nonetheless visited the count of Solms's estate in The Hague to see the cassowary that Dutch sailors had brought back from the Moluccas.[311]

When travel was too long, expensive, or dangerous, correspondence provided an inexpensive, if not wholly reliable, alternative. The naturalist who lacked time or funds to travel to distant places could, instead, receive information and material from correspondents in those places.[312] An example is the fourth volume of Conrad Gessner's history of animals, which was devoted to fishes and other aquatic animals.[313] From his study in landlocked

Zürich, Gessner was able to coordinate information not only from his travels and other books but also from his correspondents' observations and his own examination of objects they had sent.

Because plants could be transported easily, horticulturists profited greatly from such exchanges. The proximity of Vienna to the Ottoman Empire, which gave its inhabitants an almost constant sense of insecurity from 1529 to 1683, also made it ideally situated for trade in exotic Eastern flowers. Shortly after his arrival in Vienna, Clusius asked imperial representatives in Constantinople to acquire some bulbs for him and the new imperial garden.[314] By the 1580s, there was a regular cross-border trade in exotic bulbs. The best-known of these exotic imports was the tulip, first sent to the West by Ogier de Busbecq in the 1550s and described by Conrad Gessner in print in 1561.[315] From Vienna, Clusius sent bulbs throughout Europe, in particular to his friends in Nuremberg, Italy, and the Low Countries. Echoing Justus Lipsius, Marie de Brimeu, the princess of Chimay, called him "the father of all the beautiful gardens in this country."[316]

Correspondents reminded their friends of their obligations, but without being too importunate. Busbecq, in retirement in France, asked Clusius not to forget him when he divided the riches of his garden. In return, he promised to send *fritillariae,* which he had in great quantity from an apothecary friend.[317] In theory if not always in practice, such gifts were reciprocal. When Felix Platter requested an *Anemone flore rubro,* he promised other rare plants in return.[318] Anselm de Boodt, thanking Clusius for the seeds and flowers he sent, offered bulbs in return, if Clusius wanted them. Boodt, in fact, was not interested in growing plants but exchanging them: "Though I have no garden myself, I give them to my friends; and it pleases me a great deal when I can gratify them."[319]

In theory, money had no place in such exchanges. In the 1590s, Felix Platter had occasionally sold bulbs. Platter's colleague Martin Chmielecius complained to his brother-in-law Jacob Zwinger that "Platter never gives anything away. He sells everything he has, and gives nothing freely, not even to friends! He sold a red tulip bulb for a thaler; I won't pay that much. The older he gets, the greedier he becomes."[320] The next year, Zwinger's fellow student Pascal Le Coq stopped in Basel on his way to Poitiers and gave some seeds and roots to Chmielecius—but expressly not to Platter. "For it is against the duty of a philosopher to give things to sordid and envious men, who have an illiberal and asinine mind, from which they will derive filthy lucre instead of philosophical recreation."[321] Platter's behavior clearly violated the rules of exchange in the Republic of Letters, which were consciously opposed to the impersonal transactions of the market. Initiates to the republic were sometimes unaware of the rules. When Jacques Levenier introduced himself to Clusius by letter in 1597, he asked for a list of the bulbs that Clusius had available and their prices. Clusius quickly corrected him; in his next letter Levenier praised Clusius's humanity at great length, implicitly apologizing for his error, thanked him for the offer of bulbs, and asked only that Clusius not inconvenience himself by sending them. But commerce was creeping into the market, for Levenier also asked Clusius for a list of bulb-sellers ("eorum indicem qui venales plantas habent").[322] Naturalists might see their activity as a liberal art, but others were happy to profit from the growing interest in ornamental flowers.

By the 1620s, the commerce in flowers was deeply rooted in the Low Countries.[323] Late in his own life, Clusius railed against the vulgar crowd who sold flowers instead of giving them away. He had occasionally purchased them himself, but thought it shameful for a scholar to sell them: "thus is made vile this study, which used to be generous and liberal—but we must be patient, since the world is getting worse from day to day."[324] But Clusius was swimming against the tide. The first bulb catalogue, Crispijn van de Passe's magnificent *Hortus floridus,* was published in 1614. The market economy drove an ever larger wedge between dedicated naturalists and the public whom they had supplied, in the sixteenth century, with exotic florals from Asia, Africa, and the New World.

But even before this large-scale commercialization, rare and colorful garden plants tempted thieves. In 1581, while Clusius was on a trip to England, his servant absconded with many of his most valuable plants and sold them. Clusius returned to Vienna too late to assess the damage, but the following spring he realized its extent and had to send to Constantinople to replace bulbs that had been stolen. He suspected one noblewoman of having purchased the flowers from the thief: she denied it, but Clusius thought that he alone had grown some of the plants now on display in her garden.[325] Years later, in Leiden, Clusius experienced similar thefts. Once a neighbor was to blame: the man even had the audacity to announce that he was expecting plants from Constantinople, Tripoli, Alexandria, Moscow, and elsewhere, in order to allay suspicions about the new flowers in his garden.[326] Marie de Brimeu, to whom Clusius complained about his losses, offered to hire a guardian or provide a watchdog for his garden if the old botanist wished.[327] Levenier commiserated with Clusius, as he too had lost valuable plants to theft.[328] Public gardens, too, were subject to depredation. Though they were usually enclosed, they were also frequented by many more visitors than private gardens. A sign at the entrance to the Leiden *hortus* admonished visitors, "It is wicked to pluck branches, flowers, or seeds, to uproot

Figure 2.1. Leiden *Hortus botanicus*, engraving by J. C. Visscher of a drawing by J. C. Woudanus (Leiden, 1610): detail showing fenced-in beds.

bulbs or roots, or to cause the garden any injury."[329] Lest this warning prove ineffective, several beds were surrounded by a waist-high fence to discourage overly eager collectors from enriching their gardens at the university's expense (figure 2.1).

Theft, stinginess, and the growing commercial market in flowers—these developments suggest that the norm of free exchange of material was enunciated as a reaction and as a way to draw a boundary between members of the naturalists' republic and outsiders whose interest in nature was instrumental, not intrinsic. Naturalists might pay an outsider for material, but they compensated their peers not with money but with recognition in print, information, or exchanges of material. Explicit or tacit, the norm functioned throughout the sixteenth century. Much of the correspondence of sixteenth-century naturalists is concerned with such exchanges—lists of plants growing in their gardens, of seeds they had gathered in the wild, were sent back and forth, as were boxes and canisters containing carefully-packed plants. The same was true of animal and mineral objects; Gessner's *Historia animalium* and the correspondence of Jacques Dalechamps provide abundant examples.[330] Such exchanges certainly broadened the horizons of naturalists, even those like Clusius who had traveled extensively. But they served another, equally important purpose. Correspondence and travel were the warp and woof binding naturalists together in an international community.

In some cases, correspondence served to maintain earlier, face-to-face acquaintances. But letters offered an opportunity for contact between naturalists who had never met. Carolus Clusius initiated a correspondence with Joachim Camerarius on the strength of an acquaintance with Camerarius's brother and their mutual interest in wild plants.[331] During the next twenty-two years, they exchanged letters on average every five weeks. Clusius never met another of his correspondents, the Bolognese patrician Ulisse Aldrovandi, but they exchanged information and material for at least twenty-seven years.[332] The same is true of Clusius's Venetian friend Jacopo Antonio Cortuso: in 1567, Clusius, who had never seen Cortuso face-to-face, wrote a letter of introduction for his former pupil Thomas Rehdiger, who was planning to visit Venice and wished to be admitted to its learned society.[333]

Naturalists' "friendships" were sometimes matters of patronage. Clusius's friend Jean de Brancion sheltered him in Mechelen during the early phase of the Dutch revolt, when marauding Spanish soldiers sacked the town, and put his garden at Clusius's disposal.[334] His patrons Balthasar de Batthyan, Wilhelm IV of Hessen, Marie de Brimeu, and Louise de Coligny had all received his floral largesse. Yet patronage was not Clusius's primary motive for exchanging flowers. One of his long-term correspondents was the neo-Stoic philosopher Justus Lipsius.[335] Clusius gave advice and material to Lipsius, who reciprocated—in good humanist fashion—with witty correspondence and a famous punning epigram on Clusius's name.[336] Lipsius also may have influenced Leiden University's rectors in their choice of Clusius to head their new botanical garden.[337] But the canons of humanist friendship, the exchange of letters and assistance, were more important in their relationship than any concerns of patronage.

Like other humanists, naturalists conceived of their epistolary exchanges as a substitute for conversation. Clusius echoed a commonplace when he told his new friend Johann Crato von Krafftheim, in Bratislava, "Although we are separated by a great distance, and hence cannot strengthen our friendship through mutual conversation, nonetheless letters (which seem to have been invented in order to allow a conversation between absent people) can repair that inconvenience."[338] Felix Platter expressed the same sentiment, telling Rennward Cysat repeatedly that "maintaining good friendship through writing is very pleasing to me, for I think often about our conversation."[339] The link between conversation and correspondence was, in fact, natural for humanistically-educated scholars. Rhetorically trained, they tended to shape their literary productions after the models of dialogues and letters.[340] Rhetorical conventions, like Clusius's and Platter's remarks, are ubiquitous in sixteenth-century naturalists' correspondence, but they do not vitiate the honesty of the sentiments expressed in these letters.

Epistolary relationships, like any other, required maintenance. Leonhard Dold occasionally wrote from Nuremberg to his friends in Basel even though he had little to say, so as not to let their friendship deteriorate.[341] On the other hand, Conrad Gessner had harsh words for a former friend who refused respond to Gessner's letters and even the dedication of a small book.[342] Given how many letters they wrote with quill pens and ink, many sixteenth-century naturalists devoted a large portion of their time to correspondence. It was hard work to be a citizen of the Republic of Letters.

The Imagined Community

Travel and correspondence bound Renaissance naturalists into an international community that extended, and to some extent transcended, the concerns of their particular local groups. Though dispersed throughout a continent, and occasionally beyond, this community was very real. Gessner, Clusius, Aldrovandi, Bauhin—the list could go on—each wrote thousands of letters to hundreds of correspondents. Their correspondence drew on a humanist ethos that conceived of the letter as a sincere conversation between absent friends, an ethos that was inculcated and reinforced by the Latin secondary schools of sixteenth-century Europe. From this perspective the international community might best be conceived as a network; local groups were connected, through one or several of their members, to one or more nodes in a wide-flung net. Several of those nodes were main connection points—writers whose correspondence was particularly broad, or "brokers" like Giovanni Vincenti Pinelli whose main contribution to Renaissance natural history was as centers of communication. But there was no center, unlike the seventeenth-century networks that would give birth to Europe's first scientific societies.

But each link in this network, each letter exchanged between two naturalists, referred implicitly or explicitly to the other links. The notion of a community of intellectual interests—of *studium,* of devotion—lay beneath the particular exchanges that constituted part of the network. In short, naturalists imagined for themselves a broader community. They thought of themselves not only as sincere friends of other individuals but also as part of an imagined community, a province of the Republic of Letters.[343] I have used this notion loosely in the preceding pages; to conclude this chapter I would like to dwell on it, for Renaissance naturalists' conception of themselves as a united public provides the final instance of disciplinary self-awareness.

The term "republic of letters" is often used loosely to refer to the international, Latinate scholarly community of the sixteenth and seventeenth centuries.[344] However, recent historians of early modern France have pointed to a more specific notion. As Dena Goodman summarizes this scholarship, the Republic of Letters [*respublica literaria*] originated in the early seventeenth century, along with the early modern absolutist state. Its constitutional architects were Francis Bacon and, later, Pierre Bayle, whose journal, *Nouvelles de la République des lettres,* infiltrated late seventeenth-century France from neighboring Holland. According to this view, the Republic of Letters constituted a significant part of the seventeenth- and eighteenth-century "public sphere" within which absolutist political ideology could be challenged.[345] Its citizens defined themselves by their method, criticism, and by their polite, sociable culture, which French Enlightenment thinkers equated with "civilization" *tout court.*[346]

Without denying the specific Enlightenment connotations of the term *République des lettres,* it is nonetheless possible to identify a sixteenth-century ancestor. The term *res publica literaria* and its variant *res publica litterarum* have been traced back to the fifteenth century.[347] Paul Dibon suggests that the *République des lettres* of the late seventeenth and eighteenth centuries must be distinguished from an earlier, seventeenth-century *respublica literaria.*[348] I would like to take Dibon's analysis further, for the relationship between goodwill and communication that he sketches for scholars in the first third of the seventeenth century also characterizes the humanist ethos of correspondence that animated Renaissance natural history.

Sixteenth-century naturalists used the term "Republic of Letters" consciously, to refer to the world of scholarship. Moreover, the term strongly connoted a public sphere, for naturalists contrasted it with their own particular interests as scholars. In the first volume of his *Historia animalium* (1551), Conrad Gessner closed his prefatory letter with "one wish: that anyone wishing to correct me write sincerely and modestly, and that they write neither to increase their own glory nor to reprehend me, but to promote the *respublica literaria.*"[349] Half a century later, Caspar Bauhin's correspondent Johann Bachmeister expressed his desire to "adorn, increase, and promote our *respublica literaria*" by contributing to Bauhin's work.[350] In 1590, Ulisse Aldrovandi opened a letter to Carolus Clusius by saying that he had worried about Clusius's health, for his own sake and for that of the *respublica literaria,* but that his worries had been lifted when he saw Clusius's latest publication.[351]

As Aldrovandi's words indicate, the notion of the *respublica literaria* in the sixteenth century was connected with publication. The thought of publication was implicit in Gessner's reference to those who might write against him. Other correspondents connected publication with public utility: thus Ole Worm, complaining about Bauhin's delay in publishing his *Theatrum botanicum,* noted that "this work concerns public utility."[352] Earlier, Onorio Bello had mentioned to Jacob Zwinger that his letters to Clusius on the flora of Crete would become part of "public law" when they were published in Clusius's *Rariorum plantarum historia.*[353] Bello was not entirely pleased with Clusius's request, since he wished to publish descriptions himself, but he felt he owed it to the "most friendly" Clusius to agree, given Clusius's earlier kindnesses. In this instance, private duties and public service came together.

The scholarly interests of sixteenth-century naturalists—*studia nostra communia*—were thus not limited to a narrow circle of correspondents. Rather, they were the concern of an abstract literary public, a public that naturalists imagined when they composed their letters and compiled their books. The very range of sixteenth-century naturalists' interests underscores their sense of participating in a Republic of Letters. Conrad Gessner's breadth of interest and range of publications earned him the title of "the Swiss Pliny."[354] Carolus Clusius collected classical inscriptions as well as plants and seeds during his travels in Spain. He published the letters of Nicholas Clenardus, a young Flemish humanist, and contributed to Abraham Ortelius's *Theatrum orbis terrarum.*[355] Jacob Zwinger collected plants, planned a history of insects—and quizzed Andrea Cesalpino on the difference between the center of the earth and the center of the world, and whether either could move.[356] On the other side of the divide, the great philologist and chronologer Joseph Justus Scaliger had a lively interest in collecting natural objects, and Justus Lipsius was a dedicated gardener.[357] Leonhard Dold's pleasure in reading Andreas Libavius's letters, finally, points to the breadth of interest among citizens of the Republic of Letters and the centrality of correspondence within it.[358] Humanist scholars' correspondence and the thicket of epigrams they contributed to one another's books testify to the strength of the ideal. The sixteenth-century *respublica litterarum* was at least as unified as the eighteenth-century *République des lettres*—perhaps more so, given that it was restricted to an elite writing primarily in Latin.[359]

Naturalists' humanist education and their participation in the Republic of Letters explains, we shall see, how certain aspects of humanist thought and practice were adapted to serve the new discipline of natural history. Nonetheless, by the second half of the sixteenth century, natural history was a distinct province within the Republic, and naturalists saw themselves first and foremost as residents of their province, even if they made excursions into other realms from time to time. Clusius was not a great antiquarian, despite his dabbling. Lipsius, in turn, published nothing on plants, though he was happy to profit from Clusius's specialist knowledge. When Joseph Scaliger saw Bernardus Paludanus's collection, he realized the gap that separated his desultory collecting from Paludanus's *studium*.[360] When naturalists discussed their contributions to knowledge in the broadest terms, they thought of the Republic of Letters, but their core audience was initiated into the language and methods of the discipline. That was especially true of the third and fourth generation: by then naturalists had to appeal to the broader republic not only to situate themselves intellectually but also to convince their publishers to risk capital in their books.

Still, the ideal of the Republic of Letters played a powerful role in maintaining the community of naturalists. In the absence of formal institutions for exchange of humanist scholars' ideas, the affective bond of friendship and the sense of a common enterprise served to encourage scholarly investigation as a collective rather than a solitary enterprise. The scholar's study was not a place of solitude; when he entered his workroom, the sixteenth-century scholar was reminded constantly of his conversations with distant friends.[361] The circular letters and, later, learned journals of the seventeenth century developed out of such affective relationships but never supplanted them.

These affective relationships underscored the importance of *studium* in the naturalists' Republic of Letters. The republic was opposed to the corporate organization of universities, the social hierarchy of courts, and the economic organization of mercantile exchange. Within its bounds, a noble courtier like Clusius could interact freely, and on an equal footing, with Felix Platter, son of a peasant, and even Pieter Coudenbergh, an Antwerp apothecary and gardener. The early modern community of naturalists found its common ground in amity and devotion to "common studies," elements that could, within the borders of the Republic of Letters, overcome the social and corporate distinctions of early modern Europe.

Conclusion

This chapter has explored the formation of natural history in the late Renaissance. Over the course of a little more than a century, natural history developed out of medical humanism into a distinct scholarly discipline, practiced by a self-identified group of naturalists, most of whom were tightly integrated into local communities of naturalists centered on universities or courts; those communities, in turn, were closely connected through correspondence and travel. Naturalists imagined themselves as a distinct community and as part of a broader Republic of Letters.

Such was not the case in classical antiquity and the Middle Ages. What would become "natural history" in the Renaissance sense comprised three distinct traditions: natural philosophy, pharmacy, and agriculture. Renaissance naturalists drew upon these traditions in their own work, but they did more: they retrospectively invented a unitary tradition out of distinct ancient and medieval strands in order to justify and legitimate the activities of their imagined community. It is this double story—the ancient and medieval background and Renaissance naturalists' creative interpretation of it—that the next chapter relates.

CHAPTER THREE

The Humanist Invention of Natural History

The discipline of natural history was invented in the sixteenth century, but Renaissance naturalists drew upon ancient and medieval predecessors in the study of nature. Steeped in medical humanism, the first generation of naturalists turned naturally to the Roman and especially Greek classics to delineate their subject and defend their methods. Subsequent generations pursued lines of inquiry set out by ancient and medieval authors, even as those authors' works received less and less attention. Like any invention, natural history was not created *ex nihilo*. Naturalists creatively appropriated elements of the ancient and medieval tradition, turning them to new ends.

In large part, though, the ancient and medieval tradition was itself invented by Renaissance naturalists.[1] Out of a congeries of texts produced over the course of two millennia, Renaissance naturalists forged a unitary tradition, a series of naturalists like themselves who were engaged in a common project. They admitted that the tradition, like the classical tradition more generally, had been interrupted by the barbarian invasions and neglected by the majority of medieval schoolmen. Only a few voices crying in the scholastic wilderness continued to study natural history. Renaissance naturalists from Leoniceno to the Bauhins portrayed themselves, in words and pictures, as restorers and guardians of this ancient tradition.

The engraved title pages of Renaissance herbals books conveyed vividly the continuities between ancients and moderns. Rembert Dodoens's 1557 *Histoire des plantes* presented Apollo and Aesculapius, the Greek gods of medicine, along with four ancient kings and queens who had discovered the plants bearing their names (figure 3.1). Carolus Clusius added the Hebrew tradition in his *Rariorum plantarum historia* (1601), which showed the Tetragrammaton presiding over Adam and Solomon, who had named

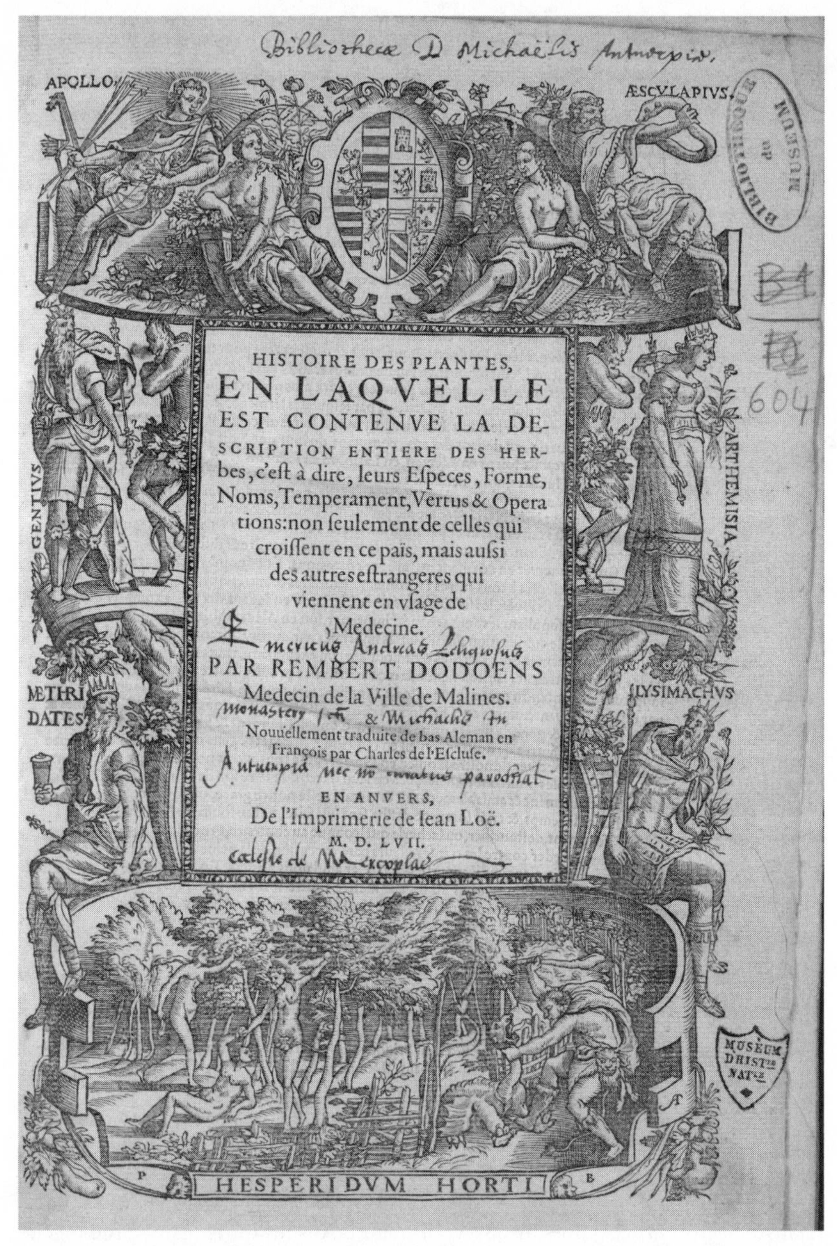

APOLLO. ÆSCVLAPIVS.

GENTIVS ARTHEMISIA.

MITHRIDATES LYSIMACHVS

HISTOIRE DES PLANTES,
EN LAQVELLE
EST CONTENVE LA DE-
SCRIPTION ENTIERE DES HER-
bes, c'eſt à dire, leurs Eſpeces, Forme,
Noms, Temperament, Vertus & Opera
tions: non ſeulement de celles qui
croiſſent en ce païs, mais auſsi
des autres eſtrangeres qui
viennent en vſage de
Medecine.
PAR REMBERT DODOENS
Medecin de la Ville de Malines.
Nouuellement traduite de bas Aleman en
François par Charles de l'Eſcluſe.
EN ANVERS,
De l'Imprimerie de Iean Loë.
M. D. LVII.

HESPERIDVM HORTI

Figure 3.1. Title page of Rembert Dodoens, *Histoire des plantes* (1557). © Bibliothèque centrale MNHN Paris 2005.

and described plants, and Theophrastus and Dioscorides, the most important ancient Greek writers on plants (figure 3.2).[2] The posthumous *Historia plantarum universalis* by Johann Bauhin and Johann Heinrich Cherler, first published in 1650–51, epitomizes the relation between ancient and modern natural history as Renaissance naturalists perceived it (figure 3.3). The book's authors reveal an elegant garden. On either side, columns represent their illustrious predecessors: Theophrastus, the investigator; Dioscorides, the expert; Pliny, the encyclopedist; Galen, the judge; they are joined by the modern naturalists Mattioli, Guilandino, and Amatus Lusitanus, the critics, and Fuchs, Gessner, and Dalechamps, the discoverers. Bauhin and Cherler's accomplishment is merely the last stage in a grand enterprise begun by the ancients under the benevolent eye of God—the modern recovery of a *prisca botanica*.[3]

Reassuring as this notion might have been to sixteenth- and seventeenth-century readers, it had little basis in history. Ancient and medieval students of nature did produce texts that were critically important for Renaissance naturalists. But they were produced for different reasons. Aristotle, Theophrastus, and their medieval follower Albert the Great considered natural history as a propaedeutic to natural philosophy: facts concerned them insofar as they contributed to a philosophical understanding of natural causes. Cato the Elder, Varro, and Columella, the Roman "scriptores rei rusticae," as well as poets like Virgil in the *Eclogues,* wrote about nature from a farmer's eye. Galen and Dioscorides treated nature as physicians concerned to understand, respectively, the causes of health and disease and the practical medical uses of plants, animals, and minerals. These three contexts could be seen as facets of a single concern with natural history only in hindsight—only after the creation of natural history as a distinct discipline.

Nonetheless, ancient and medieval texts from all three contexts shaped Renaissance natural history. The single most important influence was the fateful decision by Brunfels, Bock, and Fuchs to model their books after Dioscorides. But Renaissance naturalists assimilated other aspects of the ancient tradition. Though only a handful took up the Peripatetic philosophical approach to nature, many learned from Aristotle and Theophrastus the disciplined attentiveness required by the new science of describing. Furthermore, scattered references in ancient texts to the dignity of natural history, farming, and gardening could help persuade patrons to subsidize the work of sixteenth-century naturalists. By drawing creatively on the past, Renaissance naturalists shaped a new field within the humanist idiom of restoration.

Figure 3.2. Title page of Carolus Clusius, *Rariorum plantarum historia* (1601). © Bibliothèque centrale MNHN Paris 2005.

Figure 3.3. Title page of Johann Bauhin and Johann Heinrich Cherler, *Historia plantarum universalis* (1650). © Bibliothèque centrale MNHN Paris 2005.

The past, then, provided materials out of which natural history was fashioned. But by themselves, texts could provide neither the social framework nor the intellectual tools that characterized Renaissance natural history. For those, several preconditions were necessary. On the deepest level, Renaissance naturalists had to challenge the reasons that led most intellectuals and writers in the Middle Ages to pay little attention to the particulars of nature. More immediately, the careful study of nature had to be connected with other intellectual concerns of the day. Both of these preconditions were fulfilled by the humanist movement of the fifteenth and sixteenth centuries.

Humanists contributed to natural history in many ways. Humanist scholars and their pupils emphasized the moral and aesthetic benefits of a precise knowledge of nature. This emphasis was one aspect of the humanist model of knowledge in general, which, in the simplest possible terms, privileged facts, experience, and empirically based maxims over theory and universal truths. Where medieval philosophers concentrated on universal essences and natures, humanists focused on outward appearances and particulars—in rhetoric, in history, in philology, but also in the study of nature. In the case of nature, this approach harmonized with an aesthetic judgment rooted in the particular: humanists derived more pleasure from nature when they knew it intimately.

In this atmosphere, engagement with the problems of classical natural history texts—in particular, the knotty relationship of Pliny the Elder to the Greek writers on nature—sparked the beginnings of Renaissance natural history. As we have seen in chapter 2, natural history emerged in a specific context: the medical philology of the late fifteenth century in northern and central Italy. The archetypal humanist activity, commenting and castigating ancient texts, was brought to bear on the works of Pliny, Dioscorides, and other ancient writers on nature. Some humanists—the Venetian Ermolao Barbaro, for example—were devoted to the text and attempted to correct corrupt texts on the basis of other written works. But others, like Leoniceno, saw the text as a practical guide to the natural world. Leoniceno, the physician, wanted to reform medicine, and medicinal virtues were his major concern. But he was concerned with them in an intellectual climate that encouraged the study of particulars, and he could only acquire correct knowledge of medicinal virtues by meticulously comparing ancient descriptions with modern plants. Renaissance natural history grew out of this double movement. Its first fruits can be seen in the botanical activity of one of Leoniceno's students, Euricius Cordus.

The Invention of a Tradition

The history of natural history begins with Aristotle. Though plants and animals had been domesticated for millennia, and though earlier Greek writers and the author of Leviticus had noted and commented on them (and, if we are to believe the first book of Kings, also Solomon), Aristotle is the first writer whose extant works treat natural history as an intellectual pursuit.[4] His *History of Animals* and his successor Theophrastus's *History of Plants* contributed to the inquiry (*historia*) into nature that characterized Hellenic thought.[5] Both treated this inquiry into nature as a matter of gathering empirical data and then seeking to explain its causes.

Renaissance naturalists, too, began their historical accounts with Aristotle and Theophrastus. In his *Historia animalium*, Conrad Gessner claimed that he was merely a compiler of facts: Aristotle, on the other hand, wrote about animals scientifically.[6] In the same vein, Thomas Johnson, the tireless corrector of John Gerard's mistakes, claimed that Theophrastus had fathered botany.[7] Though Aristotle's works on natural history, usually referred to collectively as *De animalibus*, never received more than a small fraction of the attention his logical, physical, and ethical works did during the Renaissance, they engaged some of the best minds of the period, and naturalists read them, at the very least in Theodore Gaza's translation.[8] Similarly, Theophrastus, also translated by Gaza and others, was readily available in Greek and Latin to sixteenth-century readers.[9] Their methodological remarks and theoretical conclusions were part of the common knowledge of Renaissance naturalists.

While Aristotle's and Theophrastus's works presented the conclusions of their researches, they said little about how they carried out their investigations. Into this silence, some scholars have projected an entire peripatetic research school. Both Aristotle and Theophrastus included hearsay as well as autopsy among "the phenomena" of nature.[10] It is tempting to surmise that they also sponsored systematic research, assigning traveling naturalists to gather information from specific areas and to report back to their bosses in Athens. Many sober scholars have yielded to the temptation, claiming, to take one recent example, that the Macedonian "civil service" answered questionnaires for Aristotle and that former Lyceum students formed a corps of trained observers.[11] Some have insisted that the *History of Animals* presumed Alexander's Indian campaigns as a source.[12] These scholars are following a much earlier tradition, dating back to the Hellenistic era.

Aristotle and Theophrastus did conceive of their research as an ongoing project. Theophrastus consulted early drafts of the *History of Animals*, and Aristotle also referred frequently to versions of his works that are no longer extant, such as the *Problemata.*[13] They also knew about India. Aristotle occasionally referred to Indian animals like the elephant and the ant-lion, while Theophrastus devoted several chapters to the plants of India, as well as to those of Arabia and Egypt. These brief references, however, probably result from trade contacts. Theophrastus seems to have questioned travelers in Athens; whether he sent them out is another matter.

The chief evidence for the existence of a Peripatetic research program in natural history comes from ancient testimony. Pliny the Elder provided the most extreme version of the tradition in his *Natural History,* composed some four hundred years after Aristotle lived:

> King Alexander the Great being fired with a desire to know the natures of animals and having delegated the pursuit of this study to Aristotle as a man of supreme eminence in every branch of science, orders were given to some thousands of persons throughout the whole of Asia and Greece, all those who made their living by hunting, fowling, and fishing and those who were in charge of warrens, herds, apiaries, fishponds and aviaries, to obey his instructions, so that he might not fail to be informed about any creature born anywhere. His enquiries addressed to those persons resulted in the composition of his famous works on zoology, in nearly fifty volumes.[14]

The earlier writer Athenaeus, writing merely a century after Aristotle, was more circumspect, but he did write that Alexander gave his former tutor the princely sum of eighty talents to carry out his research.[15] The story grew in the telling until, as in Pliny's version, Alexander was ordering his entire realm to assist Aristotle in his researches.

These and other ancient accounts of Alexander's magnanimity toward Aristotle have been gathered and convincingly refuted by Maurice Manquat.[16] He notes that Aristotle's knowledge was the fruit of a few years' study. It was an impressive, considering Aristotle's other occupations, but it did not require a team of researchers. Moreover, it is implausible that Alexander's soldiers would have had time and resources to send specimens back. That is reading a Roman (or British) civil service back into Alexander's time. J. M. Bigwood proposes that Ctesias and other earlier writers, not Alexander's henchmen, were Aristotle's sources on the elephant and other Indian matters.[17] No textual evidence proves that Aristotle received information from Alexander. Theophrastus did write that he had spoken with soldiers who had been in India, but that was probably a group of veterans from Athens or on their way through it.

If there was a peripatetic research program in natural history beyond the individual efforts of Aristotle and Theophrastus, it was a small-scale effort carried out at the Lyceum, possibly on a scale to match the collection of 150 Greek constitutions gathered under Aristotle's direction. The Lyceum's garden may have been used to grow plants, although Theophrastus rarely referred to it in his botanical works.[18] Aristotle and Theophrastus interrogated fishers, farmers, herdsmen, and others with practical knowledge of animals, but they did not do so with funding and a staff from the great Macedonian conqueror.

But if the tale told by Pliny has managed to fool twentieth-century scholars, it was also taken seriously by Renaissance naturalists. Conrad Gessner repeatedly urged his patrons to play Alexander to his Aristotle, to supplement his limited means for travel and collection with generous stipends.[19] If Alexander were alive, Gessner wrote, he would be astounded to see that the only one of his works that survived was Aristotle's history of animals: his empire had collapsed after his death, his cities had been destroyed or had fallen into barbarism, and many of the histories that celebrated his deeds had been lost. But those books of animals have always been cherished by scholars, and they redound as much to his fame as to Aristotle's.[20] The notion that Aristotle's and Theophrastus's works had been collective enterprises suited perfectly the collective nature of Renaissance natural history, and it suggested to Renaissance naturalists another way they could imitate the ancients.

Aristotle and Theophrastus had few successors in their enterprise, collective or not. Their works on animals and plants were copied, not superseded, in subsequent generations. A true successor arose only in the thirteenth century: the Dominican natural philosopher Albertus Magnus. Albertus's works on animals and plants, conceived as commentaries on Aristotle, were conducted with the same goal: a philosophical explanation of animals and plants on the basis of both careful observation and the reports of his predecessors.[21] But Albertus too had no immediate successors.

This Peripatetic tradition—if "tradition" is the right word for Aristotle, Theophrastus, and Albertus—offered Renaissance naturalists one model for natural history, one that emphasized experience as the basis for a philosophical explanation of animals and plants but that could also discourage dwelling too long on particulars or investigating them too closely. Aristotle's trenchant remarks about observation and its problems must have prompted his Renaissance readers to reflect on their own practices, but his and Theophrastus's use of secondhand reports may have undermined his methodological precepts. Pliny's story about the resources that Alexander devoted to the Lyceum's investigation would also find an echo in the Renaissance. The books that Renaissance naturalists produced, though, were modeled not on the philosophical natural history of Aristotle and Theophrastus but, rather, on the descriptive works of medical writers and encyclopedists—and above all, on the Greek medical herbal of Dioscorides. The medical and herbalist tradition to which we now turn was the second strand woven into Renaissance natural history.

Between Theophrastus and Albertus, philosophers were little interested in the systematic study of living beings. They left the field to the physicians.[22] In the last books of the *History of Plants* and the *Causes of Plants*, Theophrastus had already devoted considerable attention to the medicinal uses of vegetable products, a subject with clear practical interest. His successors (if they may be considered such) restricted their inquiries to this domain. Within it, they naturally devoted attention to the form and habits of plants, but that attention was rigorously subordinated to the medicinal virtues of plants.

The greatest ancient herbalist was Dioscorides, whose *De materia medica* was the *summa* of ancient descriptive botany.[23] The empiricism that Dioscorides advocated in his preface was taken up as a slogan by Niccolò Leoniceno and his followers, who contrasted Dioscorides's allegedly firsthand observations with medieval physicians' slavish imitation of corrupt Arab authorities. In his descriptions, however, Dioscorides did not bother to claim observations as his own. He wanted to give a true account of plants and their medicinal virtues, unlike the false claims of his rivals. But his aims were practical, not scholarly, and he did not fussily separate what he had written from what he took from other sources. Experience, after all, could validate someone else's description. Dioscorides provided Renaissance naturalists with both a model for descriptive natural history and a challenge: his descriptions, often maddeningly vague, raised as many problems for Renaissance naturalists as they solved.

Dioscorides's work was so successful that few other ancient works on *materia medica* have survived. Galen's works on simple and compound medicines discussed medicinal virtues, but the prince of physicians explicitly renounced descriptions.[24] Instead, Dioscorides's work was copied, excerpted, and translated into Latin, becoming the basis for the herbal literature of the Latin Middle Ages. Herbalism was undoubtedly the liveliest area of studying nature in the Middle Ages. Largely craft knowledge, passed down orally, herbal lore was occasionally recorded in such diverse sources as the Carolingian *capitulare de villis* and the medical writings of St. Hildegard of Bingen.[25] One can imagine apothecaries or monastic herbalists meeting and eagerly comparing notes, but these exchanges would have been limited to knowledge of a relatively small number of traditional simples. Moreover, the herbal literature underwent a significant shift in the Middle Ages. Descriptions vanished from works on *materia medica,* leaving behind simple plant lists and collections of recipes.[26] This shift went hand-in-hand with the declining knowledge in medieval Europe of the Mediterranean flora described by the ancients: rather than write descriptions of the new plants that were being used in medicines, medieval herbalists preferred to rely on their own memory, and oral tradition, to identify the plants whose virtues formed the basis of the medieval pharmacopoeia. As a result, their detailed knowledge of plants remained personal and was confined to local contexts.

Some standard reference works, and their modern successors, were widespread. Versions of Dioscorides abounded. The fourteenth-century *Buch der Natur* by Conrad von Megenburg was also widely distributed, usually bound with medical books, suggesting that physicians, apothecaries, or literate lay healers found its contents useful for their practice, or at least for their bedside manner.[27] Other books on the nature and virtues of animals and plants were also written and read in the Middle Ages, including bestiaries. But they were written with other aims than conveying philosophical or medical knowledge of the natural world.

The most immediate and profound knowledge of the natural world, in antiquity and the Middle Ages, was undoubtedly that of the men and women who worked the land and collected its fruits. Precisely this intimacy, however, renders such knowledge elusive to the historian, except in the highly mediated form of the agricultural manual. Classical agricultural manuals, especially Palladius, were copied and read by literate nobles or their estate managers in the Middle Ages.[28] Renaissance naturalists drew on these manuals and their medieval successors. But the agricultural manuals remained on the margins of Renaissance natural history. Their economic focus ran contrary to the ethos of much Renaissance natural history, while they had relatively little to say about the form, virtues, and life cycles of wild plants. The intimate link between natural history, agriculture, and national economy that developed in the eighteenth century had only begun to develop in the Renaissance. A distinct tradition of Renaissance agricultural and gardening treatises did exist—for example, Olivier de Serres's and Thomas Hill's manuals from the late sixteenth century.[29] And a few members of the sixteenth-century community of naturalists did write about the economic benefits of the study and acclimatization of exotic plants; Pierre Belon's 1558 *Remonstrances sur le defaut du labour & culture des plantes* urged the French king Henri II to establish a garden in which foreign trees could be naturalized, to the great intellectual and economic benefit of the realm. Carolus Clusius and his publisher Plantin considered this book interesting enough to translate into Latin and publish in 1589.[30] But such concerns remained marginal to the daily activities and intimate concerns of Renaissance natural history.

The three traditions in which the study of nature was pursued—natural philosophy, pharmacy, and agriculture—rarely intersected in antiquity and the Middle Ages. There were some points of contact. Aristotle and Theophrastus learned from experienced herdsmen, hunters, and farmers, but it is doubtful that their sources knew or cared much about the Peripatetic project in natural philosophy. Dioscorides read Theophrastus, at least for his practical tips on the virtues of plants. Pliny the Elder pillaged every book he could get his hands on, extracting facts by the hundreds and depositing them in the book that gave natural history its name. Albertus Magnus, too, brought together the natural philosophical tradition, the pharmaceutical tradition of Dominican herbalists, and his own discussions with farmers. But Albertus's interests were too broad to give much depth to his interest in natural history—which was, after all, only one part of his immense commentary on all of Aristotle. After the Peripatetic school's turn away from natural history, there was no formal institutional setting in which these interests were combined and sustained. The few exceptions—Albertus himself, or the court of Frederick II, where falconers and scholars compared Aristotle with their own practical experience—were not successfully institutionalized.[31]

This fleeting overview of ancient and medieval study of nature reveals three interrelated characteristics. First, "experience" constituted both firsthand inspection and secondhand information, from written works or hearsay. Some investigators, like Dioscorides and Frederick II, distinguished between the two in theory, but in practice they did not draw a sharp line between their own observations and those taken from other sources. This, we shall see, is not true of sixteenth-century natural history, in which personal experience was increasingly preferred to secondhand reports, and those reports were carefully analyzed to determine their reliability.

Second, there was no single community of naturalists. Peripatetic philosophers, herbalists, and hunters comprised separate circles, and there was little personal contact or transfer of information between them. What exchange took place was on a literary level; Albertus Magnus's work, in which the observations of the ancients play a decisive role, provides the clearest example. The study of nature was a literary genre, not a social activity. As a result, ancient and medieval knowledge of nature was fragmentary. It was gathered together piecemeal through individual experience and study, or through initiation into a specific, clearly delimited craft tradition in the case of pharmacy. Above all, it was local and contextual knowledge, with few opportunities or reasons to go beyond the immediate and the practical. Typical of the late medieval situation was the fact that Rufus and Simon Cordo, two herbalists in the late thirteenth century, were unaware of one another's work, despite living in many of the same towns.[32]

Third, those who did pay careful attention to nature did so with aims that either discouraged lingering too long on particulars or urged too close attention to a few particulars at the cost of others. Natural philosophy discouraged studying the particular, which was no part of philosophy, and urged instead the ascent to universals, the discovery of natures or essences. Pharmaceutical botany was concerned above all with medicinal virtues, and in the Middle Ages also with the problem of *succedanea,* familiar plants that could be substituted for the exotic species found in ancient recipes. At the same time, apothecaries had no reason to bother with plants that did not have medicinal value. Farmers, pastoralists, and hunters, meanwhile, had an exhaustive focus on the particularities of the plants and animals they raised, or the game they hunted, but little reason to concern themselves with nature beyond the places where they lived.

I do not mean to suggest a teleology, as if it were natural for natural history to develop and that we must seek hindrances to explain why it did not. There is no prima facie reason that the natural philosophy, agricultural, and medical contexts for the study of nature should have come together as they did in the Renaissance. Nonetheless, it is profitable to explore, albeit briefly, more general attitudes toward nature in the high and late Middle Ages, because changes in these attitudes were a precondition for the development of natural history as a cultural form practiced by thousands of widely dispersed naturalists. Modern natural history is the product of a tissue of contingent events, but underlying the specific concerns of a Leoniceno, a Fuchs, or a Clusius is a more profound shift in attitudes toward nature.

Nature in Medieval and Renaissance Literature

Medieval Europeans did not lack an appreciation for nature.[33] Indeed, medieval literature brims with frequent references to nature and natural phenomena.[34] But these references were often governed by strict formal rules, and they reflect particular social and intellectual milieus. The vernacular world of courtly romance appreciated nature, in particular a managed nature of gardens and groves, as a setting for adventure. But this nature was painted with a palette of colors drawn from antiquity. Meanwhile, in the intellectual circles of cloister and university, the things of this world were subordinated to the spiritual and the world to come.

Medieval descriptions of nature usually followed strict conventions.[35] Poets drew on classical models to describe two idealized forms of nature: the mixed wood and the *locus amoenus,* or "pleasance." In each case, the elements of description were determined by classical precedents, and the resulting descriptions, far from being impressionistic, were highly structured.[36] The poets who created such landscapes as settings for their romances undoubtedly felt their beauty—otherwise it is hard to see why they would have chosen models to imitate.[37] But a telling detail may tell us more about what a poet read than what he saw. As Ernst Curtius wryly observed, the lions and olive trees populating the landscapes of medieval Europe did not come from Mediterranean gardens and menageries. They came from the lines of Horace, Vergil, and Ovid.[38] More prosaically, Charlemagne's *Capitulare de villis* called for the cultivation of herbs that could not grow north of the Alps, another example of such literary borrowing.[39]

The classical mixed forest, as developed in epic and pastoral poetry, involved the description of a large number of trees. Ovid and other poets described the mixed forest by enumeration: that is, by naming species and giving each a brief descriptive epithet.[40] The *locus amoenus* was also constructed by enumerating plants, particularly flowering herbs in a grassy sward, which typified beauty. As with the mixed wood, the *locus amoenus* was a literary topic.

Medieval writers often described nature following these topics. This was already the case in the allegorical use of nature for religious instruction in Saint Ambrose's *Hexaemeron* or the *Physiologus.* It was also true of Hildebert of Lavardin's nature description in "De ornatu mundi" (twelfth century), which is colorful and varied, but based on antique models rather than personal observation of nature. The Goliard poets' joy in this world was unquestionably genuine, but their descriptions of nature also drew on earlier models rather than personal observation of particular localities.[41]

The late medieval poets' appreciation of nature was thus expressed in forms drawn from their predecessors. There is nothing unusual about this, for literary tropes have a life of their own and are often unwilling to fade into oblivion. Given the normative influence of classical literature on medieval culture—both directly and through epitomes and florilegia—it would be surprising had medieval poets not quoted and adapted their predecessors. Such imitation continued into the sixteenth and seventeenth centuries; one need think only of Spenser and Milton. What is peculiar, however, is that precisely in the Renaissance—a period in which literary imitation was not only practiced but theorized[42]—this medieval approach to nature description was, as we shall see, expressly criticized for its inaccuracies.

While romance writers and poets presented an idealized, managed nature in a favorable light, theologians, as usual, were more negative, while philosophers were generally indifferent. The twelfth-century Cluniac monk Bernard of Morlaix's poem "De contemptu mundi" contrasted the faults of the present world with the pleasures of that to come. Bernard compared our state with the fleeting beauty of nature. We are born, bear burdens, and die, just as "what was a flower ran to ruin in the space of an hour."[43] But after the Final Judgment we will live in Paradise, and the good Christian should concentrate on getting there, not enjoying his present state. The fourteenth-century author of the *Imitation of Christ,* an immensely popular devotional work from the northern Netherlands, shared a similar view. Fully appreciating the Creator meant focusing one's attention on Him and neglecting the creation. The *Imitation* opens by echoing the Ecclesiastes: "All that is in this world is vanity." God should be served by forsaking the world: "Unless a man be clearly delivered from the love of all creatures, he may not fully tend to his creator."[44] Unlike later natural theologians, the author held that the study of created things diminished, not amplified, the glory of their Creator. Insofar as the things of this world are worthy of our attention, it is as signs written by the Creator. "If your heart is honest [*rectum*], then every creature will be a mirror of life, and a book of holy doctrine."[45] Creatures exist to be seen through, not seen.

Bernard's and the Imitator of Christ's radical indifference to the creation represent an extreme. But in emphasizing the allegorical interpretation of the Creation, the *Imitation of Christ* points to a more widespread attitude that plants and animals should be studied for the spiritual lessons they can teach, not out of mere curiosity. It was encouraged by the relatively brief, etymological treatments of animals and plants found in medieval encyclopedias, such as Isidore of Seville's immensely popular *Etymologiae.* Isidore saw little need to describe creatures in detail; they could be understood through their names and those names' etymology. The beaver provides a particularly memorable example:

> Beavers [*castores*] get their name from castration. For their testicles are useful for medicines, and because of this, when they see a hunter, they castrate themselves, biting off their virile parts [*vires suas*]. Cicero says of them in the *Scauriana:* "They ransom themselves with that bodily part on whose behalf they are especially sought." Juvenal writes: "who makes himself a eunuch, that he may escape—with loss of a testicle." They are the same animal as *fibri,* also called Pontic dogs.[46]

The same sensibility informed medieval bestiaries. Sometimes wrongly taken as protoscientific treatises, these devotional texts used the visible world as moral and spiritual guides, to help laypeople better remember their lessons.[47] After describing the "Cocodryllus" as a creature of the River Nile, "generally about thirty feet long, armed with horrible teeth and claws," the author of one bestiary added,

> Hypocritical, dissolute and avaricious people have the same nature as this brute—also any people who are puffed up with the vice of pride, dirtied with the corruption of luxury, or haunted with the disease of avarice—even if they do make a show of falling in with the justifications of the Law, pretending in the sight of men to be upright and indeed very saintly.[48]

The same bestiary rehearsed Isidore's story about the beaver, adding that the chaste man must "cut off from himself all vices, all motions of lewdness, and must cast them from him in the Devil's face."[49] Unlike fables—an ancient form revived in the Renaissance—which attribute human failings to animals, the bestiaries allegorized nature. There was no necessary connection between the animal behavior they described and the moral gloss given to it; rather, the allegory served to make the moral more memorable.

If bestiaries' readers learned a few facts about animals, that was incidental. If, as the twelfth-century hymnist Alan of Lille wrote, "Omnis mundi creatura / Quasi liber et pictura / Nobis est et speculum" [Every creature of the world is like a book, picture, and mirror to us],[50] then creatures are to be read for their spiritual content. Just as we do not study letters on a page for themselves, we should not study creatures as natural things. Moreover, as Hans Blumenberg has noted, Alan did not even have in mind the spiritual meaning of the creatures: he continued, "Nostrae vitae, nostrae mortis / Nostri status, nostri sortus / Fidele signaculum" [A faithful symbol of our life, our death, our state, and our lot]. We look at the book of nature only to see our own reflection.[51]

This position only apparently conflicts with medieval natural theology, which used nature to prove the existence and goodness of God. Ancient and medieval natural theology was not founded on a careful, detailed investigation of nature. Rather, it drew upon commonplaces such as astronomy, as in Cicero's *De natura deorum,* or animal stories, such as Basil the Great's *Hexaemeron,* that could be found in earlier authors. It was not based on careful contemplation of individual living things. In classical antiquity and the Middle Ages, the sense of wonder that permeated all natural theology was, as Clarence Glacken observes, usually a substitute for curious and careful inquiry into the natural world.[52]

It would be a mistake to think of this allegorical attitude, in which the created world is important only as a reflection of its Creator, as *the* medieval attitude toward nature. Albertus Magnus showed a keen interest in the things of this world. Though he subordinated natural history to natural philosophy, he clearly delighted in observing plants and animals. But Albertus was an isolated instance. And as a good medieval philosopher, he admitted that the study of particulars was not philosophy.[53] Secular poets were more interested in the particulars of the natural world than the clerical authors of Latin bestiaries and other devotional tracts. But their descriptions of nature were conventional, derived primarily from earlier literature rather than a careful observation of nature. Whether theologians or philosophers, medieval scholars' attitude toward nature was that its particulars were not essential. Spiritual contemplation, allegorical interpretation, or poetic diction were more important.[54] Conversely, the curiosity that might lead one to carefully observe natural things for their own sake was included by theologians among the vices.[55]

Even Petrarch, the prototypic Renaissance humanist, held that vain curiosity about the things of this world was irrelevant to his deepest concerns.[56] He believed that many of the bestiaries' stories about animals were false. But "even if they were true, they would not contribute anything whatsoever to the blessed life."[57] As an educated man, capable of thinking philosophically, Petrarch had no need of bestiaries as spiritual textbooks, and following Augustine, he preferred to neglect this useless branch of knowledge in favor of important things.

Petrarch did accept natural theology. He cited, at length and approvingly, Cicero's argument in *De natura deorum* and the *Tusculan Disputations* that the things of this world are so clearly designed that a creator god must exist.[58] But this argument from design relies primarily on the order of heavenly bodies, the most noble created things, not precise knowledge of more mundane creatures.[59] And unlike the natural theology of John Ray and William Paley, which revealed design in the tiniest features of insects, Petrarch (following Cicero) relied on a handful of grand examples to prove his point.[60] Once the existence of God was established through natural reason, there was little cause for further study of the world.

Even Petrarch's most eloquent expression of his experience of nature—his description of the ascent of Mont Ventoux—is tinged with suspicion of the world and his pleasure in it. Spurred on by curiosity and by a passage in Livy, Petrarch decided to ascend this mountain, in the vicinity of Avignon, for no reason other than "the desire to see its conspicuous height." After an arduous ascent, made more difficult by his search for an easy route up, he reached the top. He admired every detail. As he, his brother, and their servants prepared for the descent, he "was completely satisfied with what I had seen of the mountain."[61]

Petrarch's text, however, describes little of the natural beauty that attracted him. Rather, it underscores the spiritual implications of his journey. At the beginning, he meditated on the proper choice of a traveling companion, eventually deciding that all his friends were unsuitable. On the ascent, he searched for a long, easy way—a search that led him down three times into the valley before he followed his brother's example and took the rough, high path. In Petrarch's allegory, the blessed life is the high peak, and the longing for mean earthly pleasures keeps us away from it. At the top, after taking in the view, Petrarch opened his copy of Augustine's *Confessions.*

> I opened it with the intention of reading whatever might occur to me first: nothing, indeed, but pious and devout sentences could come to hand. I happened to hit upon the tenth book of the work. My brother stood beside me, intently expecting to hear something from Augustine on my mouth. I ask God to be my witness and my brother who was with me: Where I fixed my eyes first, it was written: "And men go to

admire the high mountains, the vast floods of the sea, the huge streams of the rivers, the circumference of the ocean, and the revolutions of the stars—and desert themselves." I was stunned, I confess.[62]

Petrarch closed the book, and on the entire downward descent he reflected on the spiritual life and his vain longing for the things of this world. Petrarch's letter is carefully crafted to underscore the metaphoric nature of his ascent. His delight in nature is, ultimately, a sign of his weakness. He did not *reject* nature; indeed, he was completely satisfied with his sightseeing. But like most medieval thinkers, he saw it as trivial in comparison with the ultimate spiritual questions.

For Petrarch, as for his contemporary, the Imitator of Christ, the attractions of nature—like the fleshpots of Avignon—were dangerous temptations, and the pleasure taken in them was a guilty one. At best, nature was the province of poets, when not of farmers. It was not an intellectually respectable subject. The only exceptions were, perhaps, natural wonders—the unicorn's horn, or the whale ribs often hung in parish churches—that testified to the wisdom and power of God and were, intellectually if not theologically, in a class with saints' relics.[63] A medieval intellectual might well consider the lilies of the field, but he would do well not to consider them too long or too deeply. He had more important concerns.

In this intellectual climate, focused on the spiritual and wary of curiosity, it is not surprising that the study of nature was the province of exceptional individuals such as Albertus Magnus, who clothed his interest in the respectable garb of philosophy, or Frederick II, who as emperor did not have to justify his pastimes. Petrarch, who in many respects felt himself to be an anachronism, was in this regard fully in the mainstream of his time. For most late medieval intellectuals, a limited, practically oriented knowledge of nature sufficed. Apothecaries and physicians needed to know a few things more closely, but they pursued their studies with practical goals in mind. The aesthetic appreciation of nature was distinct from specific knowledge of it—as is, indeed, often the case: the Romantic appreciation for natural beauty reached its culmination in the luminous mists of Turner.

The fifteenth century saw a significant change. In the generations after Petrarch, many humanists promoted the attentive study of the world, including plants and animals, in the framework of an epistemology that granted much more importance to particulars than the neo-Aristotelianism of the schools, with its focus on substantial forms, natures, and essences. Moreover, they developed an esthetic of the particular, in which appreciation of natural beauty was closely linked to proper knowledge of it. These two developments set the stage for the invention of natural history in the sixteenth century.

Humanism was a product of cities. But humanist scholars delighted in country retreats. Leonardo Bruni set his second *Dialogue for Pier Paolo Vergerio* in a country house belonging to Roberto de' Rossi.[64] Jacob Burckhardt described with great sympathy Enea Silvio Piccolomini's delight in nature, especially during his stay on Monte Amiata in the sultry summer of 1462.[65] Burckhardt was wrong to take such displays as indicative of a new appreciation of nature, "the discovery of the world," as he glossed it (following Jules Michelet). Rather, what is new with the humanists is a positive valuation of this experience. Whereas Petrarch had been torn between the seductions of the vista from Mont Ventoux and his belief in the vanity of worldly delights, later humanists embraced the world. Committed to the active life and engagement in the world, humanists felt no shame at intellectualizing the pleasure they took in nature.

Conrad Celtis, who was instrumental in bringing the humanist movement over the Alps into Germany, exalted in the world. Celtis's praise for nature and this world is couched as a reproach to the humanists' critics, who saw the *studia humanitatis* as a thinly-disguised paganism:

> Miraris nullis templis mea labra moveri
> murmure dentifrago.
> est ratio, taciti quia cernunt pectoris ora
> numina magna poli.
> miraris videas raris me templa deorum
> passibus obterere.
> est deus in nobis, non est quod numina pictis
> aedibus intuear.
> miraris campos liquidos Phoebumque calentem
> me cupidum expetere.
> hic mihi magna Iovis subit omnipotentis imago,
> templaque summa dei.
> silva placet musis, urbs est inimica poetis
> et male sana cohors.
> i nunc, et stolidis deride numina verbis
> nostra, procax Sepule.[66]

> [You wonder why I don't wag my lips in church with tongue-twisting murmurs? There's a reason: the great spirits of heaven hear the prayers within my silent breast. You wonder why I rarely tread in the temples of the gods? God is within us, and there is no reason to regard the gods in painted houses. You wonder why I greedily seek bright fields and the hot sun? Here the great image of omnipotent Jove comes to meet me, and the highest temples of god. The woods please the muses, while the city is hostile to poets and full of the unhealthy crowd. Go now, impudent Sepulus, and ridicule our gods with your foolish words!]

Celtis's reply is as stylized and exaggerated as the objections of his hypothetical critic, for humanism was rooted in cities, and Celtis himself wrote a famous praise of Nuremberg.[67] Hellenistic and Virgilian pastoral echo in the lines. Still, though Celtis probably did not court the sylvan muses during every Sunday mass, his use of the polarity between church and forest, the one with columns of stone, the other of wood, recognizes the theologians' distinction between this world and the next while embracing the former. And while the use of *mirari* is hyperbolic, it doubtless seemed strange to many contemporaries that learned scholars would want to tramp around in wet meadows and in the heat of the midday sun—and there is no reason to doubt that Celtis was sincere (*me cupidum*).

The positive emotional and aesthetic valuation of nature by fifteenth-century humanists is seen also in Francesco Colonna's *Hypnerotomachia Poliphilii.* A bizarre mixture of allegory, soft porn, and architectural fantasy, the work presents the dream adventures of the young Poliphile in a strange, unknown land. Throughout his adventures, the hero is comforted by plants, which represent an ordering principle in nature. After wandering through a gloomy forest and falling asleep in exhaustion and despair, the hero wakes in a far more pleasant place. In this grove, among all types of trees and bushes, Poliphile sees "clover, sedge, common bee-bread, umbelliferous panacea, flowering crowfoot, cervicello or elaphio, sertula, and various equally noble herbs; also many other beneficial simples, and unknown herbs and flowers, strewn about the meadows."[68]

Colonna did not merely enumerate medicinal herbs; he knew enough to put them where they belonged. After escaping from a labyrinth where he was chased by a dragon, Poliphile emerged in another pleasant place.

> I could see an unknown mountain of medium steepness, all wooded with green and pleasant foliage: acorn-bearing roburs, beeches, and oaks—holm-oaks, Turkey-oaks, winter-oaks and cork-oaks—beside the two

> kinds of ilex of which one is the yew, the other the holly or prickly one. Further on toward the plain it was dense with cornel-cherries, hazels, fragrant and flowering privets with their perfumed blossoms of two colours, showing red to the north and white to the south; hornbeams, ash, and burgeoning shrubs that looked similar to these, which were wrapped in green climbing honeysuckle and twining hops, giving a cool and opaque shade. Beneath these were the sow-bread that hinders childbirth, frilly polypodium, triple-leaved spleenwort, the two kinds of hellebore named after the shepherd, trifoliate or triangular tora, sanicula and other shade-loving herbs and woodland trees.[69]

Colonna expected his readers to take great pleasure from such enumeration, precisely as they should enjoy the detailed architectural descriptions of the book. Architectural and natural description met in Colonna's description of the Isle of Cytherée, with its garden laid out in a geometrical pattern.[70] Colonna's fantastic topiary was clearly a pleasant invention, but the plants that filled this garden were real: the "most beautiful, varied, and rare botanical species, punctiliously listed."[71] The humanists for whom Colonna wrote the *Hypnerotomachia* were supposed to enjoy the punctilious listing of plants as much as Colonna's classical conceits.

Such tropes were not confined to poetry and literary prose. Amid technical criticisms of his predecessors, Niccolò Leoniceno remarked on the joys of experiencing nature. Though he aimed to improve knowledge of the plant world, and the majority of his writing was technical, he anticipated the pleasure of wandering through the hills around Lucca in the last part of his *De Plinii erroribus*.[72] Such pleasures, though not intellectually defensible reasons for the study of nature, must have contributed to its pursuit—as indeed they did for the few medieval writers who offered themselves to such vice. The humanists' removal of the pleasures of nature, if not curiosity, from the catalogue of vices prepared the way for a much more widespread study of nature.

Even Desiderius Erasmus, the embodiment of Christian humanism, expressed an appreciation for nature that would have distressed Petrarch.[73] Though he was educated by the Brethren of the Common Life, followers of the *Imitatio Christi*, Erasmus's colloquy on the *convivium religiosum* (the godly feast) did not contrast this world with the next. Instead, he underscored the importance of nature—and, indeed, a precise knowledge of nature—to the spiritual life.

The godly feast takes place in a garden. The host, Eusebius, invites a group of friends from the city to his "suburban" estate, where he promises to feed them entirely with the produce of his property. When they arrive, Eusebius gives his guests a tour of his gardens. A small flower garden greets them at the door. A larger, more elegant garden is reserved for pleasant herbs, each type in its own plot. These plots are labeled not with the plants' names but with literary or proverbial allusions to them. For example, marjoram bears the label, "Keep away, pigs: I don't have a scent for you," referring to the belief that swine find the pleasant odor of the plant repulsive.[74] The garden contains three galleries with faux-marble columns for study, conversation, or contemplation, and its walls are painted with the images of other plants and animals. Even in winter the garden is green.[75] Beyond this *hortus conclusus* lie further gardens: one divided in two for kitchen and medicinal herbs; an open, grassy field; an orchard; and an aviary.

As literary work, the *convivium religiosum* shows clear signs of influence of the classical *scriptores rei rusticae,* and the Roman agrarian ideal, particularly the insistence on a self-sufficient household.[76] The garden as part of the ideal country estate also has a long literary heritage.[77] In its form, however—the *hortus conclusus* for pleasure, with separate medicinal and kitchen gardens, an orchard, a lawn, and an aviary—Eusebius's garden follows the pattern of the late medieval bourgeois garden.[78] It is a particularly splendid version of the patrician gardens found in south German and Dutch cities of the late fifteenth and sixteenth centuries.

Furthermore, Erasmus underscored the importance of a *correct* knowledge of nature, and natural things, to aesthetic enjoyment and spiritual contemplation. The plant labels were allusive, requiring that the reader be familiar with a range of herbal lore. Guests pointed out plants that were common in Italy but rare in northern Europe, for example aconite.[79] No olive trees grew in this garden, unlike those of medieval poetry. All of its plants really grew there; and if its owner wished to regard rarer plants, he had them painted on its walls.

Erasmus, as mentioned, was educated by the Brethren of the Common Life, to whose founder the *Imitation of Christ* has been attributed.[80] But in his colloquies, intended to serve as both moral guide and textbook for good Latin style, he placed the *godly* feast in a garden, and its participants delighted in, rather than contemning, creatures as well as their Creator. In contrasting the godly feast with others, for example the profane and the poetical feasts, Erasmus condoned such behavior.[81] He made some concessions to Christian morality: unlike Italian humanists' gardens, Eusebius's is decorated with images of Christ and the Evangelists, not with pagan herms and termini. His garden is a real one, however, not merely the metaphoric *hortus conclusus* of medieval theology, the closed garden of the Song of Songs, which was taken to represent both Mary and the Church. The plants in it are real, and Eusebius and his guests experience them as individual types with their own properties.

The place of natural history in humanist studies was underscored by François Rabelais in his parodic accounts of scholastic learning and the new studies. His gigantine hero Gargantua, studying under Ponocrates in Paris, learned every day "the nature and effect of everything placed on their table that day: bread, wine, water, salt, meat, fish, fruit, herbs, roots." Mindful of the virtues of the new learning and the vices of scholasticism, Gargantua advised his son Pantagruel, when the boy went to Paris in his turn:

> I wish you to carefully devote yourself to the natural world. Let there be no sea, river, or brook whose fish you do not know. Nothing should be unknown to you—all the birds of the air, each and every tree and bush and shrub in the forests, every plant that grows from the earth, all the metals hidden deep in the abyss, all the gems of the Orient and the Middle East—nothing.[82]

Natural history was only one of the subjects Gargantua urged on his son: he was also to learn the scholarly languages, the liberal arts, civil law and philosophy, medicine and anatomy, and the Bible. Being from the race of giants, Pantagruel had a brain capacious enough for this knowledge—but Rabelais's character seems less prodigious when we think of the accomplishments of an Erasmus or a Conrad Gessner. And compared to the seventeenth century, there was little natural history to be mastered in the 1530s.

It is important to stress that Erasmus and Rabelais treated natural knowledge as both personal and moral. Eusebius and Gargantua learned natural history because mastery of the subject, *pace* Petrarch, made them better people. All knowledge was conceived in terms of the individual; as Hans Blumenberg has observed, "in antiquity and the Middle Ages, the notion of a form of knowledge which was not related to the capacity of an individual and the realization of his or her existence did not exist."[83] The humanists' notion of knowledge as an individual good—that is, of knowledge as a preliminary to action—would later stand in tension with the notion, increasingly common among sixteenth-century naturalists, that it was impossible for one man to master all of natural history. Such knowledge was inevitably collective. But in that case, perfection was chimerical; one would have to be satisfied with knowing enough, not knowing it all, a transition that is intimately linked with Francis Bacon's notion of communally gained knowledge as power, the capacity for action. In natural history, this change was largely due to the development of a more fine-grained view of nature: by developing techniques of communal observation and description, naturalists made it impossible to learn it all.

If Erasmus's focus on proper knowledge of plants represented a turn away from the unrealistic treatment of plants in the ancient and medieval topoi, Julius Caesar Scaliger criticized those topoi explicitly. Scaliger devoted the sixth book of his posthumously published *Poetices libri septem* (1561) to a critical survey of Latin poetry. Medieval poetry was beneath his consideration, but he turned a severe eye to the ancient poets' representations of nature.

Claudianus, for instance,

> interpreted nature equally unhappily when discussing the cypress, from which, he says, torches are made for Ceres. But cypress does not have pine-wood, of which torches used to be made: and even now in the mountains around here they are made from spruces. If he borrowed this expression from the Greeks, I would have preferred that he correct them rather than follow them.[84]

Scaliger was particularly fond of the cypress. Writing of Statius, he remarked, "I have no idea where he found an alpine Cypress. I have explored almost the entire Alps, but I have never seen a cypress there." On Nemesianus's line, "*—intacto premimus vestigia musco,*" Scaliger noted dryly that "there's no moss in meadows." But Nemesianus did describe dogs admirably. Catullus also comes in for criticism: "He liked to write things from his own fancy... he also claimed that laurels were straight, and he is the only writer who does so. I have never seen one with a straight trunk, so far as I remember." Even Horace slipped:

> What is more inept than this?
> *You will strive, through banks, rivers, bogs,*
> *Where sheep were lost to theft, little goats to disease.*
> Among us, the *sheep* is the more disease-prone herd animal, although in books on husbandry it is claimed otherwise.

In Carmen 4.4, which allegorizes upon an eagle's development, Horace "describes the chick as a young adult: especially when he sends it to kill lambs and dragons. Hence there can be no mention of spring here. For the bird lays eggs at the beginning of spring, and incubates them for thirty days. Only after six months are the chicks ready to hunt. In September they are still a little weak."[85]

To Scaliger, these lapses in natural history were poetic flaws. As he noted of Claudianus, even if mistakes had been taken from literary models, it was better to correct than to follow blindly. Scaliger had a pedantic streak, but his criticisms reveal how he experienced nature. His experience did not merely involve the aesthetic pleasures of being away from the city, among beautiful flowers and plants—and he did not reach for conventional elements to describe it. It was an intellectualized form of experience. Like the participants in Erasmus's *convivium religiosum,* Scaliger observed the forms of particular plants and remembered them. His experience was not merely of a meadow, or a forest: it was of individual trees and herbs, or of species. He saw in a different way than the poets he had criticized—or at least, in a different way than their expressions led him to believe.

The union of aesthetic responses to nature with proper appreciation of natural history is found not only in pedagogical tracts and literary criticism, but also at the heart of Renaissance natural history. For Conrad Gessner, the journey from the city to its rural surroundings was not merely a means to gather material for his monumental *Historia animalium* (1551–58) and his projected *Historia plantarum.*[86] Rather, his description of Mons Fractus near Lucerne praises the mountain excursion as the purest form of earthly pleasure.[87] After the hard ascent to the summit, Gessner and his companions quenched their thirst in a cool mountain spring.

> A little below the summit there is a spring on the right, in the side of the slope, hidden in a small hollow. Its pure and frigid water refreshed us marvelously from exhaustion, thirst, and heat, once we had drunk our fill and eaten bread soaked in it. I scarcely know if a greater, more Epicurean pleasure (though it is most sober and frugal) can touch the human senses.[88]

The mountains offer pleasures for all the senses, and a significant contribution comes from plants' forms, colors, and odors. "The same plant is more odiferous and more medically efficacious in the mountains"—lest his reader forget, rapt in Gessner's praise, the purpose of the excursion. True, the lazy or ignorant will not enjoy this experience.

> But give me a man of at least average mind and body, and liberally educated—not too much given to leisure, luxury, or lust—and let him be a student and admirer of nature, so that from the contemplation and admiration of so many works of the Great Artificer, and such variety of nature in the mountains (as if everything were gathered in one great pile), a pleasure of the mind is conjoined with the harmonious pleasure of all the senses. Then, I ask you, what delight will you find in the bounds of nature that could be more honest, greater, and more perfect in every respect?[89]

For Gessner, climbing the mountain was not a means of escaping the self and spiritual concerns, as Petrarch had experienced it, but rather of finding it. And the explorer must be "a student and admirer of nature," in order truly to appreciate the experience. Lest the sedentary reader miss these pleasures, Gessner included at the end of his description a list of the various plants found in the mountain. He indicated not only their names and where they were to be found, but also whether they were pleasant: the *Rosa alpina* was a "wholly pleasantly scented bush," while the *Vitis Idaea rubris acinis* was "more acidic and less pleasant" than the variety with black berries.[90] At least in this context, the sensuous pleasure of experiencing a plant in its native place was an essential part of its description.

Leonhart Fuchs, too, held that useful knowledge of plants was complemented by the pleasure that arose from their study. In the dedication to his 1542 *De historia stirpium,* Fuchs wrote that there was no need to "expound at length the pleasure and delight that the knowledge of plants brings." Everyone knows the pleasure of walking in woods, mountains, and meadows, "garlanded and adorned with these varied, exquisite blossoms and herbs." But Fuchs, like Gessner, insisted that detailed knowledge was key. The wanderer should observe plants "with keen eyes," and his "pleasure and delight is increased" with knowledge of the virtues of plants: "For there is as much pleasure and enjoyment in learning as in looking." The ancients—not only physicians like Dioscorides and Galen but even kings and emperors—had recognized the pleasure, utility, and dignity of this subject, but the study had fallen into desuetude through the arrogance of physicians, who considered the knowledge of herbs beneath them and left it to ignorant apothecaries and superstitious old women. Fortunately a few of Fuchs's predecessors had practically resurrected this moribund study, so that now it was going from strength to strength. Fuchs's letter ended with a plea to his dedicatee, the elector Joachim of Brandenburg, to support his efforts, a plea that emphasized above all the pleasure to be gained from "contemplating"—not merely seeing—the variety of plants that God had created for human use and delight.[91]

⁂

This transformation of experience in literature and natural history, seen in the century and a half between Dante and the *Imitatio Christi* on the one hand, and Erasmus and Scaliger on the other, has its parallels in the history of art. Thomas and Virginia Kaufmann's ingenious analysis of Georg Hoefnagel's *trompe l'oeil* illustrations in his *Mira calligraphiae monumenta* has demonstrated that these images—which have nothing to do with the text but are provided for the sheer delight of accurate imitation of nature—derive from medieval pilgrimage practices.[92] Pilgrims in the late fourteenth and fifteenth centuries would collect items from different stages of their pilgrimage, for instance specially-minted pilgrim's medals, but also flowers, leaves, branches, or other parts of plants from the holy sites they visited. These would be affixed to the pages of devotional works. In some devotional books, illusionistic illustrations replaced such objects, either because it was impractical to include them in a rich book or because they had faded or decayed. Hoefnagel's illustrations, though serving a different function, are descendants of those devotional reminders.

In this case too, the medieval interest in the spiritual meanings of natural objects, objects that were important as concrete reminders of a spiritual journey, was replaced in the sixteenth century by an interest in the accuracy of naturalistic representation and delight in that accuracy. Readers of Hoefnagel's book probably enjoyed not only the representations but also the process of identifying the objects represented, just as Eusebius's guests eagerly pointed out plants that they knew. The aesthetic had taken precedence over the spiritual. Furthermore, it was an intellectualized form of aesthetic experience of nature: one in which the accuracy of representation, and the ability to recognize that accuracy, played an important role.

This form of experience was crucial for the development of a community of naturalists. By the 1520s, medical students provided a captive audience for natural history, but that alone does not explain why many of them found it an attractive pastime. Modern medical students do not usually conceive a passion for organic chemistry. The aesthetic pleasure in correctly observing nature, seen in the works of humanists like Erasmus and Scaliger, provided an important impetus for the formation of natural history as a discipline with its own community of practitioners. This intellectual appreciation of nature was itself bound up with the humanist approach to knowledge, above all the humanists' tendency to privilege empirical experience over theory, a tendency that led them to prefer natural history to natural philosophy as an idiom for talking about nature.

Cognitio Historica and Natural History

The close connection between Renaissance humanism and natural history has long been noted: in particular, Renaissance botanists' and zoologists' engagement with classical texts, a subject that will occupy us below.[93] But the affinity between humanism and botany goes deeper than the common association between Renaissance science and the ancient texts, restored by humanist critics and translators, on which scientific inquiry was founded. Humanism encouraged a specific attitude toward knowledge and the means to acquire it, one that was opposed to the natural philosophy of Aristotle and Albertus Magnus but that harmonized with the empiricism that characterized the Renaissance science of describing.

This claim may provoke anxiety or skepticism among readers who are familiar with recent arguments over the nature of Renaissance humanism.[94] The current consensus is that humanism was a cultural movement associated with the traditional humanistic disciplines of grammar, rhetoric, poetry, history, and ethics—the Roman *studia humanitatis*—and that, as teachers and scholars of the humanities, humanists neither opposed scholastic Aristotelian philosophy nor advocated neo-Platonism, Stoicism, or other philosophical systems. Individual humanists, qua philosophers, may have taken part in the culture wars surrounding Aristotelianism and its alternatives, and humanist scholarship contributed new versions of ancient texts for philosophers to deploy in their polemics, but humanism as such was simply not a form of philosophy or philosophical discourse.[95]

There is a good deal of truth to this position; certainly the humanists were not the philosophes of the Enlightenment, though many individual humanists addressed philosophical issues. But the modern position is nonetheless misleading in its implication that humanism as such was philosophically neutral, for two reasons. First, its proponents seek to define humanism by its least common denominator—by identifying characteristics that were shared by everyone whom they wish to call a humanist and excluding those that were not possessed by every humanist. In so doing, family resemblances that connect many humanists are neglected; furthermore, many humanists rejected the received form of Aristotelian philosophy even if they were not anti-Aristotelians.[96] Second, even the narrow definition of humanism as the study and practice of the humanities implies that humanists were trained in specific epistemological attitudes that are distinct from, and to some extent opposed to, philosophy as it was understood in medieval and Renaissance universities.[97]

The key epistemological attitude of humanism was a concern for the particular. Scholastic natural philosophers, even those as sensitive to the variety of nature as Albertus Magnus, privileged demonstrative knowledge of universals. They admitted that particulars could be known by means of the senses, but they denied that such knowledge was demonstrative or scientific.[98] Humanists did not deny the distinction, but their practices in the five humanistic disciplines of grammar, rhetoric, poetry, history, and moral philosophy, and in the professions of medicine and law, focused on the particular.[99] This focus privileged the particular, empirical focus of the humanist approach to nature.

Let us examine the *studia humanitatis* in more detail. Humanist grammar was perhaps the empirical discipline par excellence, and nowhere does the contrast with scholastic approaches appear more clearly.[100] Medieval grammarians had approached the subject as logicians, intent on exploring the structure of language—that is, Latin—and its relation to the world, in the process developing sophisticated theories of reference and signification that were taken up by their sixteenth-century scholastic successors.[101] Humanists, on the other hand, approached grammar empirically. In the fourteenth century, Petrarch chose to model his use of language after ancient exempla, and Lorenzo Valla's massive guide to Latin usage, the *Elegantiae,* was based on extensive investigation of ancient usage. In this work, Valla had nothing to say about language and reality; he was not antiphilosophical but rather aphilosophical.[102] Humanist grammar was not quite historicist: humanists from Petrarch to Poliziano and beyond held that classical Latin was a timeless standard against which all writing should be judged, while disagreeing about the range of authors who fell within the pale of proper Latinity.[103] Yet it was clearly empirical, and students educated in humanist schools from the early fifteenth century through the end of the sixteenth would learn to collect examples from the best writers, record them in notebooks, and model their own usage after them.[104] From the first, humanism encouraged this attention to particulars.

Humanist poetics, as we have seen in the case of the pedantic Julius Caesar Scaliger, also focused attention on particulars—not surprisingly, for as practiced by humanist teachers, poetics was a form of applied grammar. Though there were several schools of Renaissance poetics, deriving variously from Horace's *Ars poetica,* Aristotle's *Poetics,* and Plato's obiter dicta, most writers on poetics agreed that mimesis was one of the poet's chief tasks, and that verisimilitude was the criterion for successful mimesis.[105] Only the most pedantic critic could have limited poetry to mimesis, but it was a quality sine qua non. Verisimilitude could be approached in many ways. At the least, it presumed Horatian decorum, internal coherence among poetic elements, so that even if the specifics were fictive, the entire poetic work lacked incongruities: no fish-tailed women, or their poetic equivalents. The most strict critics, such as Lodovico Castelvetro and Alessandro Piccolomini, both contemporaries of Scaliger, argued that the poet should choose material only from history—for what had actually happened was necessarily similar to the truth. Poets' approach to their subjects differed from that of the historians, but their material, per Castelvetro and Piccolomini, should be the same.[106]

Few students would have encountered such esoteric poetics in their studies. Humanist schoolmasters were more likely to treat poetry, like history, as material for grammatical and ethical exposition. The poets were read as models of virtue and examples of vice, in an allegorizing light that allowed their heterodox religious and sexual mores to be passed over in silence.[107] What was important was not the work as a whole but the particular ethical precepts and grammatical rules that could be extracted from its lines.[108] Poetry, that is, served as a source for *exempla,* particular cases that carried a lesson for the reader. In this regard it, too, encouraged humanistically educated scholars to concentrate on particulars.

Textual scholarship, of course, encouraged an intensive focus on particulars, from collecting and collating manuscripts to determining the readings of specific words.[109] From the Paduan "proto-humanists" of the late thirteenth centuries, generations of humanists eagerly sought out ancient works, compiled lexica, and attempted to restore corrupt or incoherent passages. Such problematic places in the text could be attacked in two ways. Collation required the painstaking work of going through every available manuscript and finding the best reading: either the proper version or, at least, a clue to it. Conjecture, on the other hand, was flashier: often called "divination" by its critics, it involved suggesting the proper reading by drawing on a deep knowledge of classical vocabulary and usage and a finely-tuned sense of the kinds of errors that scribes made. Both kinds of criticism required a deep knowledge of the particulars of classical language and life in the ancient world, knowledge that would be displayed not only in the study but also in familiar conversation or competitive display—as happened not infrequently at the court of King Alfonso I of Naples in the mid-fifteenth century.[110]

The importance for textual criticism of detailed knowledge, along with the establishment of humanist studies in Italian universities in the later fifteenth century, reinforced the tendency, in humanist commentaries on ancient texts, to emphasize the particular. An extreme but by no means anomalous example is the *Cornucopia* of Niccolò Perotti: ostensibly a commentary on Martial, it had become, by the time Perotti finished it, a massive encyclopedia of the classical world.[111] Every verse, indeed every word of Martial's text was a hook on which Perotti hung a densely woven tissue of linguistic, historical, and cultural knowledge. Other humanist paraphrase-commentaries on poetry deployed similar, if less elaborated, apparatus, surrounding the text with a dense context, rooted in a profound knowledge of the particulars of ancient life and literature.

If grammar and poetics formed one pole of the *studia humanitatis*, the other was formed by rhetoric, history, and moral philosophy. Though the origins of humanism lay in poetry, rhetoric had taken the upper hand by the end of the fourteenth century.[112] In the chanceries of Florence, Milan, Venice, and Rome, and in the schools of Guarino da Verona and Vittorio da Feltre, rhetoric was the highest art and the orator—as defined by Cicero, Quintilian, and the author of the *Rhetorica ad Herennium*—the model of the educated man. Quintilian's *Institutiones oratoricae* served as a manual for humanist teachers, setting out the principles that they put in practice in their classrooms.[113] Because the classical orator was not only eloquent but also ethical, moral philosophy was tightly wed to rhetoric. History, too, was connected to rhetoric, in a double sense: historical examples were one of the commonplaces of rhetorical invention, but on a deeper level, the principle of *decorum* that governed rhetorical disposition encouraged attention to the particulars of historical time and place.

Humanists' pragmatic need to know historical particulars both for rhetorical argument and grammatical explication encouraged the accumulation of facts about the ancient world. The rhetorical principle of *decorum*, however, provided some humanists with the means to organize their historical knowledge in a more profound way. *Decorum* demanded that the orator who was preparing a speech, a letter, a treatise—any work intended to persuade—consider the speaker's character, the audience, the subject, the circumstances of the events described, the circumstances of the speech, that is, the various contexts with which the work was related. In a series of brilliant studies, Nancy Struever has traced the origins of humanist historical consciousness to the principle of decorum. The Florentine humanists Coluccio Salutati, Leonardo Bruni, and Poggio Bracciolini came to realize, based on meditation on *decorum* and their knowledge of historical particulars, that the past was profoundly different from the present, and thus needed to be reconstructed.[114] Lorenzo Valla's famous oration on the Donation of Constantine is a product of this combination of factual knowledge with a sense of *decorum*. Valla argued that the Donation, allegedly a document written by the emperor Constantine that granted dominion over the western Roman Empire to Pope Sylvester, was a monstrous forgery. It violated his sense of historical particulars by using words in a way that no Roman of Constantine's day would do and by committing basic anachronisms such as referring to Constantinople in a document supposedly written before the city was founded. But Valla also railed against the document's lack of *decorum:* no emperor of the fourth century could have made such a donation, and no pope as holy as Sylvester would have accepted it.[115]

Humanists also put their historical sense and rhetorical skills to work by writing histories, often as patriots or hired pens intent on proving the antiquity and dignity of their city or patron.[116] Here, too, Lorenzo Valla showed his sense of decorum. When critics questioned his use of the word *bombarda* (cannon) in his history of Ferdinand of Aragon, Valla responded that the thing did not exist in classical times, so it was wholly appropriate to use a postclassical word to designate it.[117] But Valla's historical sense did run counter to another current in humanism: the desire to model Latin on the usage of the best ancient authors. Humanists did not write histories of the ancient world (for which the surviving histories from antiquity sufficed), and the desire to purge Latin of the "barbarisms" that had afflicted it since the end of classical times sometimes clashed with the need to describe new political and social situations.[118] The "Ciceronianism" of the early sixteenth century, viciously parodied by Erasmus, which prohibited words and phrases not found in Cicero's works, was already developing in the later fifteenth century.[119] In mediocre hands, humanist historiography was quite capable of making medieval Italians sound like ancient Romans.

The Ciceronian controversy provides a curious parallel with the central question of early sixteenth-century natural history: do the ancients' descriptions of plants suffice, or are there more herbs in the world than in the books of Dioscorides and Pliny? But Ciceronianism also underscores a general tendency in fifteenth-century humanism, one that runs contrary to the deep interest in the particular that we have seen. Humanist educators and rhetorical theorists saw edification, moral improvement, as the most important goal of education. Moral philosophy provided a systematic analysis of moral principles, but poetry and history were also sources of edifying exempla, of moral philosophy taught by example. A sense of history, however, posed a twofold threat to such a pragmatic use of poetry and history. First, as Francesco Guiccardini pointed out in response to Machiavelli's use of the past, exempla are only useful if the circumstances from which they are taken are close to those in which they are employed—but the world of fifteenth- and sixteenth-century Europe was vastly different from the Roman Republic and Empire.[120] Second, the ancients were pagans, and their ethics could only be taken so far. Petrarch had struggled with this problem in the fourteenth century; his fifteenth-century successors solved it only by emphasizing those aspects of pagan ethics that were compatible with Christian morality, neglecting the others or forcing them into a Christian mold.[121] For these two reasons, the desire to wrest practical uses out of ancient texts often led to their historical specificity being suppressed.

It would be foolish to expect too much consistency in humanists' attitudes toward the ancients. Ancient works were far too emotionally charged for humanists, and their opponents, to be free of contradiction. Nonetheless, a deep feeling for historical specificity and particularity can be seen precisely in the very tension between humanists' sense of the ancient world (idealized and schematic as it was) and the demands of their own times. If scholastic philosophers naturally neglected particulars in their search for universal, general principles, humanists did so only under protest.

Sixteenth-century legal scholarship and teaching provides a clear example of the contrast.[122] Italian universities approached the civil (that is, Roman) law philosophically, emphasizing its interconnections and employing abstract principles of justice or equity to resolve its contradictions. This *mos italicus,* the "Italian style" of teaching, echoes scholastic philosophy and theology in its desire to reconcile contradictory principles by appealing to universals. The *mos gallicus,* or "French style," approached the law from a different perspective. Developed by the Italian jurist and humanist Andrea Alciato, but established above all in France, where Alciato's most influential pupils taught, the *mos gallicus* approached the law as a historical product. Contradictions existed because the law had been compiled over centuries, in very different circumstances. Moreover, the civil law was the product of a particular time and place, that of imperial Rome, and was thus not universally binding. The success of Alciato's method in France was intimately bound up with both Gallicanism and Huguenot resistance to the absolute pretensions of Renaissance French kings, founded in the Roman legal principle of "princeps solutus legibus."[123] But it also harmonized with the humanist preference for the historical and the particular.

In myriad ways, then, humanist studies and pedagogy encouraged attention to particulars rather than universals. In the case of humanist grammar and the *mos gallicus* in civil law, this preference was a methodological principle, opposed to the scholastic, universal pretensions of philosophical grammar or the *mos italicus*. But humanism also inculcated habits of thinking and working that disposed humanistically educated scholars to focus on particulars and details. Little wonder, then, that humanists took great delight in precise, detailed knowledge, including knowledge of the natural world. Philosophers and lawyers in the *mos italicus*, even if they received a humanist education, were certainly capable of developing habits of thought that ran in the opposite direction—though even there, humanist habits elicited subtle changes. Philosophers, for example, were increasingly convinced that knowledge of Greek was necessary to properly understand Aristotle.[124] In other disciplines, humanist habits met little resistance. In philology, attention to detail was essential. Medicine, too, though as a profession it predated the rise of humanism, encouraged close attention to detail, both in the diagnosis and prognosis of diseases and in the preparation of medicaments.

It should come as no surprise, then, that humanism and medicine went hand-in-hand in the later fifteenth century, as humanistically educated men entered medical schools, took their degrees, and became professors of medicine themselves.[125] These medical humanists brought intense interest and a critical eye to the texts used in medical training, often medieval Latin translations of Avicenna's *Canon*. As humanists, they had a preference or prejudice for the ancient sources of classical medicine: the Hippocratic corpus, Galen, Dioscorides, and parts of Pliny the Elder's *Natural History*.[126] Humanist philologists, even those who were not trained in medicine, also turned eagerly to those texts. Humanist inquiries and debates over these texts turned philologists' and physicians' eyes not only to texts themselves but also to the natural objects that they described. Over the course of a generation, the attempt to reconcile text and object would precipitate the nascent intellectual and aesthetic interest in nature into the beginnings of the discipline of natural history.[127] The earliest, and most vivid, controversy surrounded the text of Pliny's *Natural History*; it is to this conflict that we now turn.

Humanist Scholarship and Ancient Natural History

The years 1492–93 saw the publication of two important works on Pliny the Elder's *Natural History*. Though the work was held in relatively low esteem by medieval naturalists, it was copied and then printed several times over the course of the fifteenth century.[128] Pliny's extensive vocabulary and lapidary style, both difficult for medieval scribes, made his text one of the more corrupted ancient works to reach the humanist textual critics, and generations of scholars attempted to make it more readable and understandable.[129] It is with the earliest of these sustained attempts that we are here concerned: Ermolao Barbaro's *Castigationes plinianae* and Niccolò Leoniceno's *De Plinii et aliorum medicorum in medicina erroribus* (*On the Medical Errors of Pliny and Other Physicians*). Barbaro approached Pliny's text philologically, attempting to purge it of the errors that had accumulated over the centuries, while Leoniceno, a physician, was more concerned with the factual accuracy of Pliny's claims and did not hesitate to accuse him of blunders. Despite their differences, however, these works share both the conviction that criticizing Pliny's text was vitally important and the principle that the only proper method to do so was to compare Pliny's text both with other texts and with the objects he described. In so doing, Barbaro and Leoniceno laid the foundations of the Renaissance science of describing.

Ermolao Barbaro (1454–93) was the *Wunderkind* of late Quattrocento Venetian humanism. Scion of a patrician house, he was free to devote his energies to literary studies and state service, as befitted his standing. But he could not adapt himself to these expectations. To leave more time for his studies, he refused to marry, and at the end of his short life he was proclaimed a traitor after he accepted the pope's offer of the patriarchy of Aquileia without, as was customary, requesting the approval of the Venetian Senate. Only an untimely death freed him from this predicament.[130] Barbaro was particularly interested in Dioscorides, and he wrote a short commentary that was published posthumously.[131] During the preparation of this work, Barbaro took a half-hour daily from his busy schedule to study plants.[132] He is best known, however, for his *Castigationes plinianae* (1493), in which he claimed to have corrected more than 5,000 errors in the received text.[133]

The Augean stable left behind by the text's first editors made Barbaro's work necessary. Early editions of classical texts were printed from whatever manuscripts were to hand, which were usually recent and written in a legible humanist hand, rather than the older, more reliable but also less legible medieval manuscripts.[134] This procedure had regrettable consequences for classical scholarship, since it established until the nineteenth century the readings of texts to be corrected: only readings that did not make sense in the vulgate received critical attention.[135] In the case of Pliny's *Natural History*, however, one of the earliest classical texts to be printed, the obvious errors in the first and second editions led to an outcry, and even to the first call for press censorship—not to control dangerous ideas but to ensure that classical texts be printed only with careful editorial oversight.[136] In many cases, the readings simply did not make sense. Thus, problems in Pliny's text attracted the best philological minds of the day: not only Barbaro, but also Niccolò Perotti and Barbaro's Florentine counterpart Angelo Poliziano.[137]

Barbaro, like his contemporaries, took the received text of Pliny as his base, and corrected it only when it did not make sense.[138] He ridiculed Marcantonio Sabellico for presuming to emend a passage that was already clear—"why such ill greed or lust for correcting what is already whole?" Pliny had suffered enough under the hands of medieval scribes; even the most ambitious editor—Barbaro himself—could be content with healing those wounds. He boasted, "I have cured nearly five thousand copyists' wounds in it, or at least shown how to cure them."[139] But as any reader of Pliny's pathologically laconic prose knows, the correct reading is often hard to determine; he rarely used two words where one would do, and often employed one where two or three were necessary.

Barbaro's own view of Pliny's *integritas* must be deduced from his editorial practice. In some cases, the testimony of "old manuscripts" was conclusive: thus, where the vulgate read "Isocynamom cognominatam," Barbaro remarked, "the old reading is 'mesocynamon,' as if growing with cinnamon and cassia and intermediate between them."[140] In others, Barbaro dared conjectural emendations that did little violence to the text. Where the vulgate described both gourds and cucumbers "scandentes per artesum aspera," Barbaro eliminated the solecism with the suggestion "scandentes parietum aspera [climbing the rough parts of walls]," a reading that changed only one word and, in context, was perfectly sensible.[141]

Far more important, however, was the agreement of authorities.[142] The vulgate Pliny wrote of the wild cabbage, "folia habet bina, rotunda, parva, lenia [it has two round, small, soft leaves]." This could not be right, Barbaro argued: "I write: 'folio habet parva, rotunda, laevia [it has small, round, light leaves]' nor do I add 'bina [two],' since Theophrastus describes it as having multiple, dense leaves."[143] And where the printed text listed "pyrallis" as a kind of eared plant, Barbaro emended it to "thryallis" on the basis of Theophrastus. Where the text claimed that pine-nuts "strengthen infirmities of virile parts," Barbaro disappointed frustrated lovers, but relieved frustrated readers, by emending "virilium [of manly things]" to "virium [of strength]," following Dioscorides.[144] Even when the *vetus lectio* (old reading) made perfect sense, Barbaro often called on the authority of Dioscorides to support it. In one particularly bad case, "elebi panacea spagora Corinthymo" (a list of herbs that could be used in winemaking), Barbaro remarked caustically, "In four words there are five mistakes. Write 'elelisphaco, panace, acoro, conyza, thymo,' from Dioscorides and old manuscripts." Where the vulgate referred to "wild onions, not cultivated," the *vetus lectio*'s "onions do not grow wild; they are cultivated" was supported by Dioscorides, who described no wild onions.[145]

In a few instances Barbaro appealed to common-sense observations, not authority, to support his emendations. One is the case of gourds and cucumbers winding their way up walls. In another instance, Barbaro refuted one of Sabellico's conjectures on the words "rimosastri pallido" (describing poisonous mushrooms): "The Roman codices have 'rimoso,' but one should read 'rimosa stria'; everyone knows that mushrooms are striated and those striations look like cracks. Apuleius writes 'furrowed forehead' [*striata frons*]; those who think the text should be 'rimosastro' as in 'imitating cracks' are ridiculous." Barbaro remarked, in support of Dioscorides's description of marrubium, that "I too have seen that." In yet another instance, personal inexperience led Barbaro to suspend judgment:

> And further, where Pliny writes: *with leaves that look like sparrows if regarded from afar*.
>
> Dioscorides seems to say not "like sparrows" but "like *struthion*"; that is, like the herb that we call *radicula* and the Greeks call *struthion*, from their word for sparrow. Theophrastus also says this more clearly in his ninth book: "it has a leaf like *struthion*, with which linen is bleached." Wool is whitened with the *radicula*, as Pliny himself writes in his nineteenth book, as well as Dioscorides and others. But I have not yet seen this species of poppy with my own eyes, and Theophrastus's words are ambiguous, since his statement that linen is made white with it could pertain either to the leaf of the poppy (as Pliny understands: "Linen draws its brightness from it") or to *struthion*, so I don't dare say that Pliny is mistaken here, instead of writing with judgment in this instance.[146]

Though Theophrastus and Dioscorides, not to mention Pliny himself, hint that the emendation is called for, Barbaro refused to decide, since Pliny's description was possible. Barbaro's half-hour of daily plant study seems to have made some impression on his philological methods.

But it did not make much of an impression. For each such example, there is a counterexample in which observation might have settled a question but, in fact, did not. After emending "stannios" to "cynops," following

Theophrastus, Barbaro remarked that "In Theophrastus, I have sometimes read 'conopa,' that is 'culex' [gnat], not 'cynopa'; and Pliny uses 'culex' as the name of a herb. I have indicated the place: let the reader decide what is correct."[147] Though the quest to discover what the herb named culex was may have been in vain, Barbaro did not even try.[148] In any case, the authority of Dioscorides or Theophrastus usually sufficed to establish the correct reading. Additional observation would have been superfluous. In one case, Barbaro ridiculed modern physicians who distinguished two forms of hemlock, based on their medical properties: "they treat of this without any weighty authority," and hence they don't need to be taken seriously.[149]

A telling presupposition of Barbaro's method is that Pliny was infallible.[150] Emending the *Natural History* on the basis of Theophrastus, Dioscorides, and other, earlier writers made sense only if Pliny got it right. Barbaro said as much: after claiming to have freed Pliny from over five thousand errors committed by copyists, he added immediately, "I said 'of copyists,' lest anyone think me imprudent or believe that Pliny himself erred. In a few places, Barbaro was less cautious—for example, Pliny's remark that cinquefoil produces strawberries was a slip. But in most cases, Barbaro explained "errors" as due to the copyist's pen. The vulgate's "all of these are odorless, [also] the autumn rose that grows in brambles" was emended to "all of these are odorless, except the autumn rose that grows in brambles," since Pliny, in another passage, had described the pleasant smell of that plant.[151] Even errors that resulted from confusing words that are similar in Greek but distinct in Latin—errors that Niccolò Leoniceno offered as proof that Pliny was incompetent—were attributed, by Barbaro, to errors in the Greek manuscripts that Pliny used. When Pliny wrote that stachys resembled the leek, Barbaro at first commented, "Pliny seems to have erred in this instance, not the copyist, because the Greeks call horehound 'prasion,' while they call leek 'prason.' This hallucination, which is made also in other plants that look like horehound, was clearly the fault of a learned man."[152] But by the second confusion of leek [*porrum*] with horehound [*marrubium*], Barbaro had thought better. The vulgate's "*ballota,* which the Greeks also call black leek," was obviously a slip—but "but this was not the fault of a Latin copyist but of a Greek copyist, in whose manuscript Pliny read 'prason' instead of 'prasion,' and thus translated 'leek' instead of 'horehound.' For *ballota* is similar to horehound, not to leek," according to Dioscorides.[153] This time Barbaro passed over silently the possibility that Pliny had blundered: the only question was whether the Latin copyist or the Greek was at fault.

Barbaro's view that Pliny was generally right was based on the Roman's own testimony: Pliny himself claimed to have observed the plants of which he wrote. For a man of his energy, "this is not a difficult subject; I have contemplated almost all these plants, with the aid of Antonius Castor—the foremost authority in this subject in our age—when I saw his garden in which he grew many."[154] Indeed, Barbaro thought that Pliny was more reliable than Dioscorides when it came to Italian plants: the latter thought that *nardus gallicus* was called *saliunca* in Latin, whereas Pliny distinguished the two.[155] Barbaro saw no contradiction between this position and the thesis, sustained by his editorial practice, that Pliny had copied large parts of his *Natural History* from his predecessors, and that their work could thus be used to reverse the ravages that time and human laxity had wrought upon his.

This view of the *auctor* as copyist carried with it the notion of a closed world of experience. Pliny's encounters with plants in the garden of Antonius Castor paralleled Barbaro's own encounters in the cool summer evenings of Padua; both thought that they could learn the essentials in a short amount of time, and in both cases experience was mediated by *auctoritas.* Hence Barbaro could use Dioscorides and Theophrastus to correct Pliny's text, just as Pliny had used them to produce it. Observation and experience were not distinct from the text but intimately bound with it. Barbaro, as critic, drew on observations. But ultimately he found the text more useful.[156]

Nonetheless, Barbaro's work would be taken up by later generations of naturalists. His translation and commentary on Dioscorides were published posthumously; though the translation was quickly superseded by Marcello Virgilio's and then by Jean de Ruel's, the commentary was reprinted in Cologne in 1529 and again in Basel in 1534 as part of his collected works.[157] His commentaries on Pliny, also published in the Basel edition, proved that Pliny's text had to be reconstructed carefully. By that time, though, Pliny was no longer valuable for naturalists, or at least for botanists: Barbaro's contemporary Niccolò Leoniceno had demonstrated that the great Roman encyclopedist had made too many errors to be trusted.

Niccolò Leoniceno (1428–1524) was part of the generation of Italian humanists trained by Guarino da Verona and his contemporaries. He spent most of his distinguished career in Ferrara, where he taught moral philosophy and practical medicine; he was also a practicing physician whose *consulta* were well respected.[158] No stranger to controversy, and a master of the pen, Leoniceno expressed himself forcefully on a number of burning issues of his day—medical, pedagogical, and philosophical.[159] The study of *materia medica* was one of his concerns, and his essay *De Plinii et aliorum medicorum in medicina erroribus* intervened in a debate that touched on both humanist philology and practical medicine. Though framed as a criticism of an ancient author, Leoniceno's book draws on extensive observational activity. For Leoniceno observation was a sensible activity only within the framework of textual knowledge—and vice versa; a text could only be read by someone with experience. Leoniceno's book allows us to see the particular articulation between text and experience at the birth of Renaissance natural history.

Leoniceno's tract, first published in Ferrara in 1492, was ostensibly a lengthy letter in reply to his friend, the Florentine humanist Angelo Poliziano. Poliziano's letter, prefaced to the treatise, was itself a response to an earlier letter by Leoniceno.[160] In his prefatory letter, Poliziano lauded Leoniceno for correcting, in his university lectures, the erroneous views of Arab medical writers, which had caused so much grief to physicians—and especially to their patients. But, apparently in response to an issue Leoniceno raised in his first letter (no longer extant), the Florentine defended Pliny against the charge that he had confused hedera (ivy) and cithon (the rock rose) because their Greek names were similar. Poliziano submitted the matter to Leoniceno's judgment: "I will rejoice if a Latin author is not in the same straits as the barbarians. If I am wrong, I await your response."[161] Leoniceno responded with his treatise, which was reissued in a much expanded version in 1509.

Leoniceno swiftly brought out his big guns. His first criticism of Pliny had nothing to do with plants or medicine, but it caught the Roman encyclopedist in a ridiculous blunder: his assertion in book 2 of the natural history that the moon is larger than the earth. This example immediately established, far more persuasively than any technical mistake in medicine or botany, that Pliny could not always be trusted. Leoniceno's choice also underscored his intention to attack Pliny himself, not the medieval copyists whom Barbaro had blamed for the degenerate text of the *Natural History.* Then Leoniceno proceeded to the relationship between hedera and cithon. His response was quite different from Poliziano's. He referred first to Pliny's distinction between male and female hedera; this distinction "is made neither by Theophrastus, nor by Dioscorides, nor by any man of weighty authority about the hedera, but it is clearly described in cithus," and hence Pliny must have been deceived by the name. "We can summarize: Pliny divided the hedera into male and female; he describes both as having a flower like the wild rose," and so forth. "No author, except Pliny, writes these things about the hedera, nor are they borne out by sense itself or experience. Rather, they are said of the two forms of cithon, and in fact this is the case. These are most clear arguments that Pliny erred here, as in other places."[162] Why did he err? The Greek words for hedera and cithon, *kissos* and *kisthos,* are so similar that Pliny confused them in his notes.

In both the original 1492 publication and the 1509 enlargement, Leoniceno continued in the same vein. Pliny blundered in two ways: either he made several plants into one by confusing similar names, as in the case of cithon and hedera, or he made one plant into several by treating the Greek and Latin names as different plants.[163] In both cases, the errors resulted from two sorts of ignorance: philological and experiential. Pliny provided the perfect example of the armchair scholar who is led to ridiculous conclusions by his lack of practical experience.[164] He was confused by names, but they confused him because he did not know the things to which they referred.

Leoniceno insisted that he was concerned with the things, not merely with names: for philosophy involved disputation about things, not words, and arguments over words were vacuous for those who have important things to investigate. Hence, in defending himself and Barbaro against Pliny's advocates, he distinguished between "those things that pertain to medicine" and purely philological matters, "which I leave to literary types, whose job is to dispute about words rather than things themselves." This does not imply that words were not important to Leoniceno. Rather, their importance lay in their use as a key to the ancients' knowledge of things, "for all our knowledge depends on knowledge of terms."[165] Instead of placing Leoniceno's work on a point in the continuum between philology and science, we should take him at his word; for him, the two were part of the same enterprise.

But how did Leoniceno know that Pliny was wrong? "More reliable" authors contradicted the Roman encyclopedist. Leoniceno's criteria for determining which authors were reliable reveal the interplay between text and observation that constituted his experience of nature. In the first instance, the reliable ancient authorities observed nature firsthand. Leoniceno singled out Galen for particular praise: he traveled all the way to the remote island of Lemnos just to investigate the true nature of *terra lemnia.* Leoniceno also lauded Dioscorides for his careful investigation of the appearance and properties of plants. His description of *centaurium* is so vivid, wrote Leoniceno, that "it seems almost to be in front of our eyes."[166] It was not enough, however, to claim to have experienced nature; as we have seen, Pliny too prided himself on his firsthand experience. Leoniceno refused to believe an author's claim to firsthand experience unless his particular descriptions backed it up. He targeted Pliny because the Roman did not know enough about the material on which he wrote but borrowed extracts from other authors. Leoniceno directed the same reproach against the Arab medical writers.[167]

Hence, Leoniceno's choice of "weighty authorities" depended on the extent and quality of those authors' own experience of nature. However, in order to judge their accuracy, Leoniceno himself needed considerable knowledge of natural things. His writings provide abundant testimony of this firsthand experience of nature. Writing about Pliny's erroneous description of the ivy, he criticized the author for not observing it himself. Leoniceno noted the relative frequencies of *centaurium maius* and *eupatorium* in Italy; the latter "occurs to our eyes in meadows, gardens, thickets, plains, and hills." Even the everyday experience of peasant children was enough to convict Pliny of ignorance when he confused *aegilops* and *avena*.[168]

Explicit references to particular observations and activities dispel any suspicion that these appeals to experience were merely rhetorical figures. Leoniceno had often seen the *corneola* in swamps and on the banks of the Po. And even as an octogenarian he looked forward to studying herbs, both useful and noxious, in the hills around Lucca, "which seem to have been made for this very purpose."[169] It is difficult to tell exactly how often Leoniceno explored nature in the wild at firsthand, but his extensive knowledge of the morphology of plants indicates more than a passing exposure to them.[170]

Leoniceno's attitude toward experience is summed up in a paean to the powers of the senses—a paean supported by respectable classical authorities.

> Should Pliny, or Theophrastus's translator Gaza, have such authority that we believe them rather than our own eyes? The great philosopher Aristotle considered the judgment of the senses so compelling that where it is present, it is vain to seek a proof. Galen, prince of physicians, thought it was insane to demand proofs when the evidence of the senses is available. Avicenna, who admitted he was but the interpreter of Galen, advised that those who don't wish to believe their senses should be burnt or whipped, so that by experiencing pain they come to realize that the judgments of the senses are true. And these learned men were right. Why did nature grant us eyes and the other senses, if not that we might see and investigate the truth with our own resources? We should not deprive ourselves and, following always in others' steps, notice nothing for ourselves: this would be to see with others' eyes, hear with others' ears, smell with others' noses, understand with others' minds, and decree that we are nothing more than stones, if we commit everything to the judgment of others and decide on nothing ourselves.[171]

Allowing for rhetorical excess, this passage is a remarkable tribute to observation, fully comparable to similar passages in the botanical works of the iconoclast Paracelsus but uttered by a man who, unlike Parcelsus, observed widely.[172] Nonetheless, Leoniceno did not insist that observation was the sole arbiter in the study of medicine. Rather, it was to be used judiciously along with texts. Our senses can tell us nothing about the nature of *lysimachon,* for example, unless one first determines which plant the ancients called by that name. Similarly, the fact that the ancients' *brassica marina* is identical to the modern plant *soldana* "is clear to anyone who compares what our eyes tell us about soldana with what Dioscorides wrote about brassica marina."[173] One cannot understand ancient medical texts without knowing the things to which they refer; observation anchors texts, otherwise floating signifiers, in the universe of things. And ignorance of texts has practical consequences for physicians, whose pharmacopoeia are poorer for their neglect of ancient knowledge.[174]

Hence the proper study of *materia medica* demanded the union of textual and experiential knowledge. According to the bishop of Padua, Petrus Barotius, the proper knowledge of herbs was almost impossible in the present day, "because we do not know their earlier names." The problems of interpreting ancient texts render this science difficult, a difficulty that is augmented by how plants change as they grow and the difficulty of describing their colors and shapes. Especially in the case of plants described by the Arabic writers, the only way to evade these difficulties is to take refuge in the judgment of sense.[175] When Albrecht von Haller called Leoniceno "an erudite man, and the first in many centuries to exercise critical judgment," he recognized the importance both of textual knowledge and careful observation in Leoniceno's work.[176]

Given Leoniceno's almost naive faith in observation, one might wonder why he did not write his own book on plants rather than comment on ancient texts. Leoniceno knew that observations and text went hand-in-hand, but the next generation of botanists solved this problem by writing their own texts, excerpting and systematizing ancient reports, rather than writing commentaries.[177] One possible answer, that Leoniceno as a humanist saw commentary as the natural form of scholarly discourse, loses its attraction after a few moments' consideration. Many humanists of Leoniceno's generation wrote their own treatises. Leoniceno himself wrote a treatise *De morbo gallico,* on the disease later called syphilis, though he refused to admit that the disease was new.[178] Commentary was one humanistic genre, but it was far from the only possibility for someone like Leoniceno, who, on his own account, was concerned with things rather than words.

A more convincing reason lies in the circumstances of medical education in late fifteenth-century Ferrara. Leoniceno was above all a teacher, and as the letter from Poliziano that begins the *De Plinii erroribus* shows, his criticisms of the ancients grew out of his pedagogical activity. Hence we should look at the Ferrara curriculum, and the structure of late medieval university education, to understand why Leoniceno's treatise took the form it did. Medieval university teaching was based on the text: in particular, an ancient text, which was read and commented by the teacher in his lectures (literally, "readings"). At Ferrara's medical school, in the late fifteenth century, these texts were predominantly written by Arabs.[179] Pharmacy (*materia medica*) was taught from book two of Avicenna's *Canon.*[180] It seems most probable that Leoniceno's criticisms of Pliny took form within the framework of his teaching and an attempt to reform the medical curriculum at his university.

The title of Leoniceno's treatise, to repeat, was *De Plinii* et aliorum medicorum *in medicina erroribus.* And the "other physicians" are almost without exception Arabs. These writers drew their errors from the same source as Pliny: they copied earlier writers without understanding.[181] To Leoniceno, the Arabs' errors were far more dangerous than Pliny's. Pliny, he claimed, was valued as a literary figure, not as a source on *materia medica.* But the Arabs are another matter: their mistakes lead to deadly errors on the part of physicians who follow them slavishly.[182] Leoniceno's main target was Avicenna, as can be seen in the index to the 1529 Basel edition of the treatise. Leoniceno even promised in *De Plinii erroribus* to produce a separate treatise devoted solely to the mistakes made by Avicenna, a cruel tyrant whose followers trust their leader completely and constantly affirm that which no experience has taught them.[183] Finally, the Arabs were chiefly responsible for the confusion of names that had caused so much trouble for students of plants.[184]

The constructive part of Leoniceno's attempt at curricular reform is hinted at in his repeated praise for Dioscorides and Galen and by his activities from the 1490s to his death. In *De Plinii erroribus* he mentioned a project to provide new, more accurate translations of Galen. In 1522 he appears in the records of the University of Ferrara as having undertaken the task of translating Galen's opera, and is to be paid 400 lira per annum until the work is completed. It was an ambitious project for a man in his ninety-fourth year, and he did not live to complete it: the 1524 records note that his heirs had claimed his outstanding salary.[185] In other aspects of his career, notably his controversy over the three ordered doctrines of Galen, his pedagogical interest in ancient Greek medicine is apparent.[186]

Leoniceno's pedagogical focus explains why he did not write a treatise on *materia medica.* He was concerned with replacing bad school texts, particularly the Arabic writers, with good ones. The independent treatise was not well suited to a university, particularly not to Ferrara's medical school, where even the anatomical treatise of Mondino had acquired a patina of age and respectability. Humanist professors commented at great length on classical texts, and sometimes published their commentaries (just as scholastic philosophers commented on *their* set texts), but they generally did not compose textbooks.[187] Their scholarly and literary activity was intended to explicate and supplement the classical founts of learning and eloquence, not to replace them.

Another reason, however, lay behind Leoniceno's interest in selecting and purifying the best texts instead of writing his own. Leoniceno gave no indication in his treatise on Pliny that the ancient works might not suffice for a proper knowledge of nature—that it might be worth paying attention to plants that were not described by the ancients, or even that there were such plants. His aim was pedagogical: to determine which texts were best, to translate them into Latin if necessary, and to base medical education on them. He considered it critical to identify the plants described by those texts, but finding and describing others had no place in his activity.

The same attitude, which I will designate "humanist natural history" for the sake of concision, characterized the first generation of sixteenth-century naturalists, the students of Leoniceno and his contemporaries. One particularly important student was Euricius Cordus, who introduced this approach to Germany.[188] These humanistic naturalists went into the fields, woods, meadows, and mountains of Italy and northern Europe in order to compare ancient herbalists' descriptions with the plants they saw in front of their eyes. When they wrote treatises, it was in the humanistic forms of dialogue or commentary. Nonetheless, their concern with observation and comparison between ancient texts and modern plants would bear fruit that fell far from the tree. By the 1530s, the nascent community of naturalists had begun to reject the texts of the ancient Greeks as an adequate basis for natural history—not because they were inaccurate, but because they were insufficient. The careful investigation of nature by a large international community produced information that surpassed the slender tomes of antiquity in quantity and precision.

For that to happen, though, Leoniceno's essentially pragmatic, medical approach to the study of nature had to cede some ground to the esthetic appreciation of nature that we have already seen among many Renaissance humanists. The scope of natural inquiry had to be extended beyond *materia medica,* and the knowledge of plants had to be pursued, to some extent, as an end in itself. Erasmus's godly feast and Rabelais's instructions from Gargantua to Pantagruel reveal that this had happened by the early sixteenth century. To find out how, we will turn to the charming botanical dialogue of Euricius Cordus, which shows us how humanist natural history was put into practice.

Humanist Natural History in Practice: Euricius Cordus

The earliest field observers in Western Europe were apothecaries, and the herb gatherers who supplied them. Fresh, wild simples were prized more than those cultivated in medicinal gardens.[189] Every year apothecaries would leave the cities and towns in which they practiced to head for the hills and collect herbs. Back home, they would dry them and store them for use in compound medicines. But this activity was commercial, not intellectual. Both the location of their hunting grounds and the plants to be found there were trade secrets, to be guarded carefully—much as modern truffle hunters are jealous of their favorite spots. Herb gatherers were not practicing natural history, though some of their techniques were taken over by naturalists.

Humanistic natural history involved observing plants and comparing them with texts. As we have seen, it began with study and debate over Pliny's *Natural History.* During his sojourn in Italy, Rudolf Agricola, one of the intellectual fathers of German humanism, took the opportunity to study plants mentioned by Pliny. He could do this better in Italy, where many plants described by the ancient Roman grew, than in the north.[190] Leoniceno gained at least some of his extensive botanical knowledge from excursions around Ferrara, and he wrote with confidence about plants found throughout northern Italy.[191]

When Agricola, Leoniceno, and their fellow humanists went on excursions to observe plants, they did so with a particular goal. For these scholars, the study of nature served to better understand classical texts—themselves not purely literary but important elements of medical education and practice. The difficulty of modern botany, as Leoniceno put it, was a *confusio verborum,* resulting not only from the corruption of names by writers such as Pliny, and *a fortiori* the medieval Arab medical writers, but also the difficulty of describing plants in words.[192] As a result, humanists were unsure of the correspondence between plants and the ancients' names for them. When

Leoniceno went into the woods and meadows, he did so with Dioscorides's and Theophrastus's descriptions in mind; he wanted to find the plants to which those descriptions applied.

Whether Leoniceno took a copy of Dioscorides with him on his excursions is unknown. We know more about his German student Cordus, whose botanical dialogue, the *Botanologicon* (1534), describes an ideal botanizing excursion set in the vicinity of Marburg in the early 1530s.[193] After a brief tour of Cordus's small town garden (hortulus), he and a group of his former students set out to explore plants outside the city walls. They took with them both an edition of Dioscorides and a copy of Otto Brunfels's recently-published *Herbarum vivae eicones* (1532).[194]

For Cordus and his students, the world was also populated largely by plants that were known to the ancients, and once again their task was to connect the herbs and trees before their eyes to those mentioned in the ancients' texts. As for Leoniceno, Dioscorides was the key to this world: in typical humanist fashion, Cordus stressed that he would rather rely on the ancient Greek than on three hundred pandects or even the entire profession of physicians, although even Dioscorides was fallible.[195] The majority of the dialogue turned around the proper identification of the ancients' vegetable pharmacopoeia. For instance, argued Cordus, the plant called by moderns *pentaphyllon* is not the *pentaphyllon* of the ancients, which has a reddish root. Rather, the latter plant is the same as the modern *tormentilla.* Similarly, apothecaries name a particular plant *serpentaria* falsely, for the plant's root does not match Dioscorides's description.[196] If, for Leoniceno, the task of the herbalist was difficult because the old names were no longer known, the largest part of Cordus's task was their rediscovery.

Corresponding to this sense of a closed world of plant species was the relatively narrow range of Cordus's excursion. He and his friends went from Marburg to his suburban garden, through streets and a wood on the way there, along the bank of the Lahn on the return trip.[197] This does not, of course, represent the full extent of Cordus's own travels—earlier he had taught in Erfurt, and he referred often to his experiences in Italy as well as to plants brought to his garden from Meissen.[198] But such transfers were relatively rare and referred to plants that grew elsewhere but still formed part of the received body of knowledge. The party uprooted plants—the root was an important diagnostic element for Dioscorides's descriptions—but the only thing they collected was a handful of ash berries for a pharmacy.[199]

Humanist botanists thus experienced nature as a process of deciphering and recognition. The ancient Greek authors had given them a key to the natural world, and their task consisted in the proper application of this key to their environment. It was not, however, a simple task. As Cordus pointed out to his companions, the ancients had not bothered to describe the most common plants. They had mentioned only their medicinal properties, secure in the knowledge that everyone would be able to identify them.[200] Hence rare plants were, paradoxically, easier for moderns to recognize. Furthermore, not every correspondence was easy to determine.

In some cases the identification was easy. Cordus affirmed that the plant called by moderns *agrimonia* was the ancients' *eupatorium:* "See how exactly this plant expresses the characters of *eupatorium:* how similar are the leaves, how similar the stalks, how similar the little seeds that grow out of these and hang toward the ground, and that cling to the clothing of passersby."[201] The plant and the description were both given, one in nature and the other in Dioscorides; all the botanist had to do was to match them up.

But certain clues, Cordus had discovered, were more reliable than others. Height was untrustworthy. Cordus remarked of the plant called *verbenaca altera* by Brunfels: "Regarding its height of three-quarters of a foot, which this plant triples, I will say nothing, for I know that many herbs grow to different heights in different places." This was due in part to the nature of the location: a *hipposelinon* in Cordus's garden was not as tall as it would have been in a shady spot, where it would attain its natural height. Similarly, Cordus considered the color of flowers as variable. He identified the modern *solidago* with Dioscorides's *symphitum alterum,* though the plant before him had a white flower and Dioscorides described it with yellow flowers. "That does not make me doubt," he remarked, "since—as I said before—even the same plant often has different flowers. I saw this *symphitum* in Bremen recently with purple flowers." In certain aspects, particular plants could exhibit a wide range of variation—as also in the leaves of the true *pentaphyllon,* which were often seven-lobed in Italy.[202]

Cordus allowed less variation in other characters. Taste could clearly identify a plant and determine its medical qualities: the bistort, falsely called *serpentaria* by modern apothecaries, could not be the ancients' *serpentaria,* because the roots have different tastes. Taste was perhaps the most reliable criterion of all: at one point Cordus called it "infallible." The natural growth cycle of a plant was also important. The distinction between evergreens and plants that lose their leaves in winter was clear, and two plants with similar flowers could be distinguished by their fruit. The hallucinogenic properties of *furiosum solanum*'s root served to clearly distinguish this plant from other kinds of *solanum:* "Many characters attributed to the *furiosum solanum* are found in this herb, above all, the following: its root provides wondrous and, from the reactions of those who eat it, not unpleasant visions—a thing of which I know many examples."[203]

These distinctions between variable and fixed characters within a particular species show a clear pattern. Euricius Cordus was an experienced botanist, but these distinctions were not the products of a hypothetical "pure experience." Rather, they demonstrate how this experience was shaped by the representations of the plant world that Cordus possessed, the resources that he brought to its study, and his concerns as a physician. Cordus saw the plant world as limited in size and largely known. He brought to his investigation humanist editions of classical texts, and he was concerned with plants, like his teacher Leoniceno, primarily as ingredients for medicines or as simple drugs themselves.[204] It is no coincidence that the fixed properties of plants were largely medical properties, or those (like taste) that were used for determining those properties.[205] In this, Cordus implicitly followed Dioscorides himself, who organized his *materia medica* around the faculties of plants, not primarily their forms, and often gave relatively dissimilar plants the same names when they possessed the same virtues.[206] When his brother-in-law Ralla asked him to distinguish the different types of endive, Cordus refused for two reasons: they borrow features from one another promiscuously, and in any case they all possess the same faculties, so it was not worth the trouble.[207]

If medical and pedagogical concerns determined the irreducibly fixed characters of plants, the humanists' notion of the plant world and their resources determined how much they could vary. The necessity of identifying northern plants with the Mediterranean flora of the ancients (which Agricola had already realized might not be possible) required naturalists to allow a certain range of variation.[208] Cordus's contemporaries Ruel, Brunfels, and Fuchs all reproduced images of northern plants alongside descriptions of similar plants taken from ancient sources.[209] Such "loose" identifications were made easier by the often vague descriptive language of Dioscorides and Theophrastus and ancient works' lack of illustrations. Repeated use of comparative descriptions by the ancients—such and such a plant has leaves like oregano, only larger—eased the task even further and permitted identifications of plants which, for later botanists, were distinct but related.

The epigraph to Cordus's book epitomizes the humanist approach to natural history:

> HEUS MEDICE
> Vis varias aliter quam doctus es hactenus herbas
> scire, novus multas iste libellus habet.

[O Physician! If you wish to know different plants in another way than you have been taught, this new book contains many.]

These lines do not claim that the book will introduce the reader to new plants.[210] Rather, it will teach him to see them "otherwise" than he was taught: that is, to recognize the true correspondences between ancient names and modern herbs. Cordus emphasized this purpose again at the end of the dialogue when Ralla, an apothecary, asked him how to fill prescriptions. Should he immediately accept all of Cordus's suggested identifications? It depends, responded Cordus: is the recipe ancient or modern? and where did the prescribing physician learn his names?[211] Names changed, but the vegetable world remained the same.

But there are signs, in Euricius Cordus's work, that this world was beginning to come unraveled at the edges. Some plants, he admitted, had been unknown to the ancients. They knew neither the "scarlea," as the French called it, nor the fuller's thistle. In fact, remarked Cordus almost as an afterthought, many plants—especially *vulnerariae* and *solidagines*—had been discovered since Dioscorides's time.[212] But this had not led him to search for more.[213] Even if the world of plants was somewhat larger than it had been, it was still limited. One can compare the impact of these discoveries to that of the Atlantic islands in the fifteenth century, which widened to some degree the traditional view of the Oecumene but did not overturn it, and into which framework the Europeans' discovery of America was initially placed.[214] Only after the extent of the new land was realized did the old conception of the world fall apart. In the same fashion, the gradual discovery of new plants in the 1530s and 1540s led to a new form of natural history: one that aimed to explore and catalogue the world of plants and animals.

Conclusion: Dioscorides and Renaissance Natural History

Though naturalists after 1530 increasingly turned away from the task of identifying plants described or mentioned in ancient works to the new project of identifying and describing previously unknown ones, their approach continued to be shaped by the habits and problems of humanist natural history. Many specific instances will be considered in the next chapter. One of the most significant decisions made by Leoniceno, his contemporaries, and their students, and one whose impact on the development of natural history can hardly be overstated, was to take Dioscorides as the model of the careful naturalist—and his book *De medica materia* as the model for natural history. Though sixteenth-century naturalists quickly surpassed Dioscorides in terms of the number of plants described and the precision of descriptions, his work served as a pattern for descriptive natural history.

There were several reasons to choose Dioscorides. As we have seen, Barbaro, Leoniceno, Cordus, and other naturalists of their generation considered him the most reliable ancient writer. Hence his words were studied carefully, and his format—name, description, account of medicinal virtues, and (sometimes) tips for identifying fakes—came to seem natural to his readers. Moreover, his work, unlike Pliny's, was explicitly medical: thus, it appealed to Leoniceno and others who approached the study of nature as a branch of medicine. But unlike Galen, who also discussed plants from a medical point of view, Dioscorides described what plants looked like; Galen simply mentioned names. For humanist naturalists, whose task was to establish a set of correspondences between the written word and the natural world, Dioscorides's accuracy and his format were the ideal combination.

Furthermore, Dioscorides was completely in harmony with the humanist emphasis on particulars, surfaces, and descriptions, rather than essences or natures. He named plants, described them, and listed their uses. He neither classified them nor discussed their formal, material, efficient, or final causes. In other words, he treated each kind of plant as an individual, not as an example of a broader generic category that was being investigated. The contrast with Theophrastus is instructive: Theophrastus's *Enquiry into Plants* discussed plants systematically, by type, while his *Causes of Plants* examined them philosophically. Renaissance naturalists ate up Dioscorides. Theophrastus, on the other hand, was available in Theodore Gaza's Latin translation from 1454, but almost a century passed before the first commentary on his works was published.[215] The first Renaissance philosophical treatise on plants was Andrea Cesalpino's *De plantis* of 1583. Though Renaissance naturalists before Cesalpino read Theophrastus and Aristotle, and paid lip service to their philosophical accounts, their hearts and hands were with the minute, careful investigation of individual species that characterized Dioscorides. The successive translations and commentaries by Ermolao Barbaro, Marcello Virgilio, and Jean Ruel provided a firm basis for the Renaissance science of describing.

CHAPTER FOUR

A Science of Describing

The first generation of Renaissance naturalists, from Niccolò Leoniceno to Euricius Cordus, attempted to establish correspondences between the plants described by the ancients and those that grew around them. Encouraged by their humanist education and medical training to focus on particulars, they developed a fine-grained sense of the natural world that nonetheless allowed for a certain variety within the productions of nature. They attempted to reduce what they saw to the patterns established by the ancients, treating variations in color and size as accidental differences that could occur within the same species. Euricius Cordus's *Botanologicon* (1534) portrays, with great charm, the attempts of humanist naturalists to identify in nature the plants known to them from the ancients.

In the next generation, that task would change. Hieronymus Bock, Euricius Cordus's son Valerius, and their contemporaries and successors would continue to scrutinize creatures' appearances as carefully as possible. But they did not intend merely to establish correspondences between ancient names and modern plants. Their goal was more ambitious: to establish a new catalogue of nature, starting from the premises that the ancients did not know everything, that they had described only a fraction of the natural world, and that their modern successors could best follow their example by surpassing them, by describing the living world more precisely and more extensively. When possible, they noted identifications—hesitant or confident—between ancient and modern descriptions, but that was a secondary aspect of their effort to create a science of describing.

Their enterprise was fruitful. The herbals of the 1530s and 1540s described some eight hundred species of vascular plants; in 1623, Caspar Bauhin had catalogued more than six thousand.[1] The reasons for this success were varied. By the 1540s, a large community of naturalists had been established. Spread throughout western and central Europe and in frequent contact through travel and correspondence, this community facilitated the rapid and reliable exchange of natural history material and descriptions, allowing naturalists to compare—and especially, to contrast—specimens and species in their own areas with those that were found elsewhere. The science of describing, the new discipline of natural history, was a product of this communal enterprise.

But if the community of naturalists was able to create a new discipline, it was only by disciplining its members. From the second generation on, naturalists experienced nature in a fashion that differed subtly but significantly from that of their predecessors. They continued to observe particulars carefully, to pay attention to the surface characteristics of *naturalia,* and to delight in the variety of nature. But their experiences were shaped by assumptions about what they would see and habits for observing it. There is no such thing as "pure experience" of nature; the city dweller who takes occasional strolls in the woods will experience his or her surroundings differently than a systematic botanist. In turn, the systematic botanist's experience is not the same as that of an ecologist, who approaches the same phenomena with different habits and concerns.[2] What one observes is, to a large extent, a function of what one has been trained to observe and the vocabulary that has been elaborated to express it. From the 1530s, naturalists in Renaissance Europe were developing new habits of observation and a new vocabulary to express them. This new sensibility, or *habitus,* was not merely a set of methods or techniques; it involved a long process of self-discipline that, if successful, produced an experienced naturalist whose judgment would be accepted by his peers.

To properly exercise this sensibility, however, naturalists did require new techniques and methods—what we might call a technology of observation. As we have seen, the humanist experience of nature was not simply immediate; it involved the integration of two layers of experience, in which the naturalist tried to correlate his immediate observation of nature with ancient descriptions. Practitioners of the sixteenth-century science of describing went through a more complex process. For them, experiencing nature involved several layers. Faced with a plant or an animal, the naturalist had to remember whether he had seen it before, where and when, how many times, whether it was common or rare. He also had to know whether it corresponded with what his colleagues had described—informally, in conversation or correspondence, or formally, in printed descriptions and illustrations. Because it was possible to discover a new species, one significant constraint on humanist naturalists' experience was eliminated. Experience thus involved three layers: the immediate observation, the naturalist's own memories, and the collective experience of the community as expressed in its verbal and pictorial productions.

Naturalists developed techniques for handling each of these layers. They left the town to experience nature in the forests, meadows, hills, and mountains around them. They created gardens where they could observe species repeatedly, over the course of one or many life cycles, without the effort of travel. They invented the herbarium, a collection of dried plants that allowed repeated observation of specimens collected on one expedition. Expeditions, gardens, menageries, and herbaria allowed naturalists to deepen their experience, while field notebooks compensated for the faults of memory. At the same time, naturalists built on the model provided by Dioscorides to develop a precise form of verbal representation, allowing them to condense and communicate their experience efficiently. Finally, they seized on the possibilities offered by the realistic art of the Renaissance and the technology of woodcut illustration to communicate not only in words but also in pictures.

This chapter examines the development of this technology of observation and its interrelationship with the sensibility and discipline of Renaissance natural history. It begins with the approach taken by the second generation of naturalists in their field expeditions, then traces the development of botanical gardens and herbaria, both products of this generation. We will then turn to the myriad ways observations were represented and transmitted: individual field notes, verbal descriptions, and illustrations. Experiencing nature and representing that experience were intimately combined in Renaissance natural history, so intimately that it would be historically unsatisfying to treat them distinctly. Renaissance naturalists strove to create a kind of vicarious experience in their writings, which thus not only condensed but also recapitulated their own experience of nature. For that reason, examining technologies for experiencing and reproducing nature is the best way to approach the phenomenology of experience in Renaissance natural history: to reconstitute how common habits and concerns structured the experience of Renaissance naturalists.

Experiencing Nature

The sixteenth-century naturalist was constantly in motion—motion between the library, the cabinet, the salon (or dinner table and printers' shop), the garden, the wharf, and the countryside—and this motion between different places, this interaction between different forms of experience, was as important in constituting the sensibility of early modern natural history as the natures of the different places themselves. But from the 1530s to the end of the century, the mark of the experienced naturalist was travel in the countryside, the immediate observation of *naturalia* in their natural place. Rembert Dodoens, addressing medical students in the 1550s, admonished them of the difficulties of botany. It required "the diligent and timely examination of all plants and the careful reading of many ancient authors: that is, much work, long travel, and constant devotion."[3] Examining plants meant traveling to see them. Hieronymus Bock guaranteed the accuracy of his herbal by emphasizing the extent and difficulties of his travels. Carolus Clusius encouraged Leonhard Rauwolf to publish a botanical work, "since he is famed for his many travels, during which he must have observed many rare and exotic plants."[4] On the other hand, Clusius thought that their colleague Jacques Dalechamps traveled too little, and his work suffered for it.[5]

Clusius himself traveled widely, despite his increasingly poor health, and chafed when his duties as prefect of the imperial garden in Vienna kept him from travel.[6] At Leiden, as an elderly invalid, he had to be content with receiving exotic plants and seeds from others. In his *Exoticorum libri* he apologized for the poor quality of descriptions that were based on objects that had been sent to him. In a dedicatory poem to the book, Dominicus Baudius agreed: he compared Clusius with Ulysses—though the botanist had better motives than the Greek hero—and he mentioned the *Exoticorum libri,* where the dedication appeared, only as an afterthought, written when cruel age had crippled the scholar's body.[7]

Because Renaissance naturalists had just begun their task of cataloguing nature, travel had a different meaning for them than for their modern successors. Natural history travel since the eighteenth century has usually involved exploring areas outside of Europe. Its practitioners were usually (though not always) young, inexperienced naturalists, with precise instructions as to what to collect, and specimens were usually examined and described by other scholars after the voyage.[8] Charles Darwin spent five years collecting specimens on the *Beagle* voyage and then another six coordinating—not writing himself—the zoological analysis of his collections after his return to London.[9] Conversely, Georges Cuvier rejected Napoleon Bonaparte's offer to participate in the French expedition to Egypt because he could do better natural history in Paris, at the museum.[10]

In the sixteenth and seventeenth century, natural history was characterized by travel that was on a smaller scale than such long-term voyages but no less important for constituting the discipline. A few naturalists went to exotic lands: Pierre Belon, Pierre Gilles, Prosper Alpino, and Leonhard Rauwolf traveled to Egypt and the Near East.[11] But every naturalist could, and did, explore the areas in which he lived and through which he traveled.[12] These were the sorts of journeys that Dodoens and his contemporaries expected from their colleagues. To call them "field expeditions," while anachronistic, would not be far from the mark. Clusius was a Ulysses of the hinterland; his journeys took him through the Iberian peninsula, the Alps, and the uplands of Christian Hungary. Such voyages may not have been as spectacular as those of Columbus or Vasco da Gama, but they laid the foundations of modern natural history.

This point is worth dwelling on. Travel in the wilderness was considerably more difficult and dangerous in the sixteenth century than the present; theft and robbery were common. In 1564, Pierre Belon met his end at the hands of cutthroats in the Bois de Boulogne, not too far from the gates of Paris. Naturalists risked life and limb for their discipline. But there is another reason to emphasize that most Renaissance natural history travel was local. The voyages of discovery have often dazzled historians, who have attributed the new sixteenth-century interest in the details of nature to the discovery of America.[13] However, the discoveries are more dazzling in retrospect than they were at the time. Despite his stubborn insistence, few Europeans were convinced that Columbus had reached Asia. The knowledge that America was a new continent (or two), rather than just another Atlantic island, took decades to circulate widely.[14] By the time the natural history of the "West Indies" became widely accessible, European natural history was firmly established. Indeed, much of the natural history of the New World was made accessible to sixteenth-century readers in Clusius's condensed, annotated translations of the 1560s, 1570s, and 1580s.[15] Before being widely disseminated outside of Spain, the natural history of the New World was firmly integrated into the science of describing.

The new approach to travel emerges in the 1530s, between Brunfels's *Herbarum vivae eicones* (1532), with its rather embarrassed references to "herbae nudae" not found among the ancients, and the confident new descriptions of previously unknown plants found in Hieronymus Bock's *New Kreütter Buch* (1539). Bock underscored the danger, hunger, thirst, fear, and other difficulties he underwent in his travels; practically every chapter in his book testifies to their extent. "I have found the wild mountain Hyssop," he wrote, "in our lands on sandy cliffs and warm mountains." Often he gave precise geographical indications, for example with regard to the Hartstongue: "Our common, well-known Hartstongue grows, even without seed, in shady, moist mountains and valleys, in certain wells and on wet walls, in the Swiss mountains, in Westerwald toward the Mosel, around Veldentz." The other kind "I have found likewise in dark, damp woods, for example the Schwarzwald, the Ydar, the Waßgau and Durstberg." Another small plant he found "in the area of Worms, near the town of Altzen, in cultivated fields."[16] The geographical indications in Bock's *Kreütter Buch* testify to the extent of his travels through the "ohnwege des Teutschenlands."

Bock traveled more widely than Euricius Cordus, and for different reasons. Cordus was concerned largely with identifying Dioscorides's plants. Bock, on the other hand, spurred on by love of his fatherland's vegetation, wanted to describe every German plant he knew.[17] In the 1551 edition of his book, this came to 806 plants described in 430 chapters.[18] To be sure, some of these were foreign plants—which Bock sequestered from the native flora. The distinction between native and foreign plants was important enough for him to take precedence over the natural relations that Bock perceived among plants. For instance, Bock wrote that "if the sharp Roman caraway [cumin] were as common in German gardens as the caraway, I would have put it immediately after caraway. But since it is a foreign guest, I must place it outside of the order of caraway."[19] The majority of the plants he described, however, were native German plants or had been introduced centuries before. Their number alone demonstrates that Bock knew, named, and described significantly more plants than the 500 or so known in antiquity.

Bock had a keen eye for where plants grew and when they flowered. Of one plant, he noted: "A beautiful herb grows in April in unbuilt areas, for example by walls. . . . At the beginning of May it bears beautiful small white flowers." For caraway he even added the time when the dead parts of the plant fall off—"hence the caraway renews itself without much work from the gardener."[20] Such precise indications point to more than close observation. They demonstrate also that Bock made careful notes about each plant he saw, and that he visited places more than once to observe the temporal succession of plant life.[21]

Travel, then, for Bock, was more than an excursion among friends with Dioscorides and Brunfels to hand. When he wandered through the forests and meadows, he carried a notebook and basket to record observations and bring unusual plants back to his garden. There he could plant them and observe them more closely, and perhaps have them when needed for his medical practice. When Otto Brunfels heard about Bock's reputation for "herborizing trips and work on plants," he decided to walk from Strasbourg to Hornbach (more than forty miles as the crow flies) to meet his fellow naturalist. What he saw in Bock's house was a "varied, laborious collection of many plants, together with their descriptions, in garden and writing."[22] Even before Bock decided to write his herbal, he had been collecting not only plants but extensive notes on them.

In this respect he differs not only from Euricius Cordus and other humanistic botanists but also from his contemporaries Brunfels and Fuchs, who together with Bock are generally named the "German fathers of botany." Brunfels's and Fuchs's herbals (1530–32 and 1542, respectively) are reputed above all for their figures, which far surpass any earlier printed botanical illustrations.[23] Their texts, however, are compiled from the works of antiquity. Fuchs tried to disguise this, but Brunfels freely admitted it: his method of working, he said, was "to note down everything worthwhile that I come across while reading, and to write it down in an orderly fashion," which he did for several books, including his "Latin and German herbals."[24] Brunfels said explicitly what Fuchs, the follower of Leoniceno, may have felt in his bones: the ancients *had* known all that was worth knowing, and it was the moderns' task to restore that knowledge.

Bock's book, on the other hand, was first published in 1539, without illustrations but with new descriptions from his notes. The woodcuts in later editions were done by the Strasbourg artist David Kandel, largely from those in Brunfels's and Fuchs's works, and Bock seems to have taken little interest in them.[25] Although later editions of his herbal praised both the painter's art in general and Kandel's skill in particular, Bock added that "he who can have his own garden and gardener . . . does not need illustrated books so much."[26] Illustrations were unnecessary because the experienced naturalist could see the plants themselves before him or refer to their careful descriptions in his notes.

The untimely death of Valerius Cordus (1515–44) provides a closer glimpse of how Bock and his successors worked in the field. Cordus, Euricius's son, was dedicated to natural history, but in a different style than his father. In 1542, Leonhart Fuchs had bemoaned the father's untimely death and promised great things of the son, "unless (heaven forbid) the gods begrudge him, too, a long life."[27] By then, Valerius already had a reputation for careful, detailed study of nature. Johannes Crato eulogized, "He frequently undertook difficult journeys just to investigate one plant," and "he inquired after herbs with a wondrous devotion and endured all sorts of hardships in order to know them precisely."[28] This devotion to botanical exploration cost him his life: he fell ill and died on a trip through the Apennines to Rome after exhausting himself looking for plants.

> His death was a great loss to medicine and to his friends in Germany. For the entire duration of his short life, during which he sought out so many

> of nature's secrets, he explained the natural history of plants, metals, and fishes (he arranged to have those found in the Adriatic Sea carefully illustrated), with reference to the ancients, and added to it himself.[29]

Crato's letter evokes not only the pathos of the young man's death but also his desire to surpass the ancients by traveling through different lands and experiencing nature himself.

Valerius Cordus left behind three incomplete works: a commentary on Dioscorides, which existed in two versions; drafts for a history of plants; and a notebook of observations made in his travels. The first version of his Dioscorides commentary was printed in Strasbourg in 1551, along with a new edition of his father's *Botanologicon;* the second, along with the other works, was brought out by Conrad Gessner in 1561. The *Historia plantarum* was a hastily-written first draft, with several lacunae, and at least one manuscript copy was made before Gessner edited it and had it published. Its relation to field notes is more immediate than most published books, especially since Gessner did not alter the text.[30] The "Sylva observationum variarum" (Collection of various observations), from the title and the format, appears to be a collection "which he jotted down briefly during his travels."[31]

Cordus kept this notebook in three series, one for metals and stones, the second for plants, the third for animals, on an extended trip in 1542.[32] Many of the notes are simple indications of where particular plants grow: for instance, "Fuchs's *Hyacinthus diphyllos* with white flowers grows in the woods near Tübingen, next to the purple *diphyllos.*" In some cases, Cordus noted variations: "*Cicuta sylvestris* with larger leaves than normal, and hairy, in woods between Pferingen and Ehingen." In other cases, however, the notes amounted to a description of a plant—presumably one that Cordus had not previously seen, as in this case:

> A small bush, spread on the ground, with green, reed-like branches: myrtle-like leaves, hard, pointed, and perennial: flowers butterfly-shaped, of diverse colors, namely white, yellow, and reddish: the outermost white, the next yellow, the innermost red. It grows in mountain woods along the Eger, in Bavaria and Swabia.[33]

In such entries we see the naturalist at work.[34] Cordus described the plant before him in his notebook. Whether he also made a drawing or took a sample is unclear. His manuscript had no illustrations; those in the published version were added by Gessner.[35] Perhaps Cordus relied on such notes, and his phenomenal memory, when writing his manuscript.[36]

On the other hand, certain entries in the "Sylva observationum" imply that Cordus wrote entire descriptions on the spot. One entry records that "a peculiar kind of olive tree grows in Swabia on the Lach and along the Danube, between the villages of Donauwörth and Kelheim. Described [*descripta est*]."[37] In this case, the field notebook reminded Cordus where he had seen the plant, while the description written on the spot would have been added to the manuscript. Although it is possible that Cordus added the note "described," the organization of his *Historia plantarum* and the descriptions of these plants imply that they were made on site. The eight trees in a row which are noted "described" or "I described this" in the "Sylva observationum" are grouped together in the *Historia plantarum,* as if the manuscript descriptions were still in the order in which they had been set down in Cordus's notes.[38] Only three trees in this series are not so annotated in the "Sylva." The first is the *Cytisus albus sylvestris,* which Cordus mentioned in the "Sylva observationum" immediately after the others with the remark, "which I described shortly before."[39] Since the plant is not mentioned in the "Sylva observationum," the words "shortly before" (*paulo ante*) must refer to when Cordus wrote the description. The second exception is the *Pseudocytisus,* whose description immediately follows that of the Cytisus, and the third is the Thymelea. A further peculiarity of these descriptions, in contrast with the others in the third book of the *Historia plantarum,* is the precise indications as to where they are to be found: whereas the sundew "grows in moist and marshy sands," the larch "grows in the German Alps, abundantly near Salzburg and in the hills of Silesia and Moravia on the Bohemian border."[40] If such descriptions were not written on the spot, they were at the least set down soon after the observations had been made. They are all descriptions of trees, suggesting that Cordus described them in detail in the field because they were too large to collect and study later, at leisure.

The experience of natural history travel was thus significantly different for Bock, Valerius Cordus, and their successors than for the preceding generation. Although Bock identified classical equivalents for the majority of plants he described, he felt no compulsion to do so, and hence his distinctions were often sharper. Like Euricius Cordus, he identified *consolida maior* with *symphitum alterum,* though rather than calling its flowers white he described them more precisely as "pale whitish-yellow colored." But the smaller "kleine Walwurtz," almost identical to the larger except for size and its slightly "tougher" appearance, was for Bock a distinct plant that was clearly unknown to the ancients.[41] Bock's interest in all the plants he encountered—many of which were unknown to the ancients—led him to draw clearer lines between them. In the same fashion, Valerius Cordus referred to very similar plants as if they were distinct species: for instance, the "*Hyacinthus minor* with round, concave, purple flowers" and the "*Hyacinthus minor,* white, similar to the preceding but a bit larger. Grows in the same places, but less often."[42] By his father's criteria, these plants could be considered the same: they differed only in color and size. For the son, they were clearly distinct.

The historian of botany Edward Lee Greene called Bock and Valerius Cordus the first phytographers: naturalists who saw their task as the precise description of plants. Though the term is not contemporary, it sums up succinctly their approach.[43] In the next generation, this form of experience was extended further: numerically, as the first generation of medical students trained in the *res herbaria* came to intellectual maturity, and geographically, as they devoted more attention to the areas in which they traveled and settled. More naturalists produced a wider range of documentary evidence, which allows us to investigate the phenomenology of experience in greater detail. It is to this generation that we now turn, focusing on its most famous representative, Carolus Clusius.

Clusius (1526–1609) was the most famous phytographer of the sixteenth century. In his journey through Spain and Portugal in the 1560s, he discovered some two hundred previously undescribed plants.[44] Though Clusius did not undertake the trip to botanize (he was to accompany one of the Fugger sons to the family's branch office in Lisbon), he examined and collected plants on the way. Despite the short trip and the cursory attention he could pay to plants along the way, he found many that were rare and not to be found elsewhere—unknown to modern authors, and perhaps also to the ancients.[45]

In 1573, after accepting Emperor Maximilian II's invitation to establish an imperial medical garden, Clusius left the Low Countries and established himself in Vienna. During the next sixteen years of his life he devoted himself not only to the medical garden, which Maximilian's heir Rudolf II tore down a few years later,[46] but also to careful investigations of the plants of Austria and Hungary, often in the company of his friend and landlord Johann Aicholz. These investigations are documented not only in his Austrian and Hungarian flora but also in his extensive correspondence with the Nuremberg physician and botanist Joachim Camerarius the Younger.[47]

Clusius arrived in Vienna late in 1573. Already in the spring of 1574 he and Aicholz were planning expeditions to the neighboring mountains.

By July the first of these expeditions had been carried out. Clusius undertook these trips in part to find plants for the imperial garden. But his main goal was to discover new plants. After the first summer's herborizations, Clusius enthused: "I have found many truly elegant plants in these Alps, some of which have already been described, but briefly and negligently, some of which have been observed by no one whose works still exist."[48] Years later he repeated that most of these plants were unknown to the ancients as well as to contemporaries.[49] Though in his Spanish flora Clusius had still attempted to identify modern plants with ancient descriptions when possible, his Austrian flora was largely free of such speculations.[50]

Like Bock and Valerius Cordus decades earlier, Clusius and his companions went on their expeditions equipped with notebooks and collecting baskets. And like his predecessors, but more precisely, Clusius recorded the plants he had observed and their locations. Year after year he returned to remote locations at different times to see whether new plants had emerged and to improve his earlier notes. Alpine excursions were especially daunting: some places "are difficult to reach, so that I think we have done well if we can go there once a year. Even an unencumbered man needs six hours to reach the ridges of some of those mountains, and there are no huts to wait out storms (which occur frequently) or to pass the night." In July 1577, Clusius and Aicholz tried three times to make an alpine ascent; they reached the base of the mountains once but were driven back by heavy rains.[51]

Through repeated excursions in the same areas, combined with careful notes and catalogues of plants found there, phytographers became familiar not only with the individual plants found in the areas they visited but also with their relations—what are now called plant associations. Bock had, at least implicitly, recognized that certain plants are regularly found together, and he had carefully noted the places where most grew.[52] By Clusius's time, such relations were much better known. His exchange of notes with Camerarius allowed him to determine similarities between alpine plants found around Vienna and those in the Tyrol.[53] Their near-contemporary, the polymath Julius Caesar Scaliger, remarked on such associations in, of all places, his *Poetics*, where he criticized Statius for calling the cypress an alpine tree. He had wandered through almost the entire Italian-French Alps and had never seen one there.[54]

Hence, by the late sixteenth century, traveling naturalists had a much more finely-textured view of the vegetable world than fifty years earlier. They admitted as distinct species plants that an earlier generation would have considered the same. If, for Euricius Cordus, difference in the color of flowers was not significant, for Clusius it was a clear distinction. "A few years ago," he remarked, "who would have believed that white hellebore would put forth a purple-blackish flower unless he had seen it himself? But nowadays this plant is carefully cultivated. Here and in the surrounding hills it is so common that no other kind of white hellebore is known, and that kind that has white-tending or greenish flowers is not found except in the Styriac Alps or the meadows at their feet."[55] It was not merely the difference in color that led Clusius to consider this a different type of plant, but its geographical distribution: the distinct range of this hellebore led him to consider the difference in color not as a mere variation but as a characteristic of a population.

Wider experience of the natural world thus brought with it a substantial alteration in the way experience itself was structured. It became both more localized, as particular areas were ever better known, and—in an important sense—less immediate. As we have seen, Hieronymus Bock and Valerius Cordus took careful notes on what they observed, and Bock at least also grew plants in his garden to observe them more closely. Their successors would elaborate these techniques and develop others for deepening experience, a process that was necessary if the naturalist was to be able to determine whether a plant was really new, given the range of new discoveries and publications that characterized the second half of the sixteenth century. The garden and the herbarium allowed them to repeat their observations, either by watching other individuals of the same species go through their life cycle again, or by fixing a particular plant at the moment of observation. I will focus on botany. Zoology was a poor cousin to botany in this period, lacking as it did the firm place in the medical curriculum held by the study of plants, but it too was pursued in menageries and cabinets of curiosities.[56]

Gardens, herbaria, and cabinets were more than tools for natural history, however. With the exception of herbaria, they had existed before being adopted by naturalists, and even the herbarium developed from antecedents that had little to do with the study of nature. Their adaptation by naturalists demonstrates the close connections between natural history in its first century and the material culture of the late Renaissance. It also shows the creative ways naturalists recognized and solved the problems created by their project of cataloguing the world. Hours spent in the garden or leafing through the herbarium could substitute—with some limitations, as we shall see—for days or weeks spent in the field; they thus allowed naturalists to have a richer, if in some ways less immediate, experience, but only by removing *naturalia* from their natural setting. By so doing, gardens and herbaria helped concentrate, even as they furthered, Renaissance naturalists' concern with description.

Gardens

Botanical gardens fill a variety of functions. University botanical gardens are sites of teaching and research. Along with municipal botanical gardens, they also present an aesthetically pleasing selection of plants to a general public. Many gardens also preserve and propagate rare and exotic species. These roles emerged in the sixteenth century. Histories of botany usually emphasize the foundation of botanical gardens at Padua, Pisa, and Florence in the 1540s—followed closely by Montpellier, Leiden, Erfurt, and other universities, through Oxford in the 1630s—as a trend that emerged almost from nowhere.[57] These were the first university botanical gardens, and their foundation does indicate a new willingness, on the part of fiscally conservative university administrators, to pour a great deal of money into the *res herbaria.* In design and function, however, these gardens resembled closely their private and princely predecessors. To Conrad Gessner, observing the situation at the end of the 1550s, the "hortus publicus" in Padua was remarkable for its magnificence and variety, but it was hardly unique.[58] The Renaissance botanic garden was an organic outgrowth of late medieval gardening traditions.

The archetypal medieval garden was the monastic herb garden, which served as a model for both medieval nobles' gardens and Renaissance formal gardens.[59] The typical cloister garden was a square or rectangular plot, divided into quadrants by paths. The center, where the paths intersected, was often occupied by a well, which provided water for the monastery as well as for the garden itself.[60] These elements were already present in the ideal monastery depicted in the ninth-century Plan of St. Gall, although the garden's center there is occupied not by a well but by a tree reputed to drive away evil spirits. Cloister gardens of the high Middle Ages generally followed this pattern. Larger monasteries could devote additional space to separate kitchen and medicinal gardens, as in the Plan of St. Gall.[61]

The aristocratic pleasure gardens of the high and late Middle Ages drew on elements of the cloister garden, though their different functions led to the introduction of other design elements. In his *De vegetabilibus,* Albertus Magnus described a thirteenth-century pleasure garden. This garden contained a section for herbs and other plants; next to it was a meadow with grassy banks for sitting, a fountain, and trees to provide shade and shelter.[62] The plant garden is essentially identical to the monastic cloister or medicinal garden. Fifteenth-century miniatures, though they do not show whole gardens, display the same basic elements of garden design; if we may draw

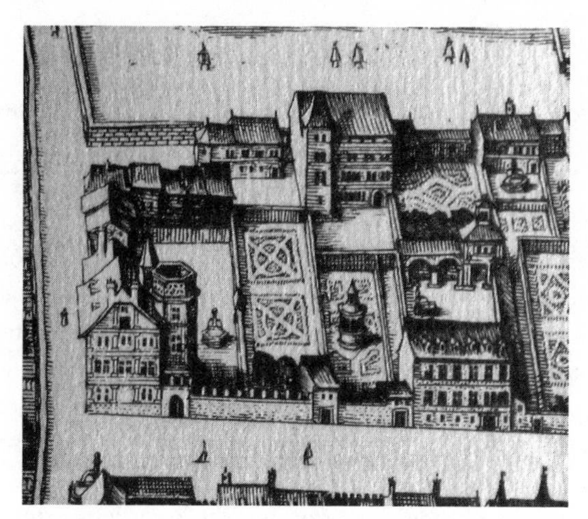

Figure 4.1. Merian, *Nova et genuina descriptio inclytae urbis Basileae* (1615): detail showing Felix Platter's houses. Photo by the author.

conclusions from them about real gardens, the typical late medieval pleasure garden was small, enclosed, and designed primarily to be seen and enjoyed from within its walls.[63]

By the beginning of the sixteenth century, private bourgeois gardens were also quite common. The courtly gardens of the late Middle Ages, themselves based initially on the monastic garden, served as immediate model. In many walled cities the *hortuli,* tiny gardens used in the summer as living space as well as for growing food plants and herbs, were considered too small.[64] Deprived of room to expand within the city itself, the bourgeois imitated noble pleasure gardens in the suburbs. Town plans from the sixteenth and seventeenth century show the proliferation of such gardens outside of cities, particularly the rich trade cities of southern Germany.[65]

Basel is typical. In both Großbasel (south of the Rhine) and Kleinbasel (north of the Rhine), the relatively compact medieval core of the city left little space for gardens. But the Vorstädte, suburbs that grew up around Großbasel in the thirteenth and fourteenth centuries and that by the fifteenth century had been enclosed by the third city wall, offered more room.[66] Houses in these areas often had substantial gardens in their courtyards, while homeowners in the inner city had separate, enclosed gardens in or around the Vorstädte. Felix Platter owned two houses near Petersplatz, with gardens between them (figure 4.1). His young colleague Jacob Zwinger, who lived inside the second city wall of 1200, possessed a garden in the Leonhardsgraben, while Zwinger's in-law Martin Chmielecius supplemented the Rhine-side back yard of his house in the Augustinergasse with a garden in the

Malzgasse. Caspar Bauhin's house had a long, narrow courtyard unsuited to a serious garden, but he owned a larger tract across an alley from his house's rear entrance.[67] The city itself was surrounded by small enclosures, though these were probably used primarily for growing food rather than as pleasure gardens.

Other sixteenth-century cities show the same pattern. In 1530s Marburg, Euricius Cordus had both a small garden (*hortulus*) next to his house and a larger one in the suburbs beyond the city gates.[68] Justus Lipsius was granted use of a garden by the city of Leiden during his tenure as professor at its university; one of Carolus Clusius's first concerns when coming to Leiden was to procure a garden for his favorite plants.[69] Many of these gardens, especially suburban ones, were probably used for agriculture rather than pleasure.[70] By the end of the Middle Ages, many cities had gardeners' guilds whose members were paid to tend such plots, whether they were for bourgeois leisure or growing vegetables.[71]

In his colloquy "Convivium religiosum," Erasmus described such a suburban garden, on a "small but well-cultivated suburban estate." The main garden, created for pleasure, was a rectangular enclosure with plants organized in beds surrounded by narrow canals. It was surrounded on three sides by galleries, open to the east, west, and north, that were painted with images of plants so that the garden would delight the eye even in winter. One of them contained an aviary. Beyond this pleasure garden was another garden, devoted to esculent and medical herbs; a lawn enclosed by a hedge; and an orchard with foreign trees.[72] Not all bourgeois suburban gardens had a lawn and an orchard, though Felix Platter had a separate plot for his fruit trees. But Erasmus's description fits well with the less-detailed images of gardens found in early seventeenth-century city plans (figure 4.2).

Contemporary book illustrations provide another source of descriptions of gardens, but these must be used cautiously. They depict, generally speaking, the largest and most splendid Renaissance gardens. The extensive gardens of Dr. Lorenz Scholtz in late sixteenth-century Breslau, for example, may reflect Erasmus's ideal, but they were laid out on a much greater scale than Eusebius's garden in the "Convivium religiosum."[73] Garden historians tend to concentrate on such magnificent gardens, largely because they are the only ones extensively, and visually, documented.[74] But even these gardens shared elements found in the more humble gardens of the less wealthy bourgeoisie. The baroque gardens of Italy and France, designed to be seen from a fixed vista as well as, or instead of, from within, are products of the late sixteenth and seventeenth centuries.

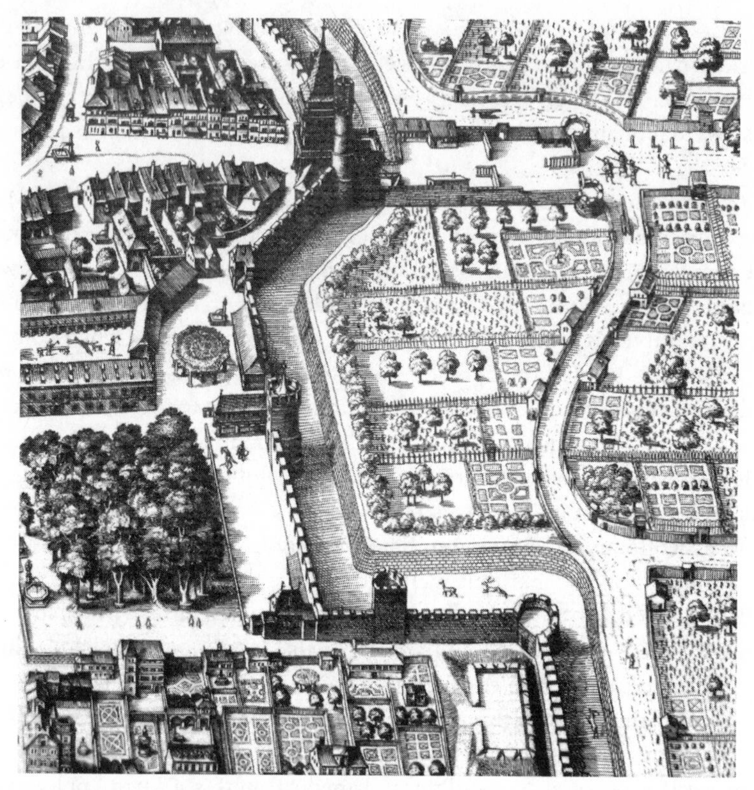

Figure 4.2. Merian, *Nova et genuina descriptio inclytae urbis Basileae* (1615): detail showing urban and suburban gardens. Photo by the author.

To Conrad Gessner, writing in the 1550s, gardens were distinguished by a short list of *differentiae.* They could be vulgar, containing only food plants; medical, with garden, wild, and foreign plants that are used in medicines; varied (*varii*), similar to the former but containing above all the more uncommon (*rariores*) plants that are grown "for admiring and contemplating nature"; elegant, such as those of women and certain opulent men (especially monks, added the Protestant Gessner), which are planted not with utility or the study of nature in mind but rather beauty; and magnificent, established by princes or republics, with waterworks, hills, groves, and fields for exercise. "These are the simple *differentiae,*" remarked Gessner; "two or more are sometimes conjoined."[75] The first four categories were based on the garden's purpose, whereas the last was determined by the size and the social standing of the owner: but in this case too the goal was of prime importance, for one point of such gardens is to impress the onlooker.

Gessner excluded the vulgar vegetable garden from further consideration. But the others all deserved a scholar's attention. Moreover, the categorical differences between different gardens ruled out legitimate rankings of them: "Hence no one should be surprised that I commend not only magnificent and more elegant gardens, but also private ones and even the most confined little gardens."[76] The treatise was published in a collection of Valerius Cordus's botanical writings, indicating that Gessner expected his readers to be naturalists, but he thought that the *studiosus rei herbariae* could learn from almost every garden.[77] What was important was not the garden's size or its owner but the plants it contained.

This was not merely a programmatic statement. The gardens that Gessner listed fall into all of his categories.[78] Of thirty-one garden owners in Germany, only nine were physicians, two surgeons, and six apothecaries. The others were ecclesiastics, teachers and professors, town officials, merchants, and one noble.[79] Furthermore, the types of gardens did not break down along the lines of profession, craft, or status that one might suspect. The gardens of the duke of Bavaria in Munich, and of the Fugger in Augsburg, were indeed of the magnificent sort, and the Antwerp apothecary Coudenbergh's garden was (according to Gessner's sources) full of medicinal plants, albeit many of these were rare.[80] But the Lindau merchant Matthias Curtius had an "outstanding garden sown with many rare plants," that is, a "varied" garden according to Gessner's scheme.[81] It should not be surprising that these categories were blurred at best, for we have already seen that the community of naturalists in Gessner's day was joined as much by social as by professional or craft bonds.[82]

In fact, naturalists used all of the kinds of gardens Gessner enumerated. Euricius Cordus, Gessner himself, Felix Platter, Caspar Bauhin, and Carolus Clusius all possessed, at one point or another, small private gardens, either behind their houses or nearby, where they could grow plants of immediate interest. Clusius was also, for two years, intendant of the short-lived imperial medical garden in Vienna. He studied plants there and appropriated some specimens when it became clear that Rudolf II had no interest in maintaining the garden.[83] Earlier, Clusius had used the gardens of his noble friend Jean de Brancion in Mechelen; he was also a familiar of Pieter Coudenbergh, the Antwerp apothecary whose garden had impressed Gessner through hearsay.[84] Caspar Bauhin's brother Johann, physician to the duke of Württemberg, worked in the ducal garden at Montbéliard.[85]

In this constellation, perhaps the most important feature that the public university gardens of the later sixteenth century possessed was their institutional continuity. Private gardens were often ephemeral. Pieter

Coudenbergh's garden was destroyed by Farnese's troops during the siege of Antwerp in 1584–85, while Brancion's garden and the imperial medical garden at Vienna did not long survive their patrons. Joachim Camerarius's *hortus medicus* in Nuremberg fell to pieces during Camerarius's final illness, especially since his friend and collaborator Leonhard Dold was also ill.[86] Caspar Bauhin, a very busy man, apparently let his garden lapse, for in 1607 his correspondent Prosper Alpino sent him seeds, with the words "I congratulate you on beginning a garden for growing plants."[87] Public gardens were not entirely free of such vicissitudes; Richer de Belleval's first garden at Montpellier was destroyed during the anti-Huguenot campaigns of the 1620s. Yet Richer quickly rebuilt the university garden even more magnificently.[88] Other public gardens, like that in Leiden, possess an unbroken institutional continuity, though they were occasionally moved to allow for expansion.[89]

In addition to being sites of medical education, university gardens served their communities as social space, just as private pleasure gardens functioned for their owners.[90] Moreover, they were planted not only with medicinal simples but also with many of the new decorative imports and splendid garden varieties of more common plants.[91] Hence it would be anachronistic to overstress the differences between public and private gardens. Both served naturalists well, from an intellectual as well as an aesthetic standpoint.

It has occasionally been claimed that Renaissance gardeners, in particular the founders and keepers of botanical gardens, were striving toward nothing less than the recreation of the Garden of Eden.[92] While there may have been intimations of Eden in the thoughts of some Renaissance gardeners, it is hard to imagine any sixteenth-century gardener taking the thought seriously: no contemporary theologian would have approved the notion that man could recreate the earthly paradise. Man might imitate and perfect the works of nature, God's second-in-command, but not the works of the Creator himself. Robert Mallet seems closer to the mark in attributing the earliest Western "nostalgia for paradise and the formal desire to restore it" to the Romantics.[93] The Latin vulgate reads: "The Lord God, in the beginning, planted a paradise of pleasure, in which he placed the man he had formed. And the Lord God produced from the soil every wood beautiful to see and pleasant to eat."[94] The text does not use the word "hortus" (garden); instead, it reads "paradisus" (paradise), and this paradise has only trees, or at least the text does not mention herbs.[95] The vernacular translations into German

(Luther) and English (King James) do use "garden" to refer to Paradise, but they do so only once: after the initial reference to God's creation, the proper noun "Eden" is used. John Parkinson's *Paradisus terrestris* (1629) may seem to be an exception, but Parkinson clearly chose his title for the play on words: *paradisus* means "park," and the full title of his work, in English, is *Park-in-Sun's Earthly Park*.[96]

Regardless of Biblical associations, sixteenth-century naturalists mentioned Eden only rarely. When Hieronymus Bock proclaimed God "der aller erst Gartner," he referred not to the trees of the Garden of Eden but to the plants of the entire world.[97] Conrad Gessner's account of gardens omitted Eden entirely.[98] Francis Bacon began his essay on gardens with Eden, but added immediately, "and indeed it [gardening] is the purest of human pleasures."[99] Bacon did not mention God again, and his attempt to produce a perpetual spring is predicated on the real climate of London, not an ideal paradise. Parkinson too insisted that his book contained only those "pleasant flowers which our English ayre will permitt to be noursed up."[100]

Even had someone claimed to plant a second Eden, Renaissance naturalists would have known that it was impossible. The challenge for sixteenth-century naturalists, as they planted and tended their gardens, was not recreating Eden: it was getting the plants to grow at all. Conrad Gessner gave the palm in gardening to the ancient Greeks, and second place to the Latins. The Germans lagged behind: in part because they were devoted more to arms, but in part "because the harsh nature of the soil and the inclement weather in areas more exposed to the north is detrimental to cultivating gardens. Hence it is especially praiseworthy that people plant gardens in our regions without being scared off by any difficulty."[101] The difficulty of the attempt underscored the achievements of those who undertook it. Eusebius, in Erasmus's colloquy, had "several foreign trees, which I am gradually accustoming to our climate."[102] Pierre Belon recognized both the difficulties of acclimatizing trees and the benefits of doing so in his proposal to establish a royal acclimatization garden in France.[103] An annotated set of manuscript plant illustrations from the Low Countries in the later sixteenth century includes frequent references to southern plants that grow only in the gardens of "devotees of botany," and then only "with great skill."[104] Even the curators of Leiden University, with their ambitious plans for a Calvinist university to rival Leuven and Paris, recognized that some plants could not be grown in the magnificent medical garden they desired to establish. When they nominated Bernardus Paludanus as *praefectus horti* (a post he refused), they specified that he should plant there "every kind of herb, plant, and sprout that can suffer the sun, air, and nature of this land to some degree [*eenichsins*]."[105]

As the adverb "eenichsins" emphasized, even those plants that could be grown often required a great deal of care. Bock described the pains that Jörg (or Georg) Öllinger took to preserve his aloe plant:

> Jörg Oellinger puts this foreign Aloe in his garden every summer, and there it grows a bit larger. When winter approaches, he digs it out and puts it in his chamber. There it loses a bit of its natural color, as if it were dying, for about three weeks, and then begins again to grow as if it were risen from the dead. But it cannot bear the winter frost in the garden, and just when it is growing the best, in autumn, the winter starts breathing down its neck. That is a reason why it cannot bloom in our lands.[106]

Gessner described the same procedure, familiar to any modern gardener, in his remarks on how to garden, adding that plants so transplanted should not be brought in suddenly but gradually accustomed to the warmer air indoors.[107] One of the early additions to the Leiden *hortus publicus* was a gallery in which such potted plants could be placed.[108] Such measures could be strikingly successful. Joachim Camerarius was justly proud of his efforts to grow *Dictamnum:* "Everyone agrees that Dictamnum grows only on the island Crete, whence it is named 'Cretan.' I assert, however, that it produces flowers for us in the summer, if it is diligently protected in the winter: in 1575 it flowered abundantly in my garden."[109] Nonetheless, they were stopgaps, means to keep certain plants alive over the winter, rather than greenhouses in a modern sense.

If some plants could not bear the northern cold, others could not tolerate warmer climes. Clusius sent *Parthenium alpinum* and *Absinthium alpinum* to his correspondents Joachim Camerarius and the Landgraf of Hessen with the instructions: "You should put these in your garden in a shady place, so that they can recuperate. From long experience I have learned that alpine plants are better conserved in gardens if they are put in an open place that gets very little direct sunlight."[110] Even so, growing these plants in lower elevations was risky. Some plants simply did not bear transplantation, such as the *Pyrora altera* found in the woods around Nuremberg.[111]

Even plants that could be transplanted had to be carefully replanted in suitable locations. As Clusius mentioned, alpine plants did best in shady areas. So did many others. Gessner stressed the importance of having several different kinds of land in the garden: "If there are many kinds of terrain in the garden (such as hills, rocks, banks, standing water, sand, etc.), many plants can grow and survive, each planted in its proper place, which otherwise would not grow or survive." Gessner managed to grow figs in Zürich by

Figure 4.3. Richer de Belleval's contrivance for a humid garden in Montpellier. Leiden University Library, Ms. Vulc. 101, drawing added to a letter of Nicolas Claude Fabri de Peiresc to Carolus Clusius, dated February 27, 1604.

planting them against a wall, but even long experience and skilful planting had its limits: he could grow marine plants in his garden, but not get them to flower.[112]

Pierre Richer de Belleval's plan for the botanical garden in Montpellier shows the creative use of landscape and construction to provide varied terrains in a small space. Richer ran a long mound from east to west in the center of the garden. Plants that required a lot of sun and warmth were planted on the southern side, while those that preferred shade or cold grew on the north.[113] For growing aquatic and marsh plants, Richer contrived an elaborate system of walls and pools, fed by a well.[114] This device allowed him to maintain a cool, damp, dark area in which such plants thrived (figure 4.3).

Despite such techniques and contrivances, gardening in the sixteenth century—as today—was a hit-or-miss proposition. A harsh winter or a wet spring could devastate gardens—and winters in the later sixteenth century were growing increasingly harsh.[115] Some plants simply refused to grow, while others flourished despite every attempt to control them.[116]

Fortunately, the well-established social network of natural history allowed brown-thumbed naturalists to replenish their gardens quickly, as we shall see below.

Gardens were places of repose, visual delight, and social exchange. But naturalists' gardens, or those they used, were also cultivated for study. Naturalists examined the plants that they received from friends and planted in their gardens. They also collected wild plants and transplanted them in their gardens to observe them more closely. In this way, they could observe the life cycles of dozens or hundreds of plants with a fraction of the effort required for field work.

What kind of plants did gardens contain? That depended, above all, on the owner's interests. In the early seventeenth century, Sigismund Schnitzer reported the princely gardens in Bamberg contained,

> in addition to common narcissus, tulips, and hyacinths, . . . *Moly flore flavo, Moly Indicum, Corona imperialis flore albo, Lilium supinum, Dens caninus, Narcissus Mathioli, Canna indica, Campana lazura, planta indica, nasturtium indicum, Fritillaria alba, Croci varii Sandalica scabiosa Hispanica fl. candido et nigra* etc. I don't have time now to list them all, but if Your Excellency likes, I can send a complete catalogue later.[117]

Carolus Clusius was particularly fond of decorative bulbaceous plants.[118] Through his efforts, as well as those of other Dutch and Belgian naturalists, such bulbs became common in the pleasure gardens of the Low Countries in the late sixteenth and early seventeenth centuries.[119] When the Parisian Jardin du Roy was first planted in the 1630s, it contained many such decorative plants, despite its nominal function as a medical garden.[120]

Of particular interest, beyond decorative plants, were those plants that a naturalist wanted to study closely. In the 1550s, Conrad Gessner, occupied by his duties as city physician in Zürich, dispatched a youth "to cross the Alps, at my expense, in order to bring me some recently uprooted Hellebore plants, as well as any other rare plants he might find, so that I can adorn my little garden and illustrate my history of plants."[121] Clusius collected and transplanted many alpine plants in his Viennese garden.[122] He also grew and described plants that his correspondents sent to him, like the *Hyacinthus obsoletior hispanicus,* sent by Francisco de Hollebeque in Spain to some of

Clusius's Belgian friends, from whom he obtained it.[123] In the case of many plants introduced via Constantinople, cultivated exemplars were the only sources of Clusius's descriptions.[124]

But if naturalists were interested in growing rare and exotic plants in their private gardens, they also cultivated more common ones. Thomas Platter may have been writing with false modesty when he told Jacob Zwinger, "I want you to take some of the seeds I have sent you and my brother and plant them in your father's garden, which is already most elegant—not because I think the plants I have collected have any beauty or rarity, but because my vulgar plants, by being placed next to your rare and exquisite ones, will increase their grace and splendor."[125] Yet the sentiment Platter expressed was genuine; in order to recognize the rare and exotic, one had to be aware of what was commonplace.[126]

Gardens intended for teaching had an even stronger reason to include common plants. Granted, experienced naturalists focused on novelties. Caspar Bartholin described the *hortus medicus* in Padua to his teacher, Bauhin, in 1608, noting that it contained "many rare and exotic plants, some of them not yet described."[127] Bartholin went on to list the plants he collected in the garden (presumably with the approval of its *praefectus*). Earlier, Gessner had praised the range of exotica in this garden.[128] It is not surprising that naturalists, writing for their colleagues and teachers, would mention above all the exotica. But surviving catalogues of sixteenth-century public gardens reveal a wider range of plants.

The *hortus medicus* at the University of Leiden is a case in point. The 1594 manuscript catalogue by Clusius and Dirk Cluyt, which indicates not only which species were planted when the garden was established but also where, provides an unrivalled resource for analyzing the kinds of plants that sixteenth-century naturalists cultivated for instruction and pleasure.[129] Many of the plants were, indeed, rare. Four of the beds were surrounded by a fence to keep out acquisitive hands. The earliest documentation for the existence of the fence is a 1610 engraving, but the area it surrounds was in fact occupied by some of the rarest plants in the 1594 catalogue: various tulips, rare Oriental hyacinths and lilies, and other ornamental plants, but also rare but not especially elegant bulbous plants like Hungarian wild garlic.[130]

But of the fourteen hundred beds in the *hortus*, many were occupied by common species. The *Pulsatilla*, unknown to the ancients but widespread in central Europe, was present, along with other common anemones; so was the common *Ornithogallum, Apium hortense, Hepatica, Absynthium vulgare,* and *Dictamnus officinalis;* the wild strawberry was to be planted as well.[131] Many of these common plants were well-known simples, used to teach medical students their forms and virtues. Indeed, some writers held that every plant held medicinal virtues, and thus deserved a place in physic gardens.[132] Yet they were not all in common use by physicians or apothecaries. It seems more likely that many plants were included for their beauty, as with the more common types of tulip; to help students differentiate between closely related species, as with the three kinds of oregano; or for both reasons, as in the case of the several *Narcissi.*

The Leiden garden was thus, in many respects, an encyclopedic collection of plants.[133] By grouping related species, the *hortus* adumbrated the same tacit notions of similarity and difference that characterized the organization of Renaissance herbals.[134] In this and similar gardens, both public and private, the scholar and the student could experience the full range of nature—or at least those plants that could tolerate the climate where the garden was located and the care its gardeners provided—without needing to travel widely. The garden could not wholly replace travel, but it provided a further means of gaining firsthand knowledge of the natural world and comparing it with descriptions.

Gardens could reduce the need to travel because plants traveled on their own. Sixteenth century naturalists regularly exchanged floral material: seeds, bulbs, roots, and even whole plants. When a naturalist could not visit other gardens, and had not found a plant in the wild, he could request it from a friend. Clusius sent seeds of several kinds of *Linum silvestre* from Vienna to his friend Brancion in Mechelen—along with dried specimens of the same plants, "so that he could see their forms immediately rather than have to wait until they grew from the seed: many plants do not reliably grow from seeds, and he was quite sickly."[135] More patient correspondents could wait until the next spring to see the plants they had been sent. Ulisse Aldrovandi congratulated in a backhanded way his young friend Jacob Zwinger on his new garden, "planted with so many exotic plants, which I at one time or another have seen, had, described, and depicted." There was one exception: "I don't know the *Lychnis* that you alone cultivate in Basel, and I would be most pleased if you could send me some seeds from it, and a few others."[136] When Mathieu de l'Obel established a medical garden for the English Lord Zouche, he wrote Caspar Bauhin asking for seeds. Bauhin's former student Sigismund Schnitzer also requested "a few rare seeds" for his private garden [*hortulus meus domesticus*].[137]

By exchanging material, naturalists spread many garden plants throughout Europe. Uncommon plants were prized in such exchanges—but the notion of rarity was relative. After Catalonian jasmine had become a common garden plant in Italy, Clusius remarked that it could now be called the humble jasmine (*Chamaejasminus*).[138] Rarity and vulgarity could even be defined relative to one garden. In 1582, Clusius sent many plants to the duke of Württemberg, only to receive shortly thereafter a catalogue of the garden including many of the plants he had sent. "I regret this, lest the Duke think I have sent him common things [*vulgaria*]."[139]

Exchanges were not limited to trades among the *studiosi rei herbariae.* Carolus Clusius had especially broad connections with horticulturists and gave freely to his friends and acquaintances.[140] Clusius had a sharp eye and a great love for garden plants, especially colorful flowers. After arriving in Vienna and exploring the surrounding Alps, Clusius remarked various kinds of Doronicum "that are different from those that grow in Belgian gardens (which, in turn, I have not seen elsewhere). The Belgian ones are now common in gardens in these parts, since I had them sent here from Belgium."[141] The tulip is one of Clusius's lasting contributions to the Dutch garden and economy: in his years in Vienna, he acted as a conduit for tulip and other bulbs between Constantinople and the Low Countries.[142] Considerable quantities of bulbs could be involved: returning to Vienna from England in 1581, Clusius brought some two hundred *Pseudo-narcissus* bulbs with him.[143] Such plants could be traded for other material. The exchange networks carefully constructed and maintained by naturalists were key to making gardens sites where nature was concentrated and multiplied.

But gardens were not always reliable sources of experience. Transplantation sometimes subtly transformed the nature of wild plants. Often, they flowered at different times, as with Clusius's fifth *Gentiana:* in the Alps, where it grew around the line of perpetual snow, it flowered in June and July. "But in milder places, where it is rarely found, and in gardens, it flowers in April and May."[144] The same was true of other alpine plants.[145] In his descriptions, Clusius specified two times of flowering for these plants, since otherwise his readers might inaccurately identify a plant found in their gardens.

Gardens could also bring forth other, more substantial alterations in plants. One of the most common was the multiplication of flowers. Writing to his friend Camerarius, Clusius observed that "the flower of the red lily is sometimes not only duplicated, but multiplied even further. In Belgium, I saw flowers having fifty or sixty leaves [i.e., petals], but the plants on which they grew brought forth simple flowers the next year. Nature sometimes plays in this fashion."[146] *Primulae* were especially prone to develop multiple and enlarged flowers when "moved into gardens and cultivated there."[147] The notion of *lusus naturae,* jokes or games of nature, was commonly used to explain such freakish alterations, be they temporary or permanent.[148]

In the year 1611, Nature was especially playful in the prince-bishop's garden at Bamberg. The court physician Sigismund Schnitzer wrote to his old teacher Caspar Bauhin with news of the oddities:

> This year we had a multiflowered lily (which Clusius mentions), with a flat stem, broad on the upper part, and a full flower, on which I counted thirty-six leaves. This summer we also had an *antirrhinum* with a very flat and broad stem on the upper part. One of the *Caryophylii* produced a flower like a Rheseda, with pale vermilion petals and calyx. A *Consolida regalis* with a flower like a green Valerian produced six or seven stems from a hollow follicle on the root, each coming out of the right side. We also saw a "clustered" juniper, similar in size and shape to a grape cluster, consisting of a multitude of berries stuck tightly together. Are these products of a luxurious nature, or are they portents? I don't think that they are peculiar species of those plants.[149]

Singularities such as these were not, in fact, considered species. But the longer-lasting garden varieties often were; even though produced by cultivation, their persistence through generations established them as separate species worthy of their own descriptions.[150] Naturalists who saw hundreds of individuals of any given plant kind could make such judgments on the basis of long experience.[151]

Nature's games were an additional reason, beyond the sheer resistance of some plants to acclimatization, that naturalists could not always trust the experience they gained in gardens. Fabio Colonna used wild plants as the basis for the illustrations in his *Ekphrasis*, since, he noted, cultivation tended to alter their form.[152] Seventeenth-century *amateurs de fleurs* attempted systematically to produce such alterations, in order to render their gardens more beautiful, but sixteenth-century naturalists were generally content to note their existence.[153] By careful comparison of wild and garden plants, they could generally determine whether a character was found in the wild plant.

Field excursions and gardens gave an annual rhythm to the *res herbaria*. Caspar Bauhin could hold the chairs of anatomy and botany at Basel because he taught the latter in the spring and summer, when plants could be inspected, and the former in the winter, when corpses could be dissected without decaying too quickly. Leonhard Dold, writing in the dead of winter, asked Jacob Zwinger about literary news—whether a new edition of Pietro Andrea Mattioli had been published, and what Cesalpino had written—before adding, "I don't have anything else to write: botany is buried in the winter."[154]

Herbaria

Leonhard Dold was exaggerating. Corresponding to the living garden was the dead one: the herbarium, or, to use contemporary terms, the *hortus siccus* or *hortus hiemalis.*[155] The herbarium was, and is, a collection of plants that have been plucked, pressed flat and dried, and affixed to paper sheets (in some herbaria, the plants were merely folded into sheets of paper without being affixed; in most, they were glued or sewn in place). In some cases these sheets were bound into volumes; in others, they were left loose. The process of making a herbarium has changed little since the sixteenth century.[156] The main differences are an increased range of drying and storing techniques for fungi, mosses, and the like, and the availability of commercial drying and pressing devices. But a sixteenth-century botanist would have no problems recognizing and using a twenty-first-century herbarium (figures 4.4, 4.5).

Detailed instructions for making a herbarium, along with a recipe for glue, were first published by Adriaan van de Spiegel in 1606.[157] This has led some historians to the untenable assumption that Spiegel introduced the practice, or at least that it is contemporary with him.[158] In fact, it seems to have been developed half a century earlier, and the name most often associated with the beginnings of herbaria is that of Luca Ghini, founding director of the botanical garden in Pisa.[159] Though Ghini never published, he was an indefatigable and much-loved teacher, and his students included some of the most prominent naturalists of the later sixteenth century. From Pisa, the technique spread throughout western and central Europe by word of mouth. More than twenty extant herbaria were begun in the sixteenth century, including those of Felix Platter, Andrea Cesalpino, and Caspar Bauhin.[160]

Still, the technique did not travel everywhere, despite its utility. Theodor Zwinger, in Basel, wrote his friend Joachim Camerarius in 1579 for instructions on making one: "I hear that you know how to dry plants. If you think I'm the kind of person to know such things, I beg you to tell me, under oath of silence."[161] Since Zwinger's colleague at the Basel University, Felix Platter, had been collecting plants in his herbarium since his days as a student in the 1550s, this request is puzzling.[162] The herbarium's spread outside of a community of dedicated botanists is hard to measure, and varied according to place: while some dried plants were de rigueur in seventeenth-century French cabinets of curiosities, the technique was unknown to Samuel Pepys until he saw John Evelyn's herbarium in 1665.[163] Yet British naturalists had been using herbaria for decades, and Linnaeus considered William Sherard's collection at Oxford among the richest in Europe.[164]

Figure 4.4. Gladiolus, page from Caspar Ratzenberger's herbarium (c. 1550–92), vol. 3, p. 573. Photo courtesy Naturkundemuseum Kassel.

Figure 4.5. *Sagittaria cuneata* Sheld, page from a twentieth-century herbarium. Photo courtesy of Utah Museum of Natural History, Garrett Herbarium.

If the history of the herbarium is easy to trace from the 1540s onward, its origins are considerably more obscure. The technique seems self-evidently useful to modern naturalists, so much so that few have stopped to ask themselves where and when it was introduced.[165] Jean-Baptiste Saint-Lager admitted that the earliest mention of herbaria dated from the 1540s, but he was convinced that the technique had a long prehistory. It was too simple not to have been invented earlier, and the earliest mentions of herbaria do not indicate that they were a novelty. Certainly it did not take a stroke of genius to discover that plants could be dried flat between sheets of paper and preserved. Saint-Lager saw the solution to the problem in the fact that paper itself was a recent introduction to Europe: papyrus and cloth were too flexible to use for affixing plants, while parchment was too expensive.[166] But thin sheets of wood or some other material could have been used for this purpose.[167]

In material terms, then, it is hard to see why the herbarium should have been invented and disseminated only in the 1530s and 1540s. But the reason becomes clearer if we consider their function. With the transition to phytographic natural history, techniques for reinforcing memory became important. Both the field notebook and the herbarium filled that function. Saint-Lager suggested that the earliest herbaria were travel souvenirs, a suggestion that is borne out by closer investigation.[168] As Thomas and Virginia Kaufmann's study of *trompe-l'oeil* in early modern illuminated manuscripts hints, the art of making a herbarium, like keeping a botanic garden, was adopted by naturalists from the material culture around them.[169]

We have already seen the origins of this technique in devotional manuals: naturalistic representations of pilgrimage souvenirs, including *naturalia*, substituted for the actual items themselves in luxurious devotional manuals. These objects functioned not only symbolically, as objects of meditation, but as souvenirs of the pilgrimage that the book's possessor had undertaken. *Naturalia* pressed in books of hours served the same function, and they were sacralized in the process of collection just as the pilgrimage sacralized space itself. The Kaufmanns speculate that *trompe-l'oeil* illustrations of naturalia were intended to serve the same function, while at the same time being persistent reminders that neither faded nor damaged the books in which they were contained.[170] In some cases, as when the stem of a plant appears to pass through a slit cut in the page of a book, pass behind it, and reemerge on the front, the intention to imitate a real object contained in similar books appears evident.[171]

Naturalia were pressed in devotional books in the fifteenth century, and other books were probably used for the same purpose. It is probable that other books were used for this purpose as well.[172] It seems plausible that some sixteenth-century naturalist (Luca Ghini or John Falconer, for example) decided to affix these dried plants to paper on which their names and other information could be recorded. The transition from devotional object to botanical tool is, however, not straightforward. It was no doubt eased by an earlier introduction in the practice of *res herbaria:* the elaborate, full-page illustrations of Brunfels's and Fuchs's herbals.

Fifteenth- and early sixteenth-century illustrated herbals had small illustrations.[173] Sometimes based on actual plants (as in the *Herbarius zu Teutsch*), more often copied from medieval manuscript illustrations, they were considerably smaller than the plants they represented, and surrounded by text. The woodcuts in Otto Brunfels's *Herbarum vivae eicones* (1532, 1536) were strikingly different and merited the book's name. Hans Weiditz, a pupil of Albrecht Dürer, was responsible for the woodcuts—and, recent scholars conclude, the plan of the book, for which Brunfels had the subordinate task of compiling descriptions to accompany the illustrations.[174] Weiditz's illustrations occupied most of a folio sheet and portrayed plants as if they had been pressed flat to best show their distinguishing features to the reader's eye.

Whether Weiditz himself worked from pressed plants is unclear. A near-contemporary image of botanical artists at work, from Leonhart Fuchs's *De historia stirpium* (1542), shows Albrecht Meyer drawing a plant that is sitting before him in a pot.[175] Yet certain of Weiditz's illustrations, such as the "Walwurtz männlin," appear to have been drawn from material flattened to fit in the available space, as the sharp bend in the stem hints (figure 4.6). Regardless of whether Weiditz worked from pressed plants, the illustrations may have given readers the idea of collecting them.

All this is at present speculation. The fact remains that naturalists in the 1540s, in several parts of Europe, possibly all inspired by Luca Ghini, began collecting plants, drying them, and affixing them with glue or thread to sheets, which they then gathered in volumes or kept in boxes. By the mid-eighteenth century, their use was so well established that Linnaeus could proclaim that "a herbarium is necessary to every naturalist."[176] That sense that the herbarium was essential, almost self-evidently so, was a product of the shift from humanist to phytographic goals in Renaissance natural history—though as we shall see later, the spread of the herbarium also led to significant changes in the aims and methods of natural history.

It is important to note that their usefulness was contested by a leading sixteenth-century naturalist. On his travels through Spain, Carolus Clusius went to great lengths to preserve plant material in as many ways as possible:

Figure 4.6. Walwurtz männlin, woodcut by Hans Weiditz, from Otto Brunfels, *Herbarum vivae eicones* (1532).

> During that trip I noted the form, place of growth, and names of many of them, to aid my memory; I sketched some of them with charcoal or red chalk; and I brought almost all of them with me, having dried them; or I sent to my friends the seeds or even the plants themselves, if they could survive being shipped (such as bulbs and tubers).[177]

But after studying what he had gathered, Clusius became pessimistic about the usefulness of herbaria in descriptive botany. In 1584, he wrote to his friend Camerarius,

> I discovered in my Hispanic flora how difficult it is to produce good illustrations from dried plants, unless the illustrator is aided by someone truly skilled in botany, even though I had an illustrator who was practiced in botanical illustration. Moreover, it is a hard job to write descriptions on the basis of dried plants, unless you have seen them growing.[178]

In his *Exoticorum libri,* he apologized more than once for the poor quality of the descriptions provided on the basis of dried plants.[179] Clusius's criticisms were those of an experienced field naturalist, used to observing a plant in its natural habitat and only too aware of the differences between a living thing and its dried husk on a page.

Nonetheless, Clusius was willing to work with dried plants in the absence of anything better. After describing a seed-pod brought back from the

Isle of Mauritius, he bemoaned the loss of Willem Marquisius's collections from the East Indies:

> If Willem Marquisius had returned here safely, we could have learned about other plants that grow there. For during the entire time the Dutch were there, he observed the plants that grew in the areas they explored and saved them between sheets of paper.

Marquisius had promised Clusius to collect plants elsewhere, but he died on Banda, and his effects were lost. Dr. Coolmans, another voyager to the Spice Islands, expired on the return voyage, but his collection of plants survived and was given by the Old East India Company to Pieter Paaw, professor of medicine at Leiden, who lent them to Clusius.[180] Coolmans did not affix his plants to their sheets, with the unfortunate result that some of them went missing, leaving only fragmentary notes on the paper; in other cases, the plants survived but without any descriptions or notes. In instructions for collectors in the East Indies that Clusius sent to the Dutch East India Company in 1602, he specified that they should not only bring back plants or their parts, preserved between sheets of paper, but also make drawings of them and notes on their growth.[181]

Clusius's approach to herbaria was purely instrumental. He did not hesitate to soak dried plants in water to restore them, as much as possible, to their original size and color. His young Provençal correspondent Peiresc once sent him a dried fungus because he did not have an artist to hand to illustrate it, but added that it would swell to its original size and take on a bit of color if soaked.[182] The same technique was described by Mattioli, whose commentaries on Dioscorides were based largely on material sent to him by friends and colleagues. Mattioli did not bother to save any of the plants he received, as his no doubt exasperated correspondent Ulisse Aldrovandi learned.[183] Both of these scholars used herbarium material as a means of extending the range of their investigations: it allowed them to describe plants that grew far away and to which they did not have immediate access. Clusius preferred live material, and encouraged his correspondents to ship living or recently plucked plants to him in place of dried specimens.[184]

Herbarium material—individual dried plants—thus played an important role in the economy of exchange of early modern natural history.[185] It was one of the means by which local naturalists were integrated into a European network. But exchanging plants did not require that they be pressed flat (though this certainly helped them to survive shipment), and their reconstitution in a bath of warm water would have been positively hindered by being glued or sewn to paper. Hence, the existence of entire herbaria from the late sixteenth century, some containing several thousand plants, points to more than their role in an exchange system. Herbaria served further purposes, both intellectual and social.

The earliest surviving herbaria were begun by medical students. The earliest sheets in Felix Platter's herbarium date from his days as Guillaume Rondelet's student in Montpellier in the 1550s.[186] It is thus contemporary with the much more modest collection of Jehan Girault, a student of Jacques Dalechamps in Lyon, who wrote in it, "this book was begun by me, Jehan Girault, on the 6th of August, 1558."[187] The herbarium of Caspar Ratzenberger, assembled in its final form in 1592 as a gift to Landgraf Moritz von Hessen, was also begun in the mid 1550s, when Ratzenberger was a student at Wittenberg.[188] Leonhard Rauwolf also studied at Wittenberg in the 1550s, where he probably knew Ratzenberger, but the earliest plants in his surviving herbaria were gathered during a journey through France in 1560–62.[189] For these naturalists, the herbarium began as a tool for learning the elements of the *res herbaria.* Ulisse Aldrovandi was not a medical student when he began his herbarium, but he was new to the study of plants, and his *hortus siccus* undoubtedly served the same purpose.[190]

For beginners, the herbarium was a Comenian teaching tool *avant la lettre.* Botanical pedagogues from Rembert Dodoens to Caspar Bauhin insisted that students must see plants not just once but several times to learn their names and characters.[191] In uniting words and things, the herbarium was a practical response to the problems that had faced Niccolò Leoniceno at the end of the fifteenth century: each page staked a claim to the identification of real plants with ancient and modern terminology.

Seen strictly as a pedagogical tool, the herbarium had clear limitations. Clusius, as already mentioned, was skeptical of their value for describing plants not already seen *in situ.* Adriaan van de Spiegel admitted that herbaria were a second-best solution to pedagogical problems, naming them "hortos hyemales," winter gardens:

> Since it is not always possible to observe plants in their place of birth—in spring, summer, and fall due to bad weather, and in winter because most plants perish then—you should inspect winter gardens (as I call books in which plants glued to sheets of paper are preserved).[192]

After herborizing excursions, Caspar Bauhin took his students to see his herbarium and pointed out the plants they had collected, so that they could recognize the differences between a living and a dried specimen.[193] The majority of plants in Platter's, Rauwolf's, and Ratzenberger's herbaria were collected by the owners themselves, who were thus in a good position to recognize such differences. In these cases, the herbarium served as a technology for extending memory long after the owner had ceased to be a student.[194]

When a naturalist encountered a plant he did not recognize, he could collect it, press it, and then show it to colleagues or compare it with books. Jacob Zwinger sent material from Padua to Basel in 1591, to be identified by his teacher Caspar Bauhin.[195] Carolus Clusius examined Leonhard Rauwolf's herbarium when he visited Rauwolf in Augsburg in 1563, identified some plants, and corrected other identifications.[196] Several surviving sixteenth-century herbaria include unnamed plants, probably collected in order to be identified. One of Caspar Ratzenberger's herbaria included several plants without names; in one case, in which three leaves and a piece of twig were conserved, the material probably did not suffice for an unequivocal identification.[197] This herbarium was assembled in the early 1590s from material that Ratzenberger had collected over the four preceding decades, so the existence of still-unidentified plants suggests strongly that even more were originally collected as unknown and then identified later, with the help of books or colleagues' advice.

Herbaria served a further purpose: as repositories for synonyms. Their owners were not content to identify the plants they contained with one name only; instead, they noted as many of the Latin, Greek, and common names as they could find. For example, Ratzenberger jotted down eleven names for one plant:

> Scandix, Pecten veneris, Herba scanaria, Cerefolium aculeatum, Nadelkrautt, Hechelkam, Nadelkoerffel, Venusstrahl, Nadel Moehren, Schnabel Moehren, Schnabelkoerffel.[198]

Felix Platter followed a similar procedure in his herbarium, though he gave fewer common names and was sometimes content with only one Latin name.[199] Such entries could serve as a learning tool, a means to master nomenclature by frequent inspection of the plants to which they referred, or merely as a reference to the relevant literature, a means of finding descriptions in whichever reference one had to hand.

For the typical naturalist in the second half of the sixteenth century, then, the herbarium served multiple functions. In his student days, it allowed him to repeatedly examine the plants that his teachers had shown him, especially in winter or inclement weather; it could be compared against living specimens in botanical gardens. If he were writing a history of plants, the herbarium supplemented notes and garden plants in producing descriptions and illustrations. If he encountered an unknown plant, the herbarium allowed him to bring it to his study rather than taking heavy reference works and learned colleagues to see it *in situ*. And it allowed him actively to seek out synonyms, thus learning the different nomenclature used by the standard literature. Finally, in addition to serving memory in these ways, it provided an outlet for the collecting urge that was widespread in late Renaissance and Baroque learned society.

Reproducing Experience

Notes, Correspondence, and Exchange

Botanical gardens and herbaria structured experience by presenting material itself to be examined more or less at will (rather less than more with gardens in northern climes). On a less immediate plane, naturalists adopted artistic and humanist literary techniques to preserve their experiences and render them accessible to others: the manuscript illustration and the commonplace book. For the phytographer needed his careful notes to be sure of what he had seen. We have already seen how Valerius Cordus took short notes, and sometimes wrote entire descriptions, in gardens and in the field. In his trip through Spain, Clusius noted down the names, forms, and places of the plants he observed, and he sketched some in charcoal or chalk.[200] A decade later he repeatedly extended his stay in Vienna while writing his Pannonian flora, because his original notes turned out to be incomplete.[201] Clusius was particularly painstaking, but others were equally reliant on their notes. On the road from Basel to Padua, the young Jacob Zwinger carefully gathered new plants, tentatively identified them, and sent them back to his friends and professors in Basel.[202] Aldrovandi took a keen interest in Zwinger's catalogue, though he misplaced it among his piles of paper; fortunately the young traveler had another copy.[203] Aldrovandi himself was accompanied on his field excursions by an artist and a secretary who carefully noted the master's words and sketched the objects he indicated.

> I often, too, wandered over the marshes and mountains, accompanied by my draughtsman and amanuenses, he carrying his pencil, and they their notebooks. The former took a drawing if expedient, the latter noted down

to my dictation what occurred to me, and in this way we collected a vast variety of specimens.[204]

Whether naturalists drew themselves, brought artists along, or acquired manuscript illustrations from others, illustrations were practically ubiquitous in the community of naturalists.[205] Felix Platter managed to obtain Hans Weiditz's original drawings for Brunfels's *Herbarum vivae eicones,* which he cut apart and pasted into his herbarium along with the plants themselves.[206] Caspar Bauhin also included pictures along with dried and pressed plants in his herbarium.[207] By 1570, Aldrovandi possessed over two thousand drawings of plants.[208] Gessner's papers, sold to Joachim Camerarius the Younger on the Swiss naturalist's death, included thousands of drawings of plants, many of them from Gessner's own hand, while Clusius gathered a comparable number of illustrations.[209] Clearly, these scholars considered manuscript illustrations to be an important resource for their studies.

Despite the beauty of many of these illustrations, they were primarily working tools. Unlike the seventeenth-century *vélins* prepared for Gaston d'Orléans that form the nucleus of the Paris Muséum d'Histoire Naturelle's collections, these sixteenth-century drawings were often covered with notes by their possessors. Gessner's "Martagon, melius quam Matth: aut alii pinxerint," displays not only the lily itself—some flowers fully open, one in the process, and one fully closed—but also its seed capsules, cut open to display the seeds within (figure 4.7). He noted that the stem was more spotted on other specimens, and that in 1565 he had such a plant that was over two cubits tall.[210] The same page has further notes on this lily, and then drawings of two kinds of wild rose with notes on their *differentiae.*

Clusius's drawings also included annotations.[211] His drawing of the martagon was divided in two parts, as if the stem had been snipped in the center, to fit the plant on a page without reducing its size excessively.[212] His Codex Fungorum, a collection of drawings of mushrooms and other fungi that vanished mysteriously from his publisher's office only to turn up, some years later, in the possession of the Leiden botany professor Arnold de Syen, contains numerous illustrations of the rather aesthetically unappealing undersides of fungi. Most are annotated lightly, often with only the name, but they are clearly working papers as well as artistic representations.[213] Clusius had intended to use them as the basis for the illustrations in his *Historia fungorum,* published as an appendix to his *Rariorum plantarum historia* (1601); when they went missing, he was forced to find inferior replacements.[214]

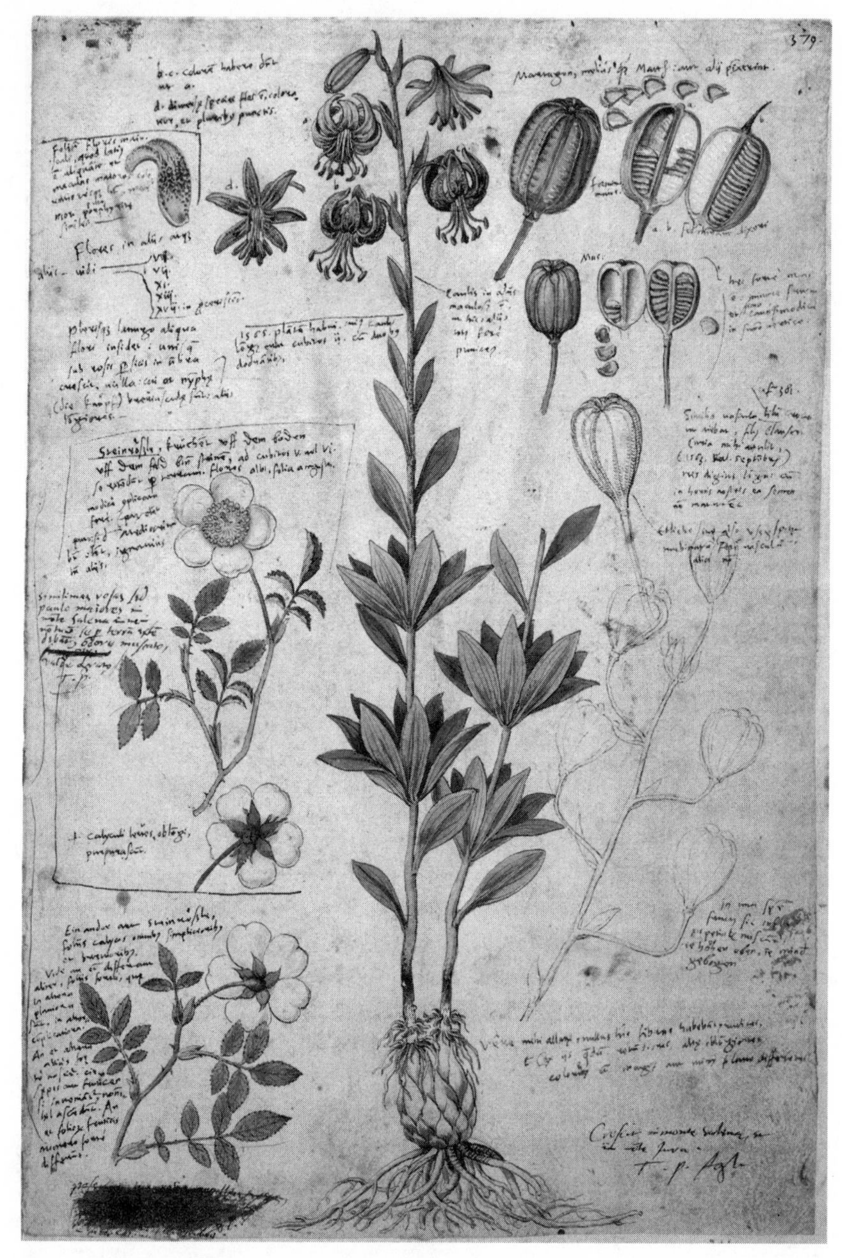

Figure 4.7. Conrad Gessner, drawing of martagon. Erlangen UB, Ms. 2386, II, 379r. Photo courtesy Universitätsbibliothek Erlangen-Nürnberg.

While few other manuscript illustrations that can be attributed to Clusius have survived, it is clear from his correspondence that he had illustrations prepared largely as the basis for woodcuts.[215]

Illustrations were not only memoranda for their author or commissioner; they were also means of communicating descriptions to one's fellow naturalists, or to those who wished to join the community. This function emerges clearly in the most spectacular surviving collection of sixteenth-century botanical watercolors, the *Libri Picturati* A.18–30, now in the Jagiellonian Library in Kraków, Poland, which contain over a thousand folia in thirteen volumes.[216] Produced in the second half of the sixteenth century, these watercolors combine aesthetic and pedagogical ends: stunningly beautiful, they are also annotated to teach the basics of the *res herbaria,* especially in its connections to *materia medica.* In their extent and state of preservation, the *Libri Picturati* are an invaluable source for the use of manuscript illustrations of plants in the sixteenth-century Low Countries.

Claudia Swan has made a strong case that the images are the six books of plants drawn from life that belonged to the Leiden apothecary Dirck Cluyt, who managed the botanical garden of Leiden University, under the nominal direction of Carolus Clusius, until his untimely death in 1598. Following Hans Wegener, generations of scholars had identified the woodcuts as the property of Clusius, but there is little evidence that he had more than a peripheral role in their production or *fortuna.*[217] Clusius *may* have written many of the annotations to the watercolors; to the untutored eye, they resemble his hand, but the paleographer S. A. C. Dudok van Heel has concluded there is nothing distinctive about them that would warrant an attribution to Clusius.[218] Complicating the matter is that at least some of the watercolors definitely date from before 1568, a period from which we have very few specimens of Clusius's handwriting. In any case, however, the annotations—regardless of their author—betray little of the interests that mark Clusius's mature botanical works, from the 1576 Iberian flora to the 1605 *Exoticorum libri.*[219]

At least two of the three hyacinths illustrated in *Libri Picturati* A.30, fol. 28, appear from the annotations to have been drawn before 1568.[220] They are identified in a "professional hand" as the first and second hyacinths of Dodoens. But another hand, probably Clusius's, has noted that "now" Dodoens, in his *Florum et coronarium historia* (1568), has given them different names.[221] Even if the second annotator had consulted Dodoens's 1568 work years after it was composed, it seems likely if not certain that the original names were given before it was available. In 1568, Dirck Cluyt was twenty-two years old—young, but not too young to have begun a manuscript collection of illustrations. His son Outgaert was even younger—only twenty—when the Leiden university students petitioned to have him succeed his father.[222] Further sets of annotations, one to the *Historia generalis plantarum* published in Lyon, 1587–88, and another to Caspar Bauhin's *Pinax Theatri botanici* (1623), suggest that the original annotations in the "professional hand" date from the 1560s, '70s, '80s, and possibly '90s. If Cluyt was indeed the owner of the illustrations, the references to the Lyon herbal might have been added by a student or his young son.

The early annotations, for the most part, imply an author from the Low Countries. They refer to geographical locations throughout Europe, but the Low Countries occur frequently, while locations elsewhere are often contrasted with where or how the plant in question grows "among us."[223] At least one annotation implies an origin in the southern or eastern Netherlands, however: the "Cochlearia folio subrotundo" (a name added by a later owner of the illustrations and taken from Caspar Bauhin) "grows wild in Holland and Friesland near estuaries and in meadows, but among us it is grown in gardens."[224] Cluyt was from Delft, in the province of Holland, which casts some doubt on the attribution to him. Until a detailed analysis of the *Libri Picturati*'s content is completed, the attribution remains plausible but not proven.

In any case, the *Libri Picturati* A.18–30 reveal one way a naturalist from the Low Countries approached the study of plants in the second half of the sixteenth century—the period in which phytographers like Clusius were compiling a detailed, precise account of the natural world around them. The *Libri Picturati* are, in many regards, untouched by this movement. The illustrations are accurate and follow the analytic style of other contemporary natural history illustrations, but the annotations are relatively simple and suggest an audience of novices with an interest in medicine. The initial references to the literature in the professional hand give pride of place to Dioscorides and Galen among the ancients and Dodoens among the moderns, not surprising given the popularity of Dodoens's 1554 *Cruydeboeck* and the 1557 translation by Clusius. The other annotations comprise, for the most part, remarks on where the plant grows and when it flowers and produces fruit.[225] Such remarks, of clear interest to medical students (who could then refer to Dioscorides, Galen, and Dodoens to find the medicinal uses of the plant), and the testimony of Leiden students that Dirck Cluyt used such a volume of pictures to teach botany in the winter, hint strongly that the collection was a teaching tool.[226] If so, its owner was quite careful with it, perhaps letting students look but not touch; after four centuries most of the illustrations are still in fine shape.

Though most of the annotations are simple, hardly the work of a phytographer like Clusius, a few exceptions hint that the collection was used, at least occasionally, by a serious naturalist. Annotations to seventeen images noted that the plants were found only in the gardens of serious botanists (*studiosi rei herbariae*), though other plants were common and some, like the *Septifolium* that propagates by runners, were serious garden pests.[227] The *Polium,* the annotator remarked, foreign to the Low Countries, could be grown there "only with great skill." Though most references to the place and time in which plants grew are in the present tense, referring to the normal course of events, a few imply specific observations of plants or attempts to cultivate them. The *Cistus vulgaris,* a native of the Mediterranean regions, "flowered for us once in May," the perfect tense underscoring the unique event. Other plants, such as the *Stramonia vulgo,* far from their native southern soil and sun, flowered "rarely" in gardens in the Low Countries. At least once the annotator used the gerundive, as if he were giving instructions to a gardener, and the note that dandelions "grow everywhere" might almost be a gardener's complaint.[228] But that does not rule out naturalists, who, as we have seen, were often avid gardeners.

The working notes, sometimes unfinished, and criticism of published botanical works provide the best clue about how the annotator approached natural history. Once the annotator expressed skepticism or reservation about where "Ruel's alfalfa" grows: "They say it grows naturally in estuaries and littorals." Presumably he had neither seen it himself nor heard of it from a fellow naturalist. Sometimes he left blank spaces to be filled in later: a "big species of clover we don't know" has blanks after "it grows" and "it flowers," suggesting that the watercolor was drawn after a dried plant or another picture, while the *Meum* has blanks instead of book and chapter references to Galen, Dioscorides, and Dodoens. Evidently he never got around to looking them up. Sometimes when he did, he took issue with what his sources said. Mattioli was his main target. The iris "doesn't go to seed around here, so I don't know whether Mattioli portrayed it correctly with seed." If in this instance he was skeptical, in others he outright contradicted the Sienese. Mattioli's woodcut of the *Chamaeiris* showed the stem as too long, contradicting his own description, while he was simply mistaken to assert that the *Bulbocastanon* was a compound medicine, not a plant. The annotator once expressed some reservation about Fuchs, but that was when Fuchs agreed with Mattioli.[229]

Whoever annotated the *Libri Picturati* had a wide-ranging knowledge and experience of natural history, even if most of his annotations imply a pedagogical focus. Later sixteenth- and seventeenth-century owners of the watercolors also used them to extend their own experience, noting down identifications (occasionally tentative) with the descriptions and woodcuts from the 1587–88 Lyon herbal and Caspar Bauhin's *Pinax Theatri botanici.* Along with Conrad Gessner's more copiously annotated illustrations, which also passed through several hands after Gessner's death, and Hans Weiditz's original watercolors for the *Herbarum vivae eicones,* acquired by Felix Platter and incorporated in his herbarium, they reveal that such collections were not merely preliminary studies for woodcuts. Instead, along with gardens and herbaria, they served to extend and refine the naturalists' experience of nature, making it available when the season, the weather, or the press of business prevented an arduous journey into the field.

In addition to annotated illustrations, Renaissance naturalists also took copious notes on what they had observed and read. Gessner, Spiegel, and their contemporaries preferred the humanist format of the commonplace book.[230] In his *Isagoge,* Adriaan van de Spiegel set out detailed rules for how to make a botanical commonplace book, or as he termed it, a florilegium. By dividing the book up properly into three parts—one on plants in general, one on plants individually, and one on questions regarding plants—the more advanced scholar could control a great amount of material and have a useful resource for writing his own books on the subject.[231]

Conrad Gessner used such a method of commonplaces to keep track of the vast amount of information contained in his numerous publications. His usual practice, at least late in life, was to keep files according to commonplaces, and to divide up his notes and correspondence into these files. By so doing, he established a comprehensive system, but at the cost of quick recall of any particular element. As he explained to the young Johann Bauhin,

> I regret that two of my letters to you have disappeared. I don't remember what I wrote in them, nor can I find your letters to which you wish me once more to respond. After responding to letters, I am accustomed to put them into my piles of paper—even to cut them up—and distribute them among my papers according to their subject. Hence it would be convenient if you could repeat, in a few words, what you wanted to ask me.[232]

Other writers suggested keeping two notebooks or files; in the first of these, the *adversaria,* one should jot down interesting thoughts or provocative literary passages in chronological order; in the second, the observations from the first should be copied according to commonplaces. By so doing, one was less likely to forget or be unable to find one's thoughts.[233]

Through herbaria, illustrations, and careful field notes, botanists of the latter half of the sixteenth century attempted to prolong their experience, to stretch it out from the fleeting moment. They were increasingly distrustful of memory. Valerius Cordus had been able to cite Dioscorides and many other ancient writers from memory, but he made careful field notes.[234] By 1600, naturalists were deeply suspicious of memory. Even one plant, claimed Spiegel, could not be accurately impressed in the memory: those who trust only in this faculty are doomed to error.[235] Botanists' memories were no worse in the 1590s than in the 1530s: rather, the number of plants they had to remember, and the number of characters used to differentiate them, had grown immensely.[236] While naturalists at the end of the century were still a hundred and fifty years away from the artificial keys that color the experience of modern field naturalists, they could no longer trust their immediate observations, and the memory of what they had seen before, to give them an accurate picture of the vegetable world they were observing. But these experiences were the foundation of further study, and they gave Renaissance natural history a lively style that vanished from phytography in the subsequent century.[237]

Illustrations and notes were still, however, an intermediate stage in the conservation and condensation of experience. They served as the basis for another type of more public fixing: the published description and illustration. While manuscripts preserved experience for their authors, published natural histories reproduced it for others. They represent the final stage in the condensation of experience. If the field, the garden, and the study were places where naturalists *did* natural history, the history of plants and animals, the herbal, and other publications were where they communicated their findings to the Republic of Letters. Having established what Renaissance naturalists did, we can now turn to what they said, treating published natural histories as the result of a process of inquiry structured by shared practices, rather than as approximations to some Platonic ideal of natural history.

Natural history descriptions were the final stage of a long condensation of observation, memory, and experience. They made sense only within a system of *differentiae* that were themselves the product of the exchange and generalization of local knowledge. The notion of complete description is chimerical. Naturalists strove, instead, to describe plants and animals as precisely as was *necessary* to distinguish congeners one from another, while still capturing the essential qualities of each species. They walked a tightrope between descriptions that were too vague, and allowed for the confusion of species, and those that were too precise, and took accidental differences to be essential.

Descriptions were complemented by woodcut illustrations. Though they lacked the precision of words or the best watercolors, woodcuts conveyed certain relationships of parts, as well as the overall appearance of a plant or animal, much more forcefully than the printed descriptions that accompanied them. The technical nature of woodcut, which required several intermediate steps between drawing an object and producing its image on the printed page, required naturalists to exercise careful control over the illustrations in their books.[238]

Both description and illustration in the phytographers' works focused on visual elements. Descriptions occasionally referred to a plant's odor or taste, but these elements were generally neglected and, in contrast to visual elements, the language used to describe them remained imprecise. The primacy of the visual rested on the increasing separation of natural history from *materia medica,* its emphasis on esthetic elements, and the necessity of being able to communicate experience on the printed page. At the same time, naturalists were not yet concerned with classification in the modern sense. For most of them, it was purely a practical issue. Like descriptions, classifications were based on *differentiae,* and they were, for the most part, unabashedly anthropocentric. Even Caspar Bauhin's *Pinax Theatri botanici* (1623), often seen as an adumbration of later systems, was more closely related to sixteenth-century concerns and methods than to the taxonomies of the later seventeenth and eighteenth centuries. For Renaissance naturalists, the individual species was important; its place in a system was a pedagogical, not a philosophical question. When they published, their main concern was how to describe what they had experienced.

Describing Plants, 1530–1630

Renaissance herbals contained descriptions modeled on their medieval predecessors and the *De materia medica* of Dioscorides. Each chapter in Otto Brunfels's *Herbarum vivae eicones* (1530–32) was mostly a summary of ancient descriptions, placed one after another, along with a woodcut drawing. His successors wrote their own descriptions, though elements were often taken from predecessors. Such descriptions generally consisted of the plant's form, place, time (of growth, flowering, and setting seed), complexion, and virtues. In some herbals, like Rembert Dodoens's *Histoire des plantes,* these elements were set off by headings; in others, they were indicated by marginal catchphrases.[239] Derivative editions of such texts were reprinted into the eighteenth century, preserving their form as well as most of the information contained in them.

In contrast, naturalists in the second half of the century modified this descriptive model for purposes that were less and less related to the medicinal aims of the herbals. A harbinger is the unfinished *Historia plantarum* by Valerius Cordus. As mentioned, Cordus's work remained in manuscript until 1561; moreover, the work was an incomplete draft, and it is unclear how Cordus would have revised it for publication himself. But as it stands, the work's descriptions focus entirely on a plant's morphology and growth. For example, Cordus's description of the black hellebore "that grows in our country" begins with the stems that grow out of the ground, then a description of the leaves that grow at their ends, followed in turn by the flowers and seeds. Cordus ended by describing the plant's root, as if he had uprooted it after watching its development for a year, then noting the places and type of soil in which it grew.[240] Gessner added a woodcut illustration of the plant, but it is unclear whether Cordus intended his book to be illustrated or not.

Cordus followed a similar pattern in his other descriptions.[241] His text was not divided up by rubrics, and occasionally he included the plant's native habitat before a morphological description, as with the ranunculae, where he stressed the different places in which similarly-formed species grew.[242] Despite slight variations, Cordus's descriptions follow a basic pattern that persists into the present.[243] However, subtle differences in this form reflect significant differences in the practices of observation. Cordus and his slightly younger contemporary Clusius usually described the root last; many of their successors, such as Caspar Bauhin, described it first. Bauhin may have had good reason to begin with the root: after all, the ancient founder of botany, Theophrastus, had commenced his own account with it.[244] But his order of description presumed that the plant itself was lying before him, uprooted, or fastened in a herbarium volume.

For naturalists before Bauhin—and, significantly, "popular" nature writers through the present[245]—a significant part of description was this vicarious element. The reader followed the description just as if he or she were observing the plant in a garden or in the wild. This can be seen in a number of natural histories published between the middle and the end of the sixteenth century. Conrad Gessner drew on a number of published accounts and correspondents' reports for his description of the "Lunaria arthritica" (or "Auricula ursi," a kind of primula), but also on his own observations. The description began with the leaves, noting their varying forms, and continued with the small, yellowish or pale flowers, "which, in our gardens, flower in April and early May, with a very pleasant odor."[246] After noting that the leaves do not fall off, Gessner described the root before turning to the alleged medicinal uses of the plant and the ease of cultivating it in a garden.

The vicarious description of *naturalia* reached its high point in the works of Carolus Clusius, which for this reason are some of the most delightful sixteenth-century natural history texts to read. Of particular importance are Clusius's two masterworks: his accounts of Spanish and Austro-Hungarian plants.[247] Each of these works was, as we have seen, the fruit of long travel and observation. Though Clusius spent less than a year traveling through Spain, his descriptions in the Spanish flora rely on observations made in gardens and the wild over a much longer period.[248] Similarly, Clusius repeatedly prolonged his stay in Vienna in order to gather material for his Austro-Hungarian work.[249] The descriptions contained in them reflect a leading naturalist's distillation of years of careful observation.

In his Spanish flora, Clusius referred frequently to his activity as "digging up" plants. He named one type of plant *Iris bulbosa,* or bulbous iris, because "though they had the flower of an Iris or Xyris, when I dug them up I discovered that they had a bulbaceous root." Another plant, the *Scilla Hispanica,* he dug up at several points: as its stem emerged, when it flowered, and when its leaves grew out. In these and other chapters, Clusius's description followed the pattern mentioned above: the description followed the plant's life history. In some cases, the history was incomplete, as with the *Pancratium,* which he had seen only when it was putting out leaves. In others, he complemented observations made during his eight month journey through Spain and Portugal with information and material from correspondents. When he saw the dragon tree near a monastery outside of Lisbon, it was not flowering; indeed, the monks claimed that it never did. But the next year a friend sent him a new shoot with ripe berries, which he reproduced in his book.[250]

The description of a plant's annual life cycle was thus an ideal for naturalists in the later sixteenth century. Clusius's eight months in Spain did not allow him to achieve this ideal in every case. His observations of plants in Austria, Pannonia, and neighboring provinces, on the other hand, were carried out over almost a decade before the book containing them went to press.[251] In this case he had ample opportunity to repeat his observations in the wild, and to grow plants from seed or bulb in his garden.

Clusius's description of the *Anemone silvestris II,* or *Anemone montana Pulsatillae folio,* provides a sterling example. His description began with the bristly top of the root, from which first leaves, then a stem bearing a flower emerged. At the end, as usual, came a description of the root. The plant grew, he noted, in high mountain ridges, among rocks with a bit of soil, and since it sent its roots into cracks in the stone, it was very difficult to uproot. "It flowers in July, occasionally in August, depending on when the

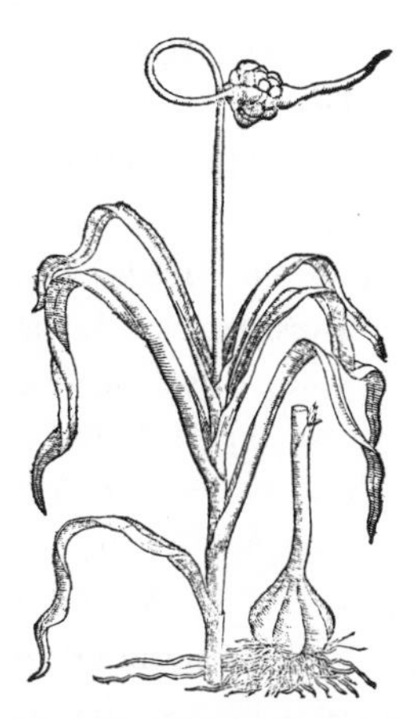

Figure 4.8. Scordoprasum, woodcut from Carolus Clusius, *Rariorum plantarum per Pannonias observatarum historia* (1583).

summer begins, that is, when the snows melt, for it is one of the first to flower." Clusius tried to cultivate it in his garden, but it was one of those alpine plants that could not bear the "milder air" of Vienna. The *Cotyledona altera taurica,* on the other hand, which was found in the Tauern (one of the Austrian alpine ranges), flowered there in July, but "in gardens, in May, like the other plants of that kind."[252]

Such vicarious descriptions seem almost like a verbal version of stop-action photography. However, they were the result of multiple observations, usually made with multiple plants in different places and times, corresponding perfectly to Dioscorides's tenet that only someone who had observed a plant at every point in its life could truly claim to know it.[253] They were a condensation of experience. Much as an illustration was intended to represent the characters of a species, not of an individual, the description had to suppress the peculiarities of individual plants. After all, as Aristotle had said, natural history deals with what happens "always or for the most part."[254] One of the more peculiar illustrations in Clusius's Austro-Hungarian flora is the second *Scordoprasum,* which twisted the top of its stem before eventually straightening out (figure 4.8). Clusius's language, here as elsewhere, focused on tendencies and expectations: "the extreme part of the stem, on which the caput is supported, is, at the beginning, accustomed [*solet*] to twist itself around with its head, so that it appears similar to a snake with a twisted neck."[255] In humanist Latin the verb *solere* was sometimes mere filler, intended to give a sentence a more sonorous ring. But Clusius was more careful in his usage. When he had seen a plant only once, like the *Leucoium bulbosum Byzantinum* or the *Sedum palustre,* he indicated precisely where and when he had observed it. On the other hand, he had observed the *Leucoium bulbosum maius serotinum* several times in the same place, and hence he could say that "It usually [*plerumque; solet*] germinates later than the others, and flowers only in May."[256]

The increasing attention to morphology in botanical description after 1550 cannot be separated from aesthetics. Clusius chose to include plants—or at least claimed to do so—following aesthetic criteria, such as elegance and curiosity. Practically every chapter in his Austro-Hungarian flora begins by praising the beauty, elegance, or rarity (often all three) of the plant in question. For instance, Clusius wrote that he could not neglect the *Ranunculus Thalietri folio,* "because it is an elegant plant, and not yet observed by anyone, as far as I know."[257] Aesthetic criteria were not exclusive, for Clusius included plants like *Polygonatum* that had no such redeeming characteristics: in fact, it stank.[258] The inclusion of stinky, ugly plants served a different purpose: the desire to enumerate the entire vegetable kingdom. This purpose shaded into natural theology, as the emblematic title-page to Clusius's collected works, the *Rariorum plantarum historia* (1601) made clear in no uncertain terms.[259]

Adriaan van de Spiegel roundly criticized the descriptive focus of natural history after Bock. As we have seen, Spiegel condemned naturalists for concentrating on plants' "form and growth" while neglecting their virtues.[260] Spiegel pointed to a significant gap between rhetoric and reality in late sixteenth-century botany. Though medical students studied plants as part of *materia medica,* the connection between these studies and medicine was not always clear. In his courses on *materia medica,* Caspar Bauhin tried to focus on those plants commonly used by apothecaries, and he asked his correspondents to let him know which plants were used for medicine in their localities.[261] A contemporary set notes from Basel for lectures on simples concentrates on those that were "commonly used in pharmacies."[262] In this regard, medical students' education was still practical. Yet public medical gardens in Italy and the north included many plants that had no medicinal purpose.

Bauhin himself was inconsistent on the relationship between botany and medicine. His chief works—the *Phytopinax* (1596), *Prodromos* (1620), *Catalogus* (1622), and *Pinax Theatri botanici* (1623)—cited prominently Galen's dictum, in "On Antidotes," that "the physician should have knowledge of all plants, if that is possible; if not, at least of many which we frequently use."[263] As a teacher, Bauhin emphasized common medicinal plants, but as an author, he expected more. The *Catalogus,* intended for medical students, listed every plant Bauhin knew that grew within a mile of Basel. Published when he was in his sixties, the work summed up a lifetime of botanical excursions with medical students. Bauhin's chief descriptive work, the *Prodromus,* included descriptions of some six hundred previously undescribed plants. Despite its citation of Galen, it contains no mention of the medicinal uses of plants. As a medical professor, Bauhin felt obliged to link his work to *materia medica,* but in as a naturalist, his main concern was the intellectual pleasure of finding and describing new plants.

Spiegel expressed the opinion of those who considered natural history a handmaid of medicine. But by the end of the sixteenth century, *materia medica* had taken a back seat to discovery and description. Herbals and commentaries on Dioscorides continued to be published and used as medical texts: for example, the University of Leiden prescribed Dioscorides as a medical textbook and ordered a copy to be placed in the portico of the botanical garden.[264] But other texts, like Bauhin's work, served the *studiosi rei herbariae,* those who were interested in nature for its own sake.

As botanical descriptions focused ever more narrowly on morphology and the place and time of growth, those aspects that interested the *studiosi,* they became increasingly detailed. The herbals provided relatively brief descriptions of morphology. In some cases they did not even bother with a description: Hieronymus Bock's chapter on common garlic (Knoblauch) began with instructions for cultivation. Though later editions of Bock's text included a picture, his only mention of the plant's form was in reference to its seed, "set on a round, smooth, rush-like stem, like the head on onion-pipes." Bock gave more detail in his description of the "Waldt Knoblauch," which he considered a separate plant. This plant

> emerges in the spring, about the same time that Aron first appears. It is not very common, and its usual place is in very dark and moist woods. It is a plant with one or two leaves, similar to those of the Meyenblumen. The leaves acquire stems like that of garlic; at the end of April they bear little white flowers, which are also like those of Meyenblumen. But they are not hollow, like the Meyenblumen, but star-shaped, and the leaves [of the flower, i.e. petals] are separated from one another. The roots of this flower are white and longish, like a small garden garlic plant that is not more than a month old.[265]

Bock presumed that his reader was familiar both with common garden garlic and the Meyenblumen. He described the Waldt Knoblauch by comparing and contrasting its features with these two plants. This method, used already by

Theophrastus, provides a powerful and economic means of describing a new object in terms of features already experienced by the observer, who draws on that previous experience to understand the novelty.

The same method was employed by Rembert Dodoens some fifteen years later in his herbal. He distinguished three kinds of garlic: cultivated garlic, wild garlic, and "bear garlic" (the same as Bock's Waldt Knoblauch). His descriptions of the first and third were also based on comparisons with known forms:

> Garden garlic has the leaves of grass or onion, between which round, hollow stems grow in the second year; on these, the flowers and seeds grow, similar to the flowers and seeds of onion. The root is round, bulbous like the bulb of an onion, composed of several nodes joined together, beneath which hang small, spindly fibers. . . .
>
> The third garlic, called bear garlic, usually has two large, broad leaves, very similar to the leaves of the "grand Muguet," between which come one or two stems, on which many small white flowers grow. The root is shaped like that of a new garlic, with a very strong smell and taste.[266]

Dodoens described the common garlic, unlike Bock, but his description of bear garlic is about as detailed as Bock's. This amount of detail, presented comparatively, characterized later sixteenth- and seventeenth-century herbals such as revisions of Dodoens's *Cruydeboeck* and John Gerard's plagiarized translation of Dodoens, the *Herball, or general Historie of plants.* It sufficed for identifying common plants that interested physicians and apothecaries.

By the end of the sixteenth century such coarse descriptions no longer sufficed for the *studiosi rei herbariae.* For example, Carolus Clusius distinguished "several kinds of Garlic or Moly, which are distinguished either in their leaves, flowers, or *capita.* Most of them, I think, have not yet been observed by others." Since these plants were found in the mountains of Austria and Pannonia, Clusius named them "Moly montanum," with brief morphological epithets to distinguish the kinds. He described the first, *Moly montanum latifolium,* as follows:

> The first has a stem that exceeds two cubits. It is surrounded by leaves to the midpoint; the leaves are much broader than those of other genera of Allium, and oblong, very close to those of the leek. At the top of the stem, which is light and reed-like, sits a head composed of many bulbs, of a darkish purple and pretty thick, among which grow pale purple

> flowers on long pedicules. These flowers are succeeded by triangular vessels that contain the seed. The root is thick and bulbous, consisting of many nuclei, with many white fibers; moreover, little round bulbs adhere to the root, similar to those that grow in the head. The whole plant has a strong smell of garlic. It flowers and bears bulbs in June and July.

Clusius continued to describe four other kinds of *Moly montanum,* one of which he divided into two "species." In describing the second, he remarked that its leaves were narrower than the first, while his description of the third noted that its flowers were "similar in shape and size to those of the second, but whitish in color, whose petals are marked by faint purple veins running lengthwise in the center and on both edges."[267] He followed this chapter with a description of two similar plants that he designated *Moly narcissinis foliis.*

Clusius's descriptions are much more detailed than those provided by Bock and Dodoens.[268] Some elements, for instance the observation that the leaves of the *Moly montanum latifolium* extend about halfway up the stem, were apparently intended to convey an overall impression of the plant's form, a verbal picture. But Clusius's descriptions were still largely contrastive—both within and across generic boundaries. He compared the leaves of the different kinds of *Moly montanum* with those of leeks, grasses, and reeds, as an economical yet accurate means of conveying their appearance. Within this chapter, he explicitly compared elements of different plants, emphasizing only those morphological elements that distinguished plants from one another. In his Spanish flora, Clusius had used the same technique: after describing the *Aristolochia longa* that grew in southern France, he identified another:

> A certain difference can be observed in the *Aristolochia longa:* for the flower of the Spanish kind (although similar to the preceding in almost every regard) is seen, by someone examining it closely, to differ a bit: the topmost part of the interior is purple, whereas it is green in the other one. Moreover, the end of the root is not slender, as in new plants of the other kind, but much more obtuse.[269]

Clusius did not bother to describe the other features of this plant: after all, they were almost identical. In this case, the two elements added to the previous description were truly *differentiae:* they were the features that distinguished these two species. Such contrastive descriptions within a genus, unlike those that compared one genus with another, did not necessarily require that the reader be familiar with the original plant, since its description was itself part of the text.

The growing precision of botanical description in the later sixteenth century was thus not due to a love of detail for its own sake. Instead, naturalists provided just enough detail for their descriptions to capture the distinctions between different kinds of plants. In Clusius's work, one can see this process within each chapter, as subsequent types of a plant are described with references to earlier descriptions. The same is true of Caspar Bauhin's lapidary morphological descriptions. In his *Prodromos Theatri Botanici,* Bauhin described three forms of *Allium.* His description of the third, *Allium montanum radice oblonga,* was extremely short:

> III. *Allium montanum* with oblong root: similar to the first, but with a taller stem and longer leaves, a little head, pale purple, with flowers consisting of six leaves and the same number of stamens.[270]

As the vegetable world became better known, it was also more populous: descriptions had to keep pace with the new distinctions naturalists made among its inhabitants.

Naturalists sometimes criticized descriptions that were too concise or vague. Clusius included the *Cotyledon altera Matthioli* in his Austro-Hungarian flora, "although its history has been presented by Mattioli and others, since in my judgment they describe it more concisely than they should."[271] Mattioli, in fact, had only repeated Dioscorides's description:

> There is also another kind of Cotyledon which some call Cymbalum, having broader, & fat leaves, like little tongues, thick about ye root, describing as it were an eye in ye middle, like as ye greater Sempervivum, binding by ye tast; a little thin stalk, & upon it flowers, & seeds like to Hypericum, but ye root greater. It is good for ye same uses that Sempervivum is.[272]

Mattioli added only that he had seen this plant in the garden of an Austrian physician and in many mountainous regions; all of Dioscorides's characteristics agreed with what he had seen.[273] Rembert Dodoens had also described this plant in the herbal that Clusius had translated: Dodoens said that it "has broad, thick, fairly round leaves, spread out around the stem, as in the large Sempervivum, from the middle of which comes the tender stem, bearing small flowers."[274] Dodoens did not even mention the eye-shaped depression in the middle of the leaves. Compare Clusius's description:

> This plant has numerous, densely-packed leaves, similar to oblong little tongues, rather thick, pointed, somewhat toothed on the edges, tending toward white, with an acid and somewhat astringent taste. From the center, or middle of the leaves, rises up a stem eighteen inches long, or sometimes only a foot, hairy on the upper part, with shorter leaves here and there. At the top of the stem, it puts out little pedicules between the leaves, on which rest two or three white flowers composed of five petals, which on the inside appear to be spotted with drops of blood. They have a pallid, shiny navel [*umbilico*] that is surrounded by ten little stamens [*staminula*]; then in little capsules very small black seed. Underneath is a thin, black, fibrous root; many other plants are attached by the root to the mother plant, forming a circle; sometimes they form a sod, as with common sedum.

This description is not only much more extensive than Dioscorides's or Dodoens's: it also serves as the basis for a contrastive description of a second variety of this plant, "with narrower and longer leaves that are not toothed but covered with a white wool, less white; its stem is shorter, thicker, and covered with many leaves, which are usually somewhat hairy on top and bottom. It has fewer flowers, scattered on the stems by the leaves."[275] By giving these two descriptions, Clusius judged that he had provided necessary information that his predecessors had omitted.

In sum, descriptive techniques in early modern natural history made sense only within a system of differences. It might have been possible to provide a verbal picture of a plant that sufficed to recreate it in the mind's eye, but naturalists did not usually attempt this. Illustrations, which are considered below, usually served the function of providing the overall *habitus* to unfamiliar readers. The goal of verbal description was, rather, to provide just enough detail to distinguish a species or variety from its congeners.

Hence the naturalist needed to be aware of a wide variety of more common plants in order to understand botanical descriptions. Adriaan van de Spiegel's botanical textbook emphasized this aspect of natural history training. "There is no part of philosophy more difficult," he admonished his readers, "than that which involves the nature of plants, on account of the incredible variety of forms, virtues, and other accidents."[276] Following Theophrastus, Spiegel devoted several chapters to the parts of plants and their *differentiae* considered in general. These told the student what qualities to look for while examining plants. But the bulk of Spiegel's first book was devoted to describing the main types of plants, "since a knowledge of these will help you greatly to recognize many species more easily, when you go to gardens or other places to learn about plants." Once a student had learned the basics, he was to go with a skilled companion to inspect plants. This *praeceptor* should point out the "middle species" of each group and its distinguishing characters; particular species should then be learned by their *differentiae,* since the entire form cannot not be committed to memory.[277]

Since Theophrastus, the natural history tradition had provided general lists of the *differentiae* of plants. Theophrastus himself, in the *Enquiry into Plants,* preceded his descriptions with a list of parts and how they differed one from another.[278] The more practically-oriented Dioscorides had not provided such a list, but Benedict Textor extracted one from his descriptions (1537).[279] Textor included *differentiae* of location, and a general discussion of similarities and differences between plants, but the bulk of the work was devoted to "differences of plants based on their parts."[280] This is the most extensive such list of *differentiae* of which I am aware; Spiegel's list in the *Isagoge,* already mentioned, is more typical of the genre.

These lists were apparently little used in practice. Conrad Gessner considered Textor's list to have been prepared "with a useful and beautiful method."[281] Students may have read them, or crammed them, to learn the proper terminology for describing plants (which was clearly Spiegel's aim in including his list; Textor's book might have been useful as an identification key). But descriptions rarely included every possible *differentia* from the lists. If one picked and chose the appropriate descriptive elements from such lists, one would have had a complete morphological description of a plant—were it not for the possibility of finding another the next week that matched every element in the description yet differed in some other way. The open-endedness of botanical inquiry, from the later sixteenth century, rendered impossible any attempt to provide a botanical identikit, a ready-made means for description.

Illustrations in Renaissance Natural History

Few Renaissance naturalists relied solely on words to describe plants. Most of their texts were illustrated with woodcuts, or, toward the end of the century, engravings or etchings. These images were praised by contemporaries as lifelike and accurate representations, and their beauty and grace undoubtedly contributed to the sales of the books that contained them. Yet they should not be considered as examples of simple naturalism; their realistic style served particular descriptive ends.[282] For Renaissance naturalists, lifelike illustrations were those that corresponded to the practices of identifying and describing plants.

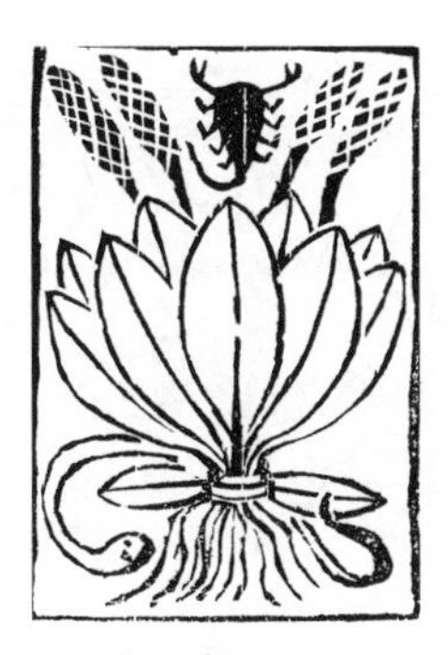

Figure 4.9. Plantago, woodcut from *Gart der Gesundheit* (1485).

Figure 4.10. Plant, woodcut from Arnaldo da Villanova, *Tractatus de virtutibus herbarum* (1499).

Early natural history illustrations, in the fifteenth and early sixteenth centuries, were crude woodcuts.[283] In some cases, they were copied after schematic manuscript illustrations, while other illustrations were evidently drawn from nature.[284] In the latter case, however, the artists or block-cutters employed by the publishers of books like the *Gart der Gesundheit* (1485) or Arnaldo da Villanova's *Tractatus de virtutibus herbarum* (1499), though capable of producing naturalistic representations, added conventional elements. If one compares these illustrations (figures 4.9, 4.10) with the much more naturalistic rendering of roses in the frontispiece to Breydenbach's *Sanctae peregrinationes* (1486) (figure 4.11), it is clear that the crudeness of botanical illustration was not due to technical limitations in the medium but rather to artistic intentions or limitations.[285] It is possible that the same artist illustrated the *Gart* and the *Sanctae peregrinationes*; if this is true, then the conventionalism of the former was undoubtedly due to the artist's following manuscript conventions.[286]

In any case, Hans Weiditz's illustrations in Otto Brunfels's *Herbarum vivae eicones* (1532) put to shame all earlier botanical illustrations.[287] Both the title of the work and Brunfels's complaints about having to include plants

Figure 4.11. Woodcut frontispiece to Breydenbach, *Sanctae peregrinationes* (1486).

with neither classical names nor medicinal virtues imply that the publisher, Johann Schott, and Weiditz bore greater responsibility for the book's basic design than Brunfels.[288] In their clear outlines, fine detail, and refusal to give in to the tendency to ornament and symmetry that characterized earlier herbal illustrations, Weiditz's woodcuts were masterpieces of realistic representation (figures 4.6, 4.12). They immediately set a new standard: just as Italians who had seen Giotto's work could no longer be satisfied by Cimabue, those who had looked into Weiditz's "living pictures" could no longer settle for the crude cuts of the *Hortus sanitatis.* Conrad Gessner was not satisfied with Brunfels's text, but he admitted that "he was the first in our time to give images of plants depicted elegantly and from life."[289]

But if modern botanical illustration sprang fully grown from Weiditz's head, like Athena from Zeus's, unlike the Greek goddess of wisdom it needed to be schooled. Certain elements of the Dürer school's realism sat uncomfortably with the demands placed on illustrations by Renaissance naturalists. The illustrations in the *Herbarum vivae eicones* represented individual plants, often with withered or insect-eaten leaves, stems, or flowers, and sometimes distorted as if drawn from a pressed specimen (figure 4.6). The same intense focus on the peculiarities of the individual characterize Dürer's own work, as seen in his 1508 *Iris.*[290] Moreover, in Weiditz's drawings and woodcuts, many parts of the plant are portrayed with foreshortening, while some parts are obscured by those in front of them. All this was highly realistic, but it was not the realism that naturalists wanted.

Brunfels's contemporary Leonhart Fuchs, whose *De historia stirpium* appeared a decade later, was convinced that illustrations were useful. But

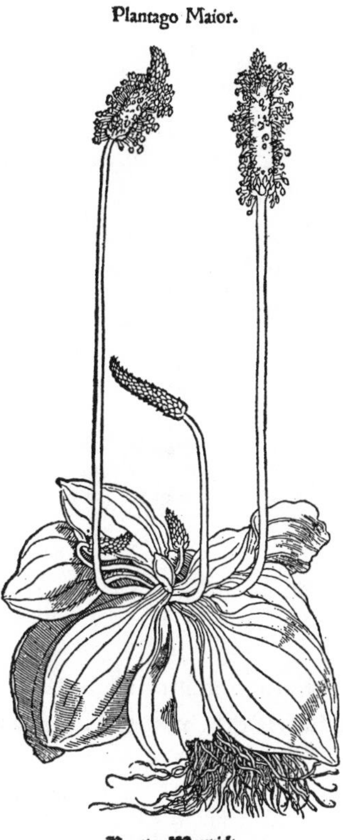

Figure 4.12. Plantago maior, woodcut by Hans Weiditz, from Otto Brunfels, *Herbarum vivae eicones* (1532).

unlike Brunfels, Fuchs felt that his illustrators needed to be carefully guided by the scholar.[291] In the preface to his work, he set out the principles that guided his illustration:

> As for the pictures themselves, every single one of them portrays the lines and appearance of the living plant. We were especially careful that they should be absolutely correct, and we have devoted the greatest diligence that every plant should be depicted with its own roots, stalks, leaves, flowers, seeds, and fruits. Over and over again, we have purposely and deliberately avoided the obliteration of the natural form of the plants lest they be obscured by shading and other artifices that painters sometimes employ to win artistic glory. And we have not allowed the craftsmen so to indulge their whims as to cause the drawing not to correspond accurately to the truth.[292]

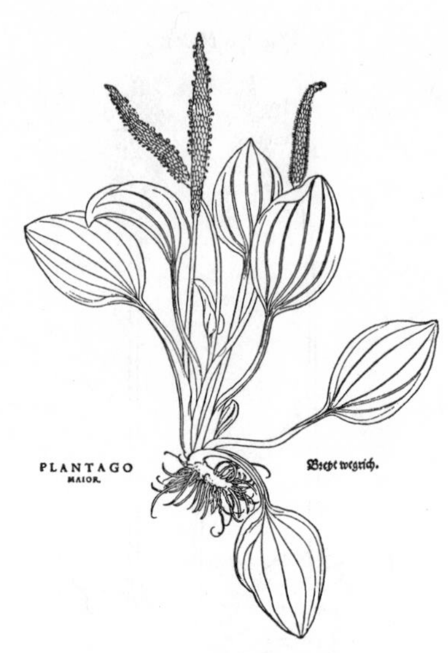

Figure 4.13. Plantago maior, woodcut from Leonhart Fuchs, *De historia stirpium* (1542).

Fuchs recognized that artistic verisimilitude, and the search for "glory," did not always correspond with the naturalist's truth. The perspectival eye of the artist, which rendered an object from a particular viewpoint with "shadows, and other less necessary things," was not that of the naturalist, who insisted that the object be represented, as far as possible, in a panoptical fashion. Fuchs's illustrations depicted plants spread out, with branches and leaves pushed to the side so as not to obscure other parts of the plant (figure 4.13).

But the naturalist's verisimilitude adopted by Fuchs went beyond issues of perspectival rendering. Fuchs illustrated plants that no naturalist could ever see—for example, plants bearing both flowers and fruits (figure 4.14). Such illustrations allowed for an economy of representation while still depicting the adult plant at various stages of its life. Whereas the artist strove to capture a particular plant at a given moment of its existence, the naturalist wanted to illustrate the plant in general, as it might be observed at different times. In extreme cases, this might require multiple illustrations: thus Clusius gave two figures of the *Scilla hispanica,* one of the bulb with stem and flowers, and another of the bulb with leaves, which emerged only in the late autumn or winter after the stem had withered away.[293]

The beautiful, large-scale woodcuts in Brunfels's and Fuchs's herbals established technical standards for subsequent work, but Fuchs's smaller

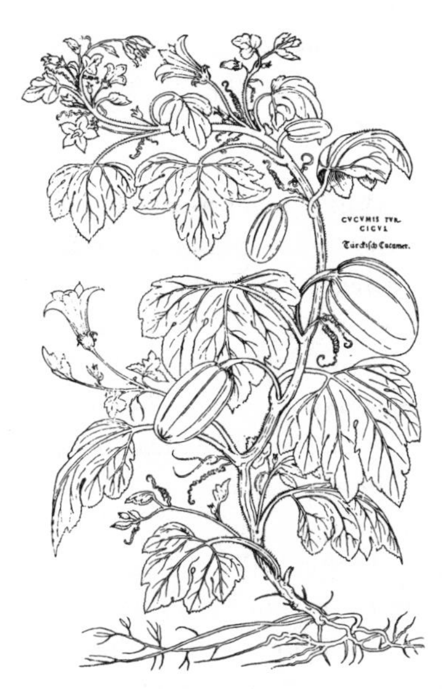

Figure 4.14. Cucumis turcicus, woodcut from Leonhart Fuchs, *De historia stirpium* (1542).

illustrations for the octavo edition of his book marked the path followed by most of his sixteenth-century successors. Large, folio-sized illustrations reappeared occasionally, but it was not until the seventeenth-century engraved florilegia that they reestablished themselves as the norm.

The usefulness of printed illustrations for naturalists was contested from the beginning. In the 1530s, before his 1542 herbal, Fuchs engaged in a polemic with the physician Sebastian Montuus over whether descriptions and illustrations could provide any real knowledge of plants.[294] Against Fuchs, Montuus argued that true knowledge of plants depended on knowing their names and their medicinal virtues, essential qualities that remained constant, whereas accidents of form—the only thing that pictures could represent—might vary and hence could not provide true knowledge. Sachiko Kusukawa argues that Fuchs's 1542 herbal, and in particular his careful supervision of his artists, was Fuchs's response to such criticism. The controversy, and Fuchs's later argument with Janus Cornarius in the 1540s and 1550s, underscores that northern European students of plants of this generation disagreed on the best way to study them and on the usefulness of illustrations.[295]

Montuus objected to illustrations on theoretical grounds; like Euricius Cordus, he held that plant forms could vary wildly while their essential virtues remained unchanged. Other naturalists of the second generation objected to illustrations on practical grounds. Hieronymus Bock did not illustrate the first edition of his *New Kreütter Buch* (1539) because he thought they were superfluous for the experienced gardener and observer:

> As far as illustrated herbals are concerned, it is clear that they are useful to a certain degree, when we do not have any living plants, or cannot get hold of recently-collected ones. But whoever has his own gardens and gardeners can plant many and various plants (as we read that, in former times, Antonius Castor did at Rome) and contemplate their living images. For such people there is no need of pictures, except for those plants that are truly foreign and that we cannot see recently collected in every place, and that completely refuse to be acclimated to our soil.

Since Bock was interested above all in German plants that he could grow in his garden in Hornbach, he did not need illustrations. Rather, Bock's publisher, Wendel Rihel of Strasbourg, paid for the young artist David Kandel to come to Hornbach and prepare illustrations under Bock's supervision.[296] The publisher bore the entire costs out of commercial interest: the market demanded illustrations.

In this regard, Bock was unusual for his time. Much more common was Fabio Colonna's belief that,

> In addition to learning a variety of subjects and languages, the investigator of nature, in order to become better, should acquire skill in painting and drawing, or at least a knowledge of it. Someone who is entirely ignorant of the art of painting cannot make true images of things whose descriptions and *differentiae* are clear in his mind.[297]

Many of the best Renaissance naturalists were skilled illustrators, and they were convinced of the value of illustration for the study of plants and animals. Not only did they keep notebooks with carefully made drawings of their subjects; they also devoted great effort to ensuring that accurately made images (*icones*) of those subjects were presented to their readers.

When Conrad Gessner edited Valerius Cordus's posthumous works, he took care to add illustrations, which Cordus had not provided.[298] Since Rihel, the book's publisher, had also published Bock and owned the blocks for the illustrations in his herbal, Gessner adopted many of those. He also added several from his own collection and copied others from earlier books.[299] Though he referred to the images as an "ornamentum," they had a more serious purpose: "Almost all will help the reader come more easily to a knowledge of unfamiliar things." On the other hand, Gessner omitted images in some cases, "for instance, where there is another picture of a plant of the same type that is almost identical to it."[300]

Gessner recognized both the advantages and limitations of woodcut illustrations for natural history. On the one hand, he inverted Bock's reason for placing little value on illustrations. Their aim was not to add to descriptions of plants already known to the reader but to aid in understanding the descriptions of those that were previously unknown. Gessner implied that pictures could successfully convey certain elements that were difficult to understand from a verbal description. On the other hand, they were useless for conveying subtle differences (especially when considerably reduced in scale, as was the case with most sixteenth-century natural history illustrations).[301] These *differentiae* could only be conveyed adequately in words—for example, the subtle distinctions between two varieties of *Aristolochia longa* described by Clusius.[302] Thus, for Gessner, the two forms of description were complementary, and each was required to complete the history of a plant.

Unlike verbal descriptions, illustrations also posed problems of control for the naturalist. As remarked above, the *Herbarum vivae eicones* was more the work of Weiditz than of Brunfels. The strong-willed Fuchs refused to let his artisans take charge of his herbal and carefully oversaw their work. The process allowed for error—or the imposition of the artisan's will—at several stages: in the original drawing, in the process of copying that drawing onto the wood block, and in the stage of cutting the block. Furthermore, the quality of the final image was determined by the paper and ink used and the skill of the press operators.[303] In order to control these elements, the naturalist had to oversee the production of his illustrations, or ensure that his publisher did so.[304]

In ideal circumstances, the naturalist was also the illustrator. Conrad Gessner's drawings served as the basis for many of the woodcuts that he prepared for his projected history of plants.[305] Gessner's illustrations included not only representations of the entire plant but often enlarged details of particular parts such as seed-pods and flowers (figure 4.7); these were faithfully reproduced in some of the woodcuts from his legacy (figure 4.15).[306] Other naturalists were not so fortunate and had to rely on trained illustrators. In such cases, as with Fuchs, they considered it essential that they guide the artist's hand. Carolus Clusius made some drawings of Spanish and Portuguese plants himself, but for most of the illustrations in his Hispanic flora, he relied on another hand: "Having found an industrious and diligent artist, I had the images of plants depicted on wood blocks, and often I was

Figure 4.15. Cyclaminus, woodcut from Mattioli, *De plantis epitome,* ed. Camerarius (1586).

beside the artist to indicate those aspects that had to be carefully observed when expressing the forms of dried plants."[307] Nonetheless, despite the high quality of the illustrations in this book, he was not entirely satisfied with the results, possibly due to careless mistakes by the artisan who cut the blocks.[308] While the book was in press, Clusius added an appendix in which he gave a "more accurate" illustration of the *Anemone tenuifolia flore multiplici,* since the one in the main body of the work was defective.[309]

If the illustrations in the books of so painstaking a scholar as Clusius could be imperfect, those in the works of less fussy writers were often much worse. Johann Bauhin loved Prosper Alpino's descriptions of plants but criticized the work of his illustrator.[310] A particularly easy target was the *Historia generalis plantarum,* published anonymously in Lyon (1586–88); Johann's brother Caspar published a scathing pamphlet pointing out that some four hundred of the illustrations in the Lyon herbal were used more than once, for different plants.[311] Clusius, too, enjoyed pointing out the inaccuracies in other naturalists' illustrations.[312] Nonetheless, illustrations continued to be de rigueur in late sixteenth-century natural histories, and those who dispensed with them, like Cesalpino, had to justify their decision.[313]

Natural history illustration was characterized by specific representational conventions. We have already seen how Fuchs insisted that certain techniques of "realistic" representation, foreshortening and shadowing, be eliminated from his pictures, and how he combined fruits and flowers on the same illustration. Further conventions were developed over the course of the sixteenth century. Gessner, as mentioned, sometimes included enlarged pictures of flowers or seeds in his woodcuts, a practice continued by

Figure 4.16. Pyra, woodcut from Mattioli, *Commentarii* (1560).

Figure 4.17. Primula veris albo flore, woodcut from Clusius, *Rariorum stirpium per Pannoniam historia* (1583).

Clusius (and in use to this day).[314] Another convention elaborated by Clusius and his contemporary Dodoens was to cut away some twigs and stems in order to reveal others more clearly. Whereas the illustrations in Pietro Andrea Mattioli's commentaries to Dioscorides appear almost decorative, and certainly require careful attention (figure 4.16), the contemporary cuts from Dodoens's and Clusius's works, both produced in Plantin's workshop, are uncluttered and clear (figure 4.17).

All of these conventions, which Treviranus considered essential elements in proper botanical illustration, serve a particular purpose.[315] They suppress certain elements of Renaissance artistic realism in favor of another kind of realism that we might anachronistically label "scientific." They were aimed at making the woodcut express, as clearly as possible, individual parts of the plant described: hence the rejection of foreshortening and modeling shadows, which might distort or obscure certain parts; the elimination of superfluous vegetative parts, which could obscure those behind them; and enlargements of certain characteristic elements. At the same time, however, these conventions distorted the overall appearance of the plant, what later naturalists would call its habit. The plantain from Fuchs's *De historia stirpium*, for instance, has its leaves pointing up (figure 4.13), rather than strewn flat on the ground, which is how the plant is found. Instances in which a plant seems literally to have been drawn from a herbarium specimen are uncommon, but the overall impression of flattening and spreading is the same.

Hence both text and image served to emphasize a focus on description of particular elements of a plant rather than its overall habit. Gessner's protestations to the contrary, they were not for the completely uneducated. Rather, they served the community of naturalists already familiar with the proper way to observe plants in the field, in gardens, and in herbaria, and already familiar with the basic forms that they would encounter. Newcomers to the discipline were to be accompanied by a teacher who could help them compare real plants with image and description.

Despite the descriptive, technical focus of sixteenth-century botanical illustrations, they were—and were considered to be—aesthetically pleasing. This is seen most clearly in the production of colored exemplars. Leonhart Fuchs referred in the text of his herbal to the colors in the illustrations, and some extant copies are colored to match Fuchs's text. Hans Weiditz's original watercolors for the *Herbarum vivae eicones* seem to have served as models for colored copies.[316] Conrad Gessner too thought that color was an important part of the illustrations to his *Historia animalium*. Though the expense of hand-coloring prevented the publisher from selling every copy with colored images, Gessner announced that his readers could order copies colored according to the original drawings.[317] In Plantin's shop, it took three months to properly color an herbal with some two thousand illustrations, and the work of coloring formed over ninety percent of the cost.[318] It is thus not surprising that most exemplars that have survived are not colored or only partially colored. Largely or completely colored books were great gifts, comparable to the richly-illustrated medieval miniatures in effect if not in cost. Plantin donated a colored copy of the Flemish edition of De l'Obel's herbal to the Duke of Prussia; the tome cost about a quarter of a university medical professor's annual salary.[319] Such deluxe editions were probably rarely owned by working naturalists.

The existence from the 1530s of colored exemplars of natural history texts thus points to a broader interest in both the *res herbaria* and in such elegantly illustrated and decorated texts as *objets d'art*. The gradual shift from woodcut to copperplate engraving or etching as a medium for natural history illustration, seen in the last decades of the sixteenth century, must be considered in light of this audience and of general trends in the graphic arts.[320] In the early seventeenth century, Clusius and his publisher had trouble, even in the Low Countries, finding a skilled wood block cutter.[321] At the same time, engravings and etchings cost more but offered more finely detailed illustrations. Fabio Colonna's *Phytobasanos* (1592) was one of the earliest botanical works illustrated with etchings.[322] Colonna drew and etched his illustrations himself, eliminating not only the artist but also the block cutter and thereby removing two possible sources of distortion. His illustrations resemble earlier and contemporary woodcuts but were more detailed. Alpino used engravings rather than woodcuts in his posthumous *De plantis exoticis* (1627), and achieved a similar fine level of detail in his illustrations.[323]

The engraved and etched illustrations in these late sixteenth-century works point toward the development, in the seventeenth century, of the *florilegium*. In these texts, engraved images took precedence over the increasingly atrophied text, until they were soon little but picture books, aimed at a market of horticulturists rather than naturalists (figure 4.18).[324] At the same time, illustrations became increasingly less useful to naturalists. Their emphasis on herbarium work—comparing descriptions with the objects themselves—made the mediated representation of the illustration less important than the unmediated observation of the plant itself, even dried. Whether the decline in quality of woodcut from the last decade of the sixteenth century, deplored by Treviranus, is cause or consequence is uncertain.[325] What remains clear is that the quality of illustrations could not measure up to herbarium material or the manuscript illustrations that naturalists continued to use. Bauhin's last works, the *Catalogus* and the *Pinax Theatri Botanici*, were published without illustrations. They point the way to the general natural histories of the later seventeenth century.

The Primacy of the Visual

The sixteenth century saw a pronounced tendency in botanical description toward an almost exclusive concentration on morphology—that is, on visual elements, a tendency reinforced by illustrations.[326] This tendency is

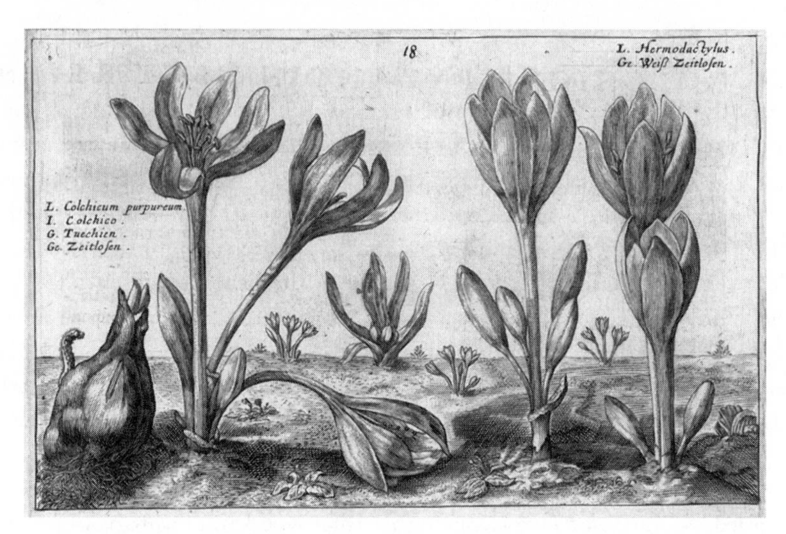

Figure 4.18. Crocus, engraving from Crispijn de Passe, *Hortus floridus* (1614).

inseparable, in its historical context, from the separation of natural history and *materia medica,* and from the necessity, in an international community, to have precise yet communicable descriptions.

In the de facto separation of natural history from *materia medica* in the later sixteenth century, Hieronymus Bock's work represents the path not taken by his successors. Conrad Gessner, in 1552, had singled out Bock for his attention to nonvisual elements: "He diligently expressed tactile qualities, and smells and tastes." But even Gessner's praise of Bock concentrated on his careful visual inspection of plants and his reproduction of that visual experience: "He gave almost all these descriptions not from commentaries or histories, but as an eyewitness and observer of each." Gessner added that Bock listed a number of medical faculties learned not only from books but from his own experience and from interrogating rustics and illiterates. He did not shy away, as Brunfels had wished, from indicating plants whose classical names and descriptions were unknown to him, since "learned physicians easily learn the virtue and nature of any remedy that is unknown to them and never before seen, from its odor, flavor, and other modes of judging and experiencing."[327] Gessner implied, though he did not state, that Bock's emphasis on nonvisual qualities was related to such techniques.

For Gessner, writing only thirteen years after the first edition of Bock's *New Kreütter Buch* such descriptions were still very much present. But his own work, he admitted, was concerned more with names and forms, not medical faculties.[328] As we have seen, Adriaan van de Spiegel, looking back from the beginning of the seventeenth century, saw Bock as the last great writer on *materia medica.* Spiegel's textbook emphasized the importance of smell and taste for determining the medicinal qualities of plants. Echoing Euricius Cordus, Spiegel held that "there is no surer index of manifest qualities than taste."[329] This proclamation introduced fifteen pages of practical advice on how to determine those qualities by smelling and tasting plants.

Spiegel's lead was not followed by his contemporaries and their successors—at least, not in their natural histories. Caspar Bauhin's descriptions focused even more exclusively on morphology than Clusius's descriptions in the previous generation. Perhaps it is no surprise that the *Isagoge* was Spiegel's only contribution to botany; afterwards, he turned to anatomy and other areas of medicine.[330] He must have felt increasingly isolated from the concerns of scholarly naturalists.

If the focus on finding and describing new plants led to a concentration on their morphological *differentiae,* this focus was reinforced by the necessity of being able to produce detailed, communicable descriptions. Lists of *differentiae,* from Textor's in 1537 to the most recent, demonstrate the existence of a wide range of precise terms for describing the shape and arrangement of parts of plants.[331] Color terms were much more vague when compared with the range of terms available to later naturalists. This may have indicated a lack of interest in describing color, as William Stearn holds.[332] But in the absence of color reproduction techniques, with which samples can be reproduced in mass, it is difficult to attach precise meanings to color terms.[333] Odors are even more difficult to describe without fixed references.

When early modern naturalists described color, odor, and flavor, they were thus limited to a fairly small, imprecise range of terms. According to Clusius, the *Linum silvestre II* had "very bitter tasting" leaves, and its flowers were "a very elegant yellow." The leaves of the *Soldanella alpina* were "green on top, and a more dilute green on bottom; they taste at first drying, then sharp and unpleasant." In some cases, Clusius could indicate differences in color only by comparison: so, for instance, the *Carduus mollis humilis* had leaves covered with "a white wool (but not quite like the plants previously described)."[334] His terms for describing odors were extremely limited, usually restricted to "pleasant" or "unpleasant," with "bitter," "sharp," "drying" as indicators of flavor.[335]

Clusius stretched his color terminology to the limits in describing the *Tulipa praecox.* "The variety of color in the *praecox,*" he observed, "is enormous; nor do I remember seeing a greater variety in any other flower, except possibly the poppy. It can be entirely yellow, or red, or white, or purple, and sometimes two or more of these colors are found mixed in one flower." Red tulips, in turn, could be

> more saturated, or more dilute and pleasant to see, sometimes inelegant and worn-out. The tips of the petals are either wholly yellow, or yellow with a light sooty color sprinkled in it: in some of them there is a black spot in the middle of the yellow tip; in others this is a big spot, so that there is just a yellow line surrounding a black tip. Sometimes these tips are so large that they occupy half of the petal.[336]

The use of adjectives of saturation, and words like soot, allowed Clusius to indicate relative degrees of color intensity, but not to specify colors absolutely. Someone who had not seen a tulip would be hard-pressed, from this description, to imagine its colors precisely. On the other hand, Clusius's description of the arrangement of yellow and black on the petals was relatively precise. When combined with his woodcut illustrations, this description provided enough information to understand the plant's form.

The primacy of the visual, then, was above all a primacy of shape, form, and arrangement.[337] Color was less precisely described, and though colored exemplars of a book could partially offset this lack, they were prohibitively expensive for most naturalists. Color, and to an even greater extent, odor and flavor, were elements of experience that tended to fall out of descriptions. They had to be experienced firsthand, in the wild, the garden, or—if these were not available—the herbarium or collection of paintings. As such, they underscore the limits to how far experience could be condensed and reproduced in the words and images of Renaissance natural histories.

Conclusion: From Local to Universal Knowledge

The reader who has made it to the end of this chapter has traveled a long distance, from the empirical orientation inspired by Renaissance philology and medical humanism to the complex technology of observation developed by sixteenth-century naturalists. In field expeditions, gardens, and herbaria, naturalists repeatedly observed nature, building up through their memory and notes a complex tissue of experience that they then reduced to its starkest lineaments in the words and pictures of their books. In condensing experience, as we have seen, naturalists from Valerius Cordus and Hieronymus Bock to Carolus Clusius and Caspar Bauhin also schematized it, reduced it to its essentials, stripped it of much incidental detail—or, as an aesthetically oriented critic might add, impoverished it. Such is true of all representations, of course, at least in fields of inquiry whose dominant form of representation is metonymic, fields whose practitioners seek to distinguish between the essential and the superfluous aspects of their objects of study and to represent only the former. But Renaissance naturalists carried this process to an extreme, providing an increasingly detailed account of the morphology of the natural world—especially of plants—while systematically eliminating or reducing the presence of color, odor, taste, and other qualities that had earlier been a vital part of description, when the scholarly study of plants was almost wholly a branch of *materia medica.*

The immediate experience of nature remained far richer, of course, than the schematic representations in the pages of Renaissance natural histories would suggest. It is important to keep in mind that published texts were not the end product of the process of natural history research; rather, they were themselves employed as tools by naturalists seeking to make sense of their particular experience. But the increasingly visual and formal technology of observation disciplined naturalists to pay more attention to morphology and less to other aspects of nature. Hieronymus Bock had noted frequent, regular associations between certain plants. Conrad Gessner had observed during an alpine ascent that the regular succession of seasons (or as we would say, climactic zones) changed as one climbed higher. Neither of these suggestions would be developed systematically by their successors. Their technology of observation shaped habits of observation, and was in turn shaped by those habits, in a continuing dialectic that focused attention above all on defining and describing new species or varieties of plant.

As a further consequence of the use of herbaria, gardens, and published descriptions and the constant exchanges between naturalists, local knowledge was becoming universalized. The flora and fauna of a particular region, explored in depth by local naturalists or travelers like the younger Cordus and Clusius, was rendered accessible, in specimen or description, to naturalists everywhere. This, in turn, only fostered the primacy of morphological descriptions, for they were the most easily communicable and the easiest means to distinguish—or to create—differences. By the end of the sixteenth century, an increasing number of specimens and descriptions from North Africa, the Levant, the Americas, even the Far East, were also entering the circulation system of Renaissance natural history, where eager naturalists attempted to compare and contrast them with local products.

This universalization of local knowledge carried with it serious practical and cognitive consequences. Because descriptions were often contrastive, providing only enough detail to distinguish a species from others known to the describer, it was sometimes hard to determine whether two descriptions referred to the same or to different species. Furthermore, naturalists gave different names to the same—or, at least, arguably the same—plant or animal. These were both practical concerns for naturalists from at least the 1550s. The challenge they posed resembled, in certain regards, the earlier enterprise of determining which ancient descriptions corresponded to modern plants. But there were two significant differences. On the one hand, naturalists could communicate more detailed observations or specimens, which made it easier in principle to resolve disputes. On the other, the sheer number of new plants described made it harder in practice to keep up. By the 1560s, writers from Clusius and Gessner to the anonymous annotator of the *Libri Picturati* A.18–30 had to list not only the vernacular synonyms of the plants they discussed but also, in many cases, the synonyms given by different modern naturalists.

This proliferation of names and species had serious cognitive consequences, as Scott Atran's brilliant study of modern natural history demonstrates.[338] The trained human mind, it seems, can keep track of some five hundred or so distinct natural kinds, without needing a system to classify them. That limit was equaled by the herbals of Fuchs and Bock, and by the 1570s it had been surpassed. The Lyon herbal listed over two thousand different plants. By the 1620s, Caspar Bauhin would be able to enumerate over six thousand. The technology of observation, the method of the science of describing, had created this problem. The next and last chapter of this study turns to the naturalists who recognized this problem and attempted to resolve it.

CHAPTER FIVE

Common Sense, Classification, and the Catalogue of Nature

By the end of the sixteenth century, naturalists had developed a sophisticated technology of observation and description, a technology that allowed them to transmit local knowledge through precise description and communication. The very success of these techniques transformed the practice of natural history. In the first generation, Renaissance naturalists compared classical texts with specimens from the field. Those of the second and third catalogued and described nature, in increasing detail, on the basis of personal observation. Naturalists of the fourth generation, on the other hand, were able to turn collections and herbaria into independent sites of research, while beginning to leave field observation to underlings or students.

At the same time, the dizzying pace of discovery, description, and publication of new species created problems of organization. By the late sixteenth century, many naturalists felt dismay at the confusing babble of names for the same plant; with no authoritative nomenclature, every naturalist felt entitled to name a new discovery himself. A few scholars were beginning to think more systematically about taxonomic systems: Andrea Cesalpino adopted an Aristotelian approach to classifying plants in his *De plantis libri* (1583), while Adam Zaluziansky used a more pragmatic organizing principle in his introductory *Methodus herbariae* (1592). The great age of systems lay in the future; most Renaissance naturalists organized plants along common-sense lines. But their research methods created problems that were beginning to overwhelm common sense.

These methods themselves could be applied fully only in Europe or in those few other places where European naturalists traveled. In the last third of the century naturalists began to travel abroad: Leonhard Rauwolf's notes on Near Eastern plants in his *Real Description of a Journey to the East* (1582) and Prosper Alpino's *De plantis Aegypti* (*Plants of Egypt,* 1592) brought first-hand observations to the studies of Europe. But most of the plants and animals of the world beyond Europe were known to European naturalists only through travelers' tales or odd, fragmentary specimens. Even the publications of Spanish and Portuguese physicians on the natural history of the "East and West Indies" were unsatisfactory, for their authors were not members of the Renaissance community of naturalists. The natural history of New Spain written by Francisco Hernández remained in manuscript in Seville, locked away by cautious Castilian bureaucrats worried about divulging commercial secrets.[1] Faced with this problem, Renaissance naturalists did as best they could when describing exotics. They were willing to publish incomplete, even potentially misleading descriptions when they could do no better; they aimed, after all, to describe the world as fully as possible.

Only in the seventeenth century, with the expansion of the Dutch and English commercial empires to rival those of the Iberian powers, did European naturalists or their agents begin to have extensive, firsthand experience of the world beyond Europe. In response, and in concert with their new interest in taxonomic systems, they began to purge dubious or anomalous exotics from the catalogue of nature. Renaissance naturalists had been skeptical of monsters and mythical creatures, but they were willing to follow trustworthy authorities if their reports were plausible. With few facts about exotics, they were not so fussy as they were with European species; their desire to catalogue the world prevailed over their skepticism. Their seventeenth-century successors, overwhelmed by exotica as their predecessors had been by the European flora and fauna, took a harder line. They would eliminate whatever had not been seen by a trained eye and described by a skilled pen. In so doing, they carried the science of describing to its logical conclusion.

The Unexpected Consequences of the Science of Describing

Herbaria: From Memoranda to Documentation Centers

In the middle of the sixteenth century, collections had become a significant part of natural history. Cabinets of *naturalia,* botanical gardens, and herbaria were sites where natural objects could be gathered and experienced, albeit separated from their context. Any naturalist could make a herbarium, though some, like Clusius, used them only as a last resort. Collectors of herbaria used them to fix and deepen their own experience. The herbarium extended their memory, allowing them to experience a plant without returning to where they had found it.

By the end of the century, however, some scholars were using their herbaria for a further purpose. For Caspar Bauhin, the herbarium served not only to preserve experience but, more importantly, to expand it. As set out in his *Phytopinax* (1596), Bauhin's method would hardly have struck his older contemporaries as unusual. Not only did he read books and inspect plants where they grow; "furthermore, I have chosen the plants as carefully as possible, dried them, stored them (at the moment, I have several thousand of them), and compared them with the authorities."[2] At this point in the 1590s, Bauhin's herbarium contained his own specimens, which he preserved in order to more effectively compare them with his botanical library. This herbarium material allowed Bauhin to achieve his goal in the *Phytopinax* and its successor, the *Pinax Theatri botanici* (1623): to bring order to botanical nomenclature by establishing a list of synonyms for each plant described in the existing literature.[3]

But Bauhin soon adopted a different approach to his herbarium. Between 1596 and 1623, he drew extensively on material that was collected and sent to him by others. Bauhin continued to collect and dry plants for himself.

> But not content with this, to obtain a more true and certain knowledge of them, I asked medical students going to France and Italy to send back plants of every kind, even the most common ones like grasses and clovers. They did this most liberally, and I have included their names in a list after the preface (since they are now doctors). Moreover, I have written (and continue to write) to famous men, both doctors and apothecaries, in Germany, France, Italy, the Low Countries, Denmark, and Crete, asking them either to send plants or to give their opinion about them for the public good.[4]

In relying on this correspondence network to send him material, Bauhin turned the herbarium from personal memoranda, a living commonplace book, into what Justin Stagl has called a "documentation center."[5] By 1623, Bauhin's herbarium contained "more than four thousand plants, in addition to fruits, seeds, woods, and spices."[6] Each plant was labeled with its synonyms. Bauhin compiled the *Phytopinax* and *Pinax* by going through his herbarium, noting the synonyms, and arranging the lists systematically. He may also have drawn on the herbarium of his Basel colleague Felix Platter, which was reorganized and bound following Bauhin's scheme.[7]

Bauhin's method and social circles epitomizes the change from the third to the fourth generation of Renaissance naturalists. In relying on a network of correspondents to provide him with herbarium material, Bauhin lessened the importance of travel to a knowledge of natural history. When his herbarium became a primary research tool, the plant in its natural habitat became almost secondary. Certainly for his own interests, nomenclature and organization, knowing the forms of as many plants as possible was much more useful than having observed them in their native habitat.

Bauhin and Platter were aware of the problems posed by working from dried plants. The plants in Bauhin's herbaria were not attached to their sheets, possibly allowing him to soak them to restore shape and some color.[8] Felix Platter took another approach: in some cases, he manipulated the dried plants in order to preserve more of their natural appearance. For example, the blue flowers of *Campanula* turn brown after the plant is uprooted. To show their natural color, Platter replaced some of them with artificially-cut flowers made from *Delphinium* flowers, which have preserved their color for centuries. Such extreme manipulations were rare, but Platter often included parts from several different plants on the same page, meticulously noting the additions.[9]

When Platter trimmed *Delphinium* flowers and pasted them next to his *Campanula* pressings, he admitted that herbarium specimens did not always faithfully represent living plants. In this case, the experience of a living plant was better conveyed by an artificial contrivance than by the plant itself. Caspar Ratzenberger may have called his herbarium a "herbarius vivus" (living herbal), but the plants in it were dead.[10] Combined with the botanical garden and field expeditions, which gave the scholar experience of living plants, the herbarium increased the scope of experience. As Bauhin realized, plants could be compared and contrasted with one another much more easily with a herbarium than any other tool. But as the size of herbaria increased and the herbarium specimen became the basis for natural history descriptions, the lively language of sixteenth-century phytographers, accustomed to observing the living plant, itself faded into the dry prose of the herbarium botanist who described the specimens sent to him by his students and correspondents.[11] These scholars experienced plants, at least in their scientific work, as a set of static features, not as living creatures to be enjoyed as well as examined.

The shift is already clear in Caspar Bauhin's work, despite his efforts to compare herbaria material with plants in the wild and the garden. In the *Prodromos Theatri Botanici* (1620), Bauhin provided descriptions of some six hundred previously undescribed plants. His descriptions are marvels of precise concision. But they also implied a new relationship between the naturalists' observations and their objects. While Clusius and earlier naturalists had been careful to describe the place and time of plants' growth, Bauhin often provided a purely morphological description, as with the *Narcissus angustifolius albus minor:*

> From a bulbous root, covered with a blackish layer and with a few fibers, emerge two or three oblong, narrow, leek-like leaves. Amid these is a little stem, a palm long, on top of which sit two flowers, which are small, white, and provided with an oblong, narrow calyx. It is found not far from Montpellier in rocky, mountainous areas.[12]

Aside from the brief indication of the plant's location, possibly provided by a correspondent, Bauhin described only the plant's form. Compare this telegraphic description with Clusius's chapter on the *Narcissus serotinus montanus:*

> In March it puts forth four or five oblong, narrow leaves, similar to those of the Narcissus. Amid these emerges in mid-April a foot-long stem, which is round, smooth, and spongy on the inside. At the top of the stem, from a thin membrane, is produced first one, then another flower, consisting of six white leaves spread in the form of a star, with a flat, pallid calyx (whose edges are adorned with red fringes) in the middle of the flowers, and six little yellowish stamens of unequal length. It is scented and similar to the flowers of the *Narcissus medio purpureus.* The flowers are succeeded by triangular capsules, rather thick, which contain uneven black seeds. The root is bulbaceous, consisting of many white layers; the outermost, which encloses the others, is dark, and there are many dark fibers emerging from its lowest part.
>
> It grows a "long mile" above the Carthusian monastery of the Virgin on the Throne at Gaminga, in meadows next to a mountain lake, on the estate called "Seehof." The Carthusians' leaders ensured that we were always received there most humanely and splendidly when we were on our way to the high ridges of Herrenalben and Durrenstein, or when we returned from the mountains to Gaminga. It flowers in May.
>
> I recall that a similar plant grew in Belgian gardens when I lived there.[13]

I have quoted each chapter in its entirety, to convey the difference in style and content between the two writers. Clusius provided a history of the plant's growth, beginning with the leaves shooting up from the ground, followed by the stem and flowers, finally the seed pods. He described the bulb last, as if he had uprooted the withered plant in the autumn to examine it. And—part of the charm of Clusius's book—he mentioned not only the place where he found it but the hospitality of the Carthusian monks toward himself, a confirmed Calvinist.

Bauhin's descriptions lack this vicarious element. Bauhin began with roots, then proceeded to describe stem, leaves, flowers, and sometimes fruit, but on the basis of the dried plant lying before him on his worktable. Bauhin's choice of verbs reflects his use of the herbarium: where Clusius often used versions of *crescere,* to grow, Bauhin restricted himself to *prodire, abire,* and other forms of *ire,* to go; sometimes even the preposition *unde,* whence, without a verb. The vicarious tradition left its traces in these verbs—and the contrast should not be exaggerated—but by the early seventeenth century, the vicarious tradition of description was on the way out, if only because Bauhin often described plants that he had not himself collected. Of thirty-five grasses described in the first book of his *Prodromos Theatri botanici,* at least thirteen were sent to Bauhin by correspondents.[14] The *Gramen caninum maritinum spicatum* was described from a dried specimen, and the illustration was apparently made from one that had been twisted around to fit on the paper.[15]

Bauhin's dry, clinical descriptions were almost exclusively morphological. He described plants' forms but devoted little attention to where and when they grew. Not that he completely neglected these aspects of description. He recorded, for example, that the *Gramen nodosum spica parva* "grows in moist heaths around Beford (a day's journey from Basel), where my late brother Johann Bauhin collected it." But he rarely devoted more than a few short lines to the *loci* of a plant. Not surprisingly, when he did so it was usually for plants he had observed himself: so, for instance, the *Erysimo similis hirsuta* "grows in rocky places, ruins, and walls, as for instance here in Basel by the walls and at the castle of Monchenstein, and it flowers at the beginning of spring. Also at Montpellier in shadowed ruins, but there it is smaller; less frequently in sandy areas."[16] Even such a long account of *loci* was much shorter than the morphological part of the description. For Bauhin, description was largely, if not exclusively, a matter of form. He saw form as the primary means for putting the botanical house in order.

In many respects, Bauhin's descriptions and his working methods closely resemble those of Carolus Linnaeus, the eighteenth-century founder of modern systematics. Like Linnaeus, Bauhin established a herbarium, sent students to collect specimens for him, determined species, and either established their synonyms in the published literature or described them if they were new. Bauhin worked on a smaller scale, but this aspect of his work would have been quite familiar to the Swedish naturalist. Moreover, Bauhin's descriptions are, on a superficial level, quite similar to Linnaeus's. But Linnaeus's descriptions differed from Bauhin's in two significant respects. First, they were more detailed and precise about plant anatomy, in particular the anatomy of fructification. This might seem to be merely a difference of degree; naturalists' descriptions had already become increasingly precise in the century from Brunfels and Bock to Bauhin. But Linnaeus's descriptions were not only more precise than Bauhin's; they also depended on determinations of generic and specific characteristics that were completely foreign to Bauhin's understanding of nature. Like Linnaeus, Bauhin and his Renaissance predecessors described species contrastively; but their descriptions, unlike Linnaeus's, did not depend on a taxonomic scheme. They had not yet decided to reduce nature to a system.

Taxonomy and Classification: Renaissance Folkbiology

Though classification often gets much attention in histories of natural history, it was not a major problem for Renaissance naturalists.[17] Some were largely indifferent to it: Carolus Clusius, for example, was content to describe the plant he called *Ranunculus Thalietri folio.* "I leave to others," he wrote, "the name that should be imposed on it, and the class in which it should be placed."[18] Other naturalists were more concerned with classification as a practical problem. But Andrea Cesalpino, who considered the matter within the framework of Aristotelian philosophy, was atypical in his desire to find a philosophical basis for classifying plants.

Of course, naturalists recognized the existence of a natural order. In the introduction to his history of plants, Hieronymus Bock rejected the medieval herbals' use of alphabetical order to arrange their entries:

> In describing these plants, I have followed the principle of joining in my book the plants that nature seems to have joined by a similitude in form. But I have separated each plant into its own chapter. I did not want to follow alphabetical order, as the old herbalists were wont. In my opinion, that method is very inconvenient, for those who use it in their works must necessarily place here an herb, there a tree, there a bush, in a great confusion of things. But how can plants that seem to have an affinity with one another be recognized, if they are described in this order?[19]

Bock intermingled two principles of classification in this passage. First, he preserved the basic division of plants into trees, shrubs, "under-shrubs," and herbs, as set out by Theophrastus.[20] Second, he adumbrated the existence of a general "similitude in form" that united some plants. Many of the series in which Bock described plants are considered natural groupings by modern systematists.[21]

These two principles were adopted by other sixteenth-century botanists. They regularly separated trees, shrubs, and herbs, and treated them separately. Dodoens, like Bock, started with herbs and put trees last; others, like Clusius and the editors of the *Historia generalis plantarum,* began with trees.[22] For Clusius, the nobility and rarity of certain trees established their right to be treated first; in this respect, he was following popular opinion, which ranked plants in a way that reflected the social hierarchy.[23] Clusius also grouped plants that he perceived as having natural relations; in some cases, he did so even when it subverted his other organizing principles. So, for example, he included the *Telephio hispanico* among the *coronariae,* plants with elegant flowers, because of its root.

> Since many *coronariae* have tuberous, globulous, or knotty roots, I could not separate from them those plants that have similar roots, even if their flowers are not particularly elegant—especially since there aren't very many of them, and they don't seem to rate their own class.

Likewise, the *Carduus mollis* belonged with the thistles, despite its lack of spines, "because of its similarity with certain common softer thistles." Elsewhere he simply placed one plant after another because they were "similar in form and temperament," like the *Primula* and the *Auricula ursi.*[24]

But, as Clusius implied when he stated that some plants did not rate their own class, the division into four major groups, and arrangement by similarity within those groups, were not the only principles at work in sixteenth-century classifications. Most of these schemes involved a pronounced anthropocentric element. Clusius did not think it was worth creating a new division in his work to accommodate a few plants, so he put them in with others. Bock himself separated "foreign" from "German" plants, as, for instance, with the caraway.[25] Some classifications were essentially arbitrary: the most extreme example is Conrad Gessner's use of alphabetical order in his *Historia animalium.* Though Gessner did not justify that choice in his prefatory remarks, it seems likely that he thought it unnecessary to adopt a more philosophical order when there were so few animals to discuss. Nonetheless, even Gessner did not hew entirely to the alphabet, for his articles often included related species under the same heading.

The authors of the *Historia generalis plantarum* adopted unabashedly an anthropocentric classification. Historians have criticized the unscientific nature of this classification, but that is to misunderstand its purpose.[26] Its authors had a sense of natural order in the plant world; after discussing flowers in general, they moved on to individual plants "with flowers that please"—"not all of those mentioned, of which some have already been described, and some remain to be described in further books, in order not to separate similar plants."[27] But the upper-level classifications of the book, beyond its basic division of the plant world into trees, shrubs, and herbs, were determined by wildly varying criteria. For instance, book 4 describes grains and legumes, and book 6 is devoted to umbelliferous plants. But book 8 contains "odorous plants," while book 11 has "plants that are found in shadowy, wet, damp, and rich places." In book 13, one finds "plants that climb on other plants"; in book 14, "thistles and other spiny and needled plants"; in book 17, venomous plants, and, in the final book, foreign plants.[28] It certainly does seem a mishmash, if the intention was to found a logical system of classification.[29]

But this was not the authors' intent. Rather, they chose to divide their work into books according to certain *differentiae* that were particularly notable in the plants in question. In large part it was an anthropocentric classification, designed for convenience sake. In the introduction to book 11, they wrote, "If some plants are found in the previous book that should be here, or vice-versa, it is not surprising or absurd, since plants are joined to others not only according to where they grow, but also from a similarity or affinity in their leaves or other parts, or their virtues, or some other character." In the case of the fourteenth book, on spiny and needled plants, the salient *differentia* was "very different from the previous ones," but at the same time it is "so conspicuous, that it forces itself on us, and it does not leave unscathed those who collect or touch these plants carelessly."[30] In this case, the naturalist had strong practical reasons for recognizing these plants as a class!

At times, the authors seem to have deliberately varied their main *differentiae* for pedagogical ends. Thus, "many of the *differentiae* that are derived from the parts of plants, including those that are most necessary for the naturalist to know, are taken from the diversity of roots, as Theophrastus diligently explained."[31] From this standpoint, the hodgepodge nature of the text was a virtue: it kept constantly before the reader's eyes the variety not only of plants but also of their *differentiae.* The Lyon herbal contained some two thousand plants, and claimed to be universal—that is, to include all, or almost all, plants that were known and described.[32] Hence, its system was adopted with the goal of ordering what was known, rather than, as with seventeenth- and eighteenth-century taxonomies, of accommodating new discoveries.

But what of Bauhin's *Pinax*? Bauhin did arrange plants in what seems, superficially, like a modern taxonomic system. He divided the *Pinax* into books and sections, each section containing "genera" divided into a number of "species," according to Agnes Arber, who also credited Bauhin with consistently using binomial nomenclature. Arber admits that many of Bauhin's groupings were unnatural (from the standpoint of modern systematics), like the similar if more far-fetched groupings of the Lyon herbal. But he did hit upon natural relations among Compositae and Cruciferae, and he even used the term "Umbelliferae." Though Bauhin's success in this matter was rather hit-or-miss, Arber excuses him, noting "how much difference of opinion exists among systematic botanists, even to-day, upon the subject of the relations of the families to one another," and she implies that taxa are themselves only conceptual schemes with no basis in nature.[33]

Nonetheless, it is deeply anachronistic to treat Bauhin as a Linnaeus *avant la lettre.* The differences between them mark a deep reorientation of natural history in the seventeenth century. I do not mean to imply a discontinuity. Indeed, Linnaeus himself referred frequently to Bauhin's *Pinax,* and thus indirectly to the century of Renaissance natural history that lay behind Bauhin's work. Bauhin's research methods, developed on the basis of his sixteenth-century predecessors' works, continue to be used by naturalists even now. But Linnaeus's descriptions, though similar to Bauhin's in form and concision, rely on concepts of genus and species that Bauhin did not possess, while Linnaeus's taxonomy was intended not to organize what was known but to provide a system that could accommodate the as-yet-unknown.[34]

On its title page, the *Pinax* did claim to list plants "methodically according to their genera and species." Bauhin used these terms consistently to refer to what modern taxonomists would identify with the same names, though they might disagree about Bauhin's grouping of species into a genus. But Bauhin was using these terms in an Aristotelian sense. In Aristotelian logical division, "genus" (*genos*) and "species" (*eidos*) are relative concepts; each genus is divided into species that, in turn, can be considered genera for purposes of future division. Only the "indivisible species" (*atomon eidos*), the species that is represented only by individuals, cannot be considered a genus relative to other species.[35] Bauhin used "species" in this sense and "genus" to refer to the relatively small groups comprising a small number of very similar species.

This usage, which Bauhin shared with Clusius and other sixteenth-century naturalists, was consistent but pretheoretical: that is, it was not based on an explicit species concept.[36] Bauhin drew on common-sense notions of natural kinds or "folk taxa," notions that folkbiologists have identified in traditional societies the world over.[37] In the 1930s, the ornithologist Ernst Mayr noted that natives of New Guinea named and identified kinds of birds that generally corresponded to species determined locally by Western ornithologists.[38] Recently, Jared Diamond and K. David Bishop have established that the Ketengban people of New Guinea identify with unique names almost every bird species that the Western literature identifies; there were few exceptions, and in only one case did the Ketengban fail to discriminate between similar species.[39] Unlike Westerners, who generally rely on fine-grained morphological distinctions to identify species (continuing the approach of Renaissance naturalists), the Ketengban use behavioral and ecological data in their identifications. But the correspondence between scientific species and these low-level folk taxa is overwhelming. Bird species appear to be especially easy to discriminate (possibly because we like to watch birds), but the level of correspondence between low-level folk taxa and scientific species is, in general, astonishing to anyone with nominalist inclinations.[40]

Most educated Westerners are familiar with the modern taxonomic hierarchy of kingdom, phylum (for animals) or division (for plants), class, order, family, genus, and species (though some might be surprised to learn that these days the two old biological kingdoms have become balkanized; there are now five). The scheme was developed by Linnaeus, though earlier naturalists had recognized groups above the level of the genus; the phylum is a more recent addition.[41] Each particular grouping at any of these levels—Mammals (a class), Canidae (a family), or *Canis familiaris* (the domestic dog, a species)—is a taxon. In this taxonomic hierarchy, every species can be located within a higher-level taxon all the way from genus to kingdom. Folk taxonomy is organized in different ways. Despite a great deal of variation in the way that people conceptualize and theorize taxonomies, folkbiologists have identified four taxonomic levels that are found in most folk taxonomies: life-forms, generic taxa, specific taxa, and family fragments. As we will see, these taxonomic levels characterize not only traditional non-Western societies but also Renaissance natural history.[42]

Life-forms are the broad division of animals and plants into mutually exclusive categories—for instance, Theophrastus's division of plants into trees, shrubs, "sub-shrubs" or bushes, and herbs, a division that was followed by Renaissance naturalists. Life-forms are often exhaustive, though this may be due to the existence of one or a few residual categories into which large numbers of generic or specific taxa are grouped. (Through the eighteenth century, European naturalists lumped the invertebrates into the residual categories of "insects" and "worms.") Generic taxa and specific taxa are the low-level groupings that generally correspond with scientific genera and species. According to Brent Berlin, the generic level is primary: folk genera are the natural kinds that any knowledgeable observer can easily identify, while folk species require skill and attention to distinguish. In practice, folk usually identify specific taxa contrastively, whereas generic taxa tend to be identified in terms of overall habit or gestalt. (For instance, "oak" would be a generic taxon, as would "maple," while "silver maple" and "sugar maple" are specific taxa.) And folk genera that are divided specifically usually include a "prototype" folk species from which the others are differentiated.[43] Generic and specific taxa, taken together, do not usually exhaust the natural world as known to experts in a society, but they are, in Atran's words, "virtually exhaustive": they name all the biologically or culturally salient species in a locale and folk experts *expect* that any unfamiliar natural kind *should,* in principle, belong to a folk genus. Folk genera do not necessary correspond to scientific genera: bats, for example, are a folk genus for Westerners, but for taxonomists they are an order comprising several families.[44] Family fragments, on the other hand, are far from exhaustive: that is, not every generic or specific taxon belongs to a family fragment, and only some life-forms include family fragments. Moreover, family fragments can be covert: that is, experts may recognize their existence without giving them a name.

Most folk genera are monotypic: they are not further differentiated into folk species. But in agricultural societies, approximately twenty percent of folk genera are polytypic, containing two or more folk species. Most foraging societies, on the other hand, have none or at most a few polytypic folk genera.[45] To Berlin, these data suggest that folk taxonomies result from the interaction of biological facts—the existence of reproductively isolated local populations of animals and plants, which Ernst Mayr terms nondimensional species (in contrast to the Darwinian species, defined as a portion of an evolutionary lineage)—and cultural interest in the natural world, including the production of new varieties of animals and plants through domestication and the greater opportunity afforded to specialists in agricultural societies to distinguish folk species within folk genera.[46] That is, agriculturalists pay more attention to the distinctions among natural kinds than foragers (beyond those species on which foragers subsist), but the relatively constant distribution of monotypic and polytypic folk genera suggests that the distinctions recognized by agriculturalists really exist in nature.[47]

Folkbiology explains two puzzles about Renaissance natural history: why Renaissance naturalists were often indifferent to classifications, and why those who proposed classificatory schemes felt little or no need to justify them. (Atran's research is especially important in this regard.[48]) Bauhin and his predecessors used "genus" and "species" in a pretheoretical, folk sense, and their works were organized according to tacit principles of folk taxonomy.[49] They identified genera by overall habit and then defined species contrastively within those genera. And when they grouped genera based on an overall sense of resemblance, which Arber characterized as a kind of groping toward science, they were identifying family fragments. Bock rejected an alphabetical classification because he perceived these similarities, not because he had a theory of taxonomic relations. Clusius placed the *Carduus mollis* among the thistles for the same reason. Jacques Dalechamps organized one of his ornithological albums similarly, by broad categories of obviously related birds, following the lead of Pierre Belon.[50] Finally, the Theophrastean categories of trees, shrubs, bushes, and herbs are plant life-forms: their validity as taxa would first be seriously questioned in the 1590s by Adam Zaluziansky, but despite protests in the middle of the century by Joachim Jung, most naturalists retained the distinction through the seventeenth century.[51]

But Renaissance natural history was a peculiar kind of folkbiology, for historical and methodological reasons that we have already examined. Rooted in ancient Greek natural history and *materia medica*, the natural history of the first generations represented an amalgam of the local knowledge of Italian and northern scholars with that of Aristotle, Theophrastus, Dioscorides, and other ancient writers on nature. This led initially to the problem that motivated Niccolò Leoniceno and Euricius Cordus, and to a lesser extent Leonhart Fuchs and Hieronymus Bock: how to identify ancient names with modern plants. In the longer term, it led to a realization that the ancient plants were, in fact, often distinct from similar modern species. This realization drove the phytographical research of the middle and late sixteenth century—and that research, which universalized formerly local botanical knowledge, exceeded the strictly local scope of folkbiology.[52]

Late Renaissance natural history went beyond folkbiology in three respects. First, the technology of observation developed by Renaissance naturalists largely eliminated the contextual, ecological information that folk experts use to identify and classify creatures. Botanical gardens removed plants from their original environment, while herbaria and collections preserved dead plants and animals from which most ecological clues had vanished, except when they were noted down. As Atran has noted, this is how scientific biology, including scientific taxonomies, originated: "by de-contextualizing nature, by curiously tearing out water lilies from water so that they could be dried, measured, printed, and compared with other living forms detached from local ecology and most of the senses."[53] The modern naturalists Jared Diamond and K. David Bishop remark on the difference between their own methods for identifying and determining species, based on morphology, and those of their Ketengban informants, who used the birds' behavior, location, and calls as well as their appearance to identify them. This difference emerged in the sixteenth century and it has characterized natural history ever since.

This methodological shift both resulted from and contributed to the second and third divergences between Renaissance natural history and folk-biology: the increasing number of species known to the community of naturalists and the concomitant increase in the number of polytypic genera with which naturalists had to deal. Naturalists' field research, communication, and description pushed their knowledge beyond the bounds of common sense. In the 1551 edition of his herbal, Hieronymus Bock had described 806 plants in 430 chapters.[54] Less than forty years later, the Lyon herbal contained some 2000. Bauhin's *Pinax* listed 6000, 10 percent of them described for the first time by Bauhin alone.[55] Six thousand plants is well beyond the upper limit found in any folk taxonomy. In the research conducted through the 1980s, only one folk taxonomy contained more than 2000 named plants; the average was 500. Even more significantly, Berlin found that the typical upper limit of folk genera was around 500.[56] If each of Bock's 430 chapters can be considered a folk genus, his work falls comfortably within the norm; the large number of polytypic genera results from the broad range of his investigations. Bock's work also fits the folkbiological norms in his use of smell, other sensory data, and medicinal virtues in describing plants.[57] But the Lyon herbal was already at the limit of folkbiology, and its descriptions were chiefly morphological. In the next generation the balance tipped: Bauhin's 6000 plants, defined morphologically based on his herbarium, were outside the realm of common sense, but for him the folk categories sufficed. He had fifty years of experience, after all.

Others were not so sure. It seems no coincidence that the first Renaissance naturalists to think seriously about the problem of species, taxonomy, and classification were active in the last quarter of the sixteenth century. There can of course be no proof, but it seems likely that Andrea Cesalpino, Adam Zaluziansky, and Adriaan van de Spiegel were driven to their considerations on plant systematics by a nagging sense that they just could not keep all those species straight. Zaluziansky perceived the problem especially acutely because he was writing for students who were overwhelmed by new material. The first modern theoretical approaches to defining and systematizing biological taxa were produced by the careful labor of describing new kinds of plants.

Andrea Cesalpino's most important work, his *De plantis libri XVI* (1583), was a cross between Aristotelian philosophy and long years of experience with plants.[58] Cesalpino had studied with Luca Ghini at Pisa and had succeeded Ghini as director of the botanical garden there; he also taught and wrote on Peripatetic philosophy. He introduced the book with a brief account of the history of botany as the humanists understood it: founded by the Greeks, botany had declined in the Middle Ages but had been brought back to life by Barbaro, Ruel, and others. But no one had yet organized the mass of new plants that the moderns had brought to light; like soldiers in an unruly camp, they were disorderly and rowdy. Without some kind of order, no one could keep track of every plant; the consequences included mistakes and "sad altercations." "All science consists in collecting what is similar and distinguishing what is dissimilar"; that is, organizing the natural world into its species and genera, and that is what Cesalpino promised to do in his book.[59] He would take what three generations of naturalists had learned and reduce it to a proper philosophical method.

Theophrastus had had the right idea, Cesalpino admitted, but he had not carried it very far; in practice "he accepted commonly known genera, most of them based on how the plant was used." Dioscorides, on the other hand, had set a terrible example by organizing his work according to medicinal virtues. Jean Ruel was the only modern to follow Theophrastus, but Ruel scarcely did more than his Greek model. Like Bock, Cesalpino dismissed alphabetical order: though it was intended to serve the memory, it in fact made it harder to keep track of plants because it completely subverted their natural resemblances. Only a natural order that divided plants into their genera and species could truly serve the memory. These genera and species would have to be identified and, if necessary, given names. Such a procedure would even assist physicians, for similar plants often possessed related medicinal virtues.[60]

But what made an order natural? Cesalpino insisted that the only philosophical way to classify plants was psychological: psychological in the Aristotelian sense, based on the faculties of the plant's soul.[61] Aristotle had identified four broad faculties of the soul: vegetation, including nutrition and reproduction; sensation; local motion; and thinking or understanding.[62] (These are not different souls; rather, they are different faculties of a unitary soul, as Aristotle pointed out in *De anima*.[63]) Plants, which by definition lacked sensation, motion, and reason, thus had only two natural functions or operations: nutrition (or growth) and reproduction. For the most part, the organs of a plant correspond to these two operations.[64] (Cesalpino mentioned that some climbing plants *seem* to have sensation, for they appeared to seek out and then climb supports, but he did not grant them true sensation.[65]) Cesalpino's natural classification would thus be based on the organs of nutrition and reproduction: the root and shoot on the one hand, and the flowers and fruit on the other. Cesalpino has been praised by historians of botany for his forward-looking choice of floral structure as the key to classification, but his reasons for doing so were rooted in the Aristotelian science of the sixteenth century.[66]

By focusing on the operations of the vegetative soul, Cesalpino thought that he could eliminate distracting accidents that were unrelated to the plant's substance. The exact relationship between substance and accident had been long debated in the Peripatetic tradition, but for Cesalpino, the difference was, in theory, simple. Every quality that differentiated species was substantial; all others were accidents. Species, in turn, were to be differentiated on the basis of morphological differences in the organs that served the soul's functions. Human beings were a limiting case. The human species was defined by reason, and—Cesalpino claimed—the faculty of reason does not admit of further differentiation. Hence all differences between humans—even giants, Ethiopians, and "other monstrous forms"—are accidental; all are members of the same species.[67] With plants, the situation is different. Roots and shoots, flowers and fruits vary in many ways, and Cesalpino knew from experience that not every difference constituted a new species. But which differences were accidental and which constituted true *differentiae* that could determine species and genera?

Cesalpino solved this problem with another deceptively modern move. He defined species reproductively. Plants that grow in different places or are cultivated might appear to differ from other members of the same species, but "everywhere like gives birth to like, according to its nature, and of the same species." But this principle raises difficulties. Some domestic varieties differ greatly from their wild counterparts, but if domestic seeds are sown, wild varieties often grow. Trees are especially susceptible to changes produced by the environment.[68] Half a century earlier, Euricius Cordus had observed the same variability and thought nothing of it; his notion of what constituted a species had been loose enough to accommodate such differences. But Cesalpino adhered to a notion of species that was much more rigid than Cordus's—more rigid even than Aristotle's.[69] He had to be able to differentiate between variable and fixed characters.

His solution was to concentrate on the organs of fructification. The root and shoot were simply too variable; if the shape of the root were taken as constituting a genus, for example, the turnip and cyclamen would have to be joined, even though their natures differed in almost everything else, while the turnip and rape would have to be separated. Cesalpino could not overcome his instinctive sense of similarity and difference. The organs of nutrition were unreliable except to determine the first genera of plants: if they were woody and harsh, the plant belonged in the class of trees; if they were soft, the plant belonged with the herbs. (Cesalpino rejected Theophrastus's additional life-forms of shrub and sub-shrub.) Further distinctions needed to be based on the flowers and fruit, which were the most variable parts of the plant and thus most apt to distinguish different species. But accidental differences affected the organs of fructification, too; colors and odors might vary within a species. Only three aspects of the organs mattered: number, location, and shape. "Nature sported with those *differentiae* in many ways in forming fruit, on which the various genera of plants are constituted."[70] Everything else might differ accidentally, but these three were true *differentiae.*

Cesalpino allowed that the stem, leaf, and other parts might be considered when defining genera, but only insofar as they appeared to serve the plant's reproduction—such as when leaves protected flowers or fruit. And accidental qualities had to be strictly excluded. "Whatever does not contribute to the constitution of the entire plant or of the fruit is an accident, like colors, odors, flavors, and other things of the sort: they frequently change according to cultivation, or the diversity of the place or the heavens."[71] Whereas Euricius Cordus had decided that morphological qualities such as color or height might vary but that the flavor of a plant was a reliable key, Cesalpino strictly excluded everything but certain aspects of morphology.

By concentrating on morphology, Cesalpino opened up possibilities for a classification that would go beyond the tacit principles that govern folk taxonomies.[72] In that important sense, his classification differs from those offered by predecessors like Bock, who grouped plants that he felt were similar, or the Lyon herbal, whose authors organized plants according to *differentiae* that Cesalpino would have rejected as accidental. Cesalpino's system—flawed as it would seem to his seventeenth- and eighteenth-century predecessors—opened up the possibility of a scientific taxonomy.[73] He did not reject folkbiology; one of his strengths as a taxonomist was his refusal to create genera that violated his sense of which plants were closely related. But he attempted to find a rational basis for this sense of affinity, to justify his pretheoretical judgments. In this he was unusual; most Renaissance naturalists did not see any need to go beyond common sense when they organized their histories of plants. From the perspective of later centuries, Cesalpino was ahead of his time; from his contemporaries' point of view, he was imagining problems that did not really exist.

Nonetheless, in significant ways Cesalpino's work is marked by the science of describing that his contemporaries had elaborated. Cesalpino offered what he considered to be strong Aristotelian arguments for concentrating only on specific aspects of plant morphology. But Aristotle himself had admitted ecological criteria among the *differentiae* of animals. Furthermore, Cesalpino rejected qualities like medicinal virtues even though he admitted that they were constant and that related plants often possessed similar or identical virtues. He insisted that such qualities, "even when they exist themselves in plants, nonetheless are not *differentiae* that constitute its substance"—an assertion that seems to beg the question.[74] It seems likely that Cesalpino's austere morphology was encouraged, at the least, by his predecessors' and contemporaries' intense concentration on form and their comparative neglect for ecological information and qualities that could not easily be described in pictures or words. In that regard, Cesalpino shows how Renaissance naturalists unwittingly made scientific classifications possible.

The other sixteenth-century naturalist to theorize about classification did so on pedagogical and disciplinary grounds. Adam Zaluziansky, a Bohemian physician, was a follower of Peter Ramus, and in his *Methodi herbariae libri tres* he applied Ramus's principles of logical division to the vegetable world.[75] Zaluziansky apparently did not know Cesalpino's work, for he claimed that "no one, to my knowledge, has ever defined the genera of plants, especially the higher genera." Like Cesalpino, he criticized existing classifications, claiming that they were hopelessly confused. But where Cesalpino had traced the confusion to bad philosophy—focusing on accidents like taste, not substantial *differentiae*—Zaluziansky blamed it on disciplinary confusion.

> These days, in the history of plants and the arrangement of their species, some completely omit the genera, while others arrange them according to the alphabet, dietary or medical use, or ornamental value. In other words, they confuse plants by applying to them alien genera from Grammar or Medicine. Botany [*herbaria*] should not be crammed with references to the alphabet, foods, medicines, or garlands. Its proper subject is the nature and growth of plants—not teaching the alphabet, giving food or medicine, or making chaplets. Its material is the nature of plants: that is, what root, stem, branch, leaf, iris, aloe, lemon, and so forth are.

Zaluziansky acknowledged that plants were used in many different aspects of life—topiary, building, gardens, agriculture, and silviculture—and that the ancients had considered "herbaria" to be the major part of medicine.[76] But those practical aspects of the subject had to be based on proper theoretical knowledge of the nature of plants in themselves.

Zaluziansky considered theoretical botany under two heads: causes (*aetiologia*) and history. Causes were discussed in the first book, which addressed plants in general, above all their parts, moving from elements first to consimilar parts (those that when divided remain similar, like the "flesh" of plants) and then to dissimilar parts like the root and the stem. Following Ramus's method, the divisions become progressively more refined until Zaluziansky arrives at the flowers and the fruit.[77] In this book, he employed Ramus's method not to classify plants but to organize a discussion of their properties in general; it is a pedagogical tool.

In the second book, on the history of plants, Zaluziansky turned from plants in general to specific species.[78] Once more he organized his material by division. Plants are imperfect or perfect. Imperfect plants are fungus or moss, while perfect plants are short or tall. Short perfect plants, in turn, have nerves in the stem or not.[79] The book eventually reaches intermediate genera, such as hyacinths, onions, lilies, and thistles, that were regularly used by sixteenth-century naturalists, but within these groups Zaluziansky continued to make binary divisions until individual species were reached. As a result, his "history" is not a description of individual species but a progressive list of *differentiae* that apply to any given species included in the book.

In this regard Zaluziansky's history resembles nothing so much as the artificial keys found in modern field manuals.[80] They too present a hierarchical series of characters; by determining in turn which of those characters a specimen presents, the reader is directed first to the family and then to the genus in which it is located. But the key is not a scientific classification; it is a pedagogical or practical tool. It provides an artificial guide to the natural classification according to which the manual's descriptions are organized. An artificial key is useful insofar as it is based on characters that are easy to identify and its alternatives are unambiguous. Modern taxonomists might spend their entire careers defining or revising one family of plants or animals, deciding exactly which species to include and in what genera to group them. The key is for everyone else—especially for learners.

Zaluziansky wrote for the same audience. Like Cesalpino, he emphasized that his predecessors, by neglecting method, had left the study of plants in a confused state. But unlike Cesalpino, he did not offer any philosophical justification for his choice of *differentiae.* He excluded practical uses from his dichotomous key not because they were accidental but because they belonged to other disciplines. Most of his *differentiae* were morphological; in this regard he continued the practices established by his predecessors and contemporaries. But he included those, such as color and smell, that Cesalpino had strictly excluded. The careful reader might find this puzzling, for earlier Zaluziansky had condemned Pliny and Dioscorides for illegitimately defining species in terms of their odor.[81] But that was a theoretical complaint; on a practical level, Zaluziansky found that color and smell could help beginners. And beginners were his audience: describing thistles, whose variety was immense, he specified, "In this genus we have adopted a useful brevity in the style of a precept, repeating only what is necessary." In keeping with the book's pedagogical aims, Zaluziansky devoted a mere ten pages to the third book, on "the exercise of botany."[82] The student should learn to analyze plants according to the method of books 1 and 2, with his teacher's help and frequent demonstration; he should also learn the various uses of plants. But these, as a matter of practical experience, are beyond the present work's scope. It conveyed the basics.

If Cesalpino strove for a philosophically defensible classification that took into account his folkbiological intutitions, Zaluziansky established a practical method based only on pedagogical utility and disciplinary purity. Though he has occasionally been praised for being the first to emancipate botany from medicine, the praise is misplaced: predecessors like Clusius had in practice already distinguished botany from medicine, while Zaluziansky based his distinction on a Ramist notion of disciplinary autonomy and freely admitted that utility—medical or otherwise—was the main reason to study plants.[83] Zaluziansky's concern with the problems faced by beginners was shared by Adriaan van de Spiegel, whose *Isagoge in rem herbariam* (1606) also stressed the pedagogical utility of clearly defining the characteristic features of a genus in terms that a beginner could understand. Spiegel too offered a rough-and-ready key to plants, without the strict division that characterized Zaluziansky's work. If beginners were taught the basics, then that would suffice; most medical students—still the main audience for formal lectures and demonstrations in botany—would need no more.[84]

To conclude: classification was not a significant problem for Renaissance naturalists. Most organized their works according to folktaxonomic intuitions about the relationships among plants without giving the matter much thought. Those few who did explicitly address the organizing principles of their works did so either out of a professional commitment to Aristotelian method or a practical concern for the needs of students (in Zaluziansky's case, combined with the trendy appeal of Ramist division and tables). Nonetheless, the research methods and goals of Renaissance natural history—above all, the documentation centers of the late sixteenth and seventeenth centuries—did remove plants (and to a lesser degree, animals) from the ecological contexts in which they were usually encountered by folk. The universalization of local knowledge permitted by those methods presented late sixteenth-century naturalists with many more polytypic genera than their predecessors had to confront; at the same time, they had to deal with a number of species that by the 1580s equaled, and by 1623 had definitely surpassed, an individual's cognitive capacity. The resulting problems were perceived at the time as chiefly pedagogical; only in the later seventeenth century would more than a few naturalists begin to engage with classification as a serious problem.[85] They would attempt to solve it by going beyond Renaissance naturalists' goal of organizing what was known: by inventing systems that had a place for what was still unknown.

Evaluating the Unknown: Renaissance Naturalists and Travelers' Tales

For Renaissance naturalists, the problem with information and objects brought from distant lands, within Europe or beyond, was not classification. Rather, they were concerned primarily with the accuracy and reliability of that information, and the precise provenance and nature of those objects. Trained to observe specimens in the field, study them in gardens and collections, and describe them precisely, naturalists like Gessner and Clusius were often frustrated by the poorly-preserved plants and animals that arrived from distant lands, often in fragments, and with the incomplete and contradictory stories that accompanied them. Nonetheless, they were unwilling to omit from their histories anything that could add to the catalogue of nature, incomplete as it might be. When they could not themselves observe a species firsthand, they collated and cross-checked the information they could gather from specimens, travelers, printed books, and manuscripts. Unlike their ancient and medieval predecessors, they identified their sources precisely and weighed them carefully. The accounts that resulted, imperfect as they were, added to the store of human knowledge and spurred further investigation.

Many of these new, exotic plants and animals came from the "Indies," East and West—from Africa and Asia, where Portuguese sailors had established themselves in late fifteenth and sixteenth centuries, and the Americas, claimed by Spain and Portugal and disputed by many other European nations. But many came from Europe itself. Europe's margins were little known to naturalists. At the Council of Trent, the Swedish exile Olaus Magnus awoke deep interest with his stories of the strange creatures and customs of the northern lands.[86] Pliny the Elder had described Scandinavia as another world; his sixteenth-century successors felt much the same. When Adriaan van de Spiegel echoed the sixteenth-century commonplace that the variety of plants is infinite, he summed up with a twist on a Greek proverb reported by Pliny: "Europe, Asia, Africa, and the Indies seem always to produce something new."[87] Naturalists and their correspondents could and did examine the new species of western and central Europe, but exotics were another matter—whether they came from Africa, Asia, America, or the far reaches of Europe itself.

The majority of new species described by Renaissance naturalists were from well-known lands, but exotics held a grip on the European imagination in the sixteenth and early seventeenth centuries.[88] Often grouped together on account of their very exoticism,[89] these creatures challenged received views of nature. Their importance in contemporary debates cannot be denied. Exotica raise a fundamental question for our understanding of early modern natural history. How did practitioners of this science, in which personal experience was becoming the standard against which knowledge was defined, judge truth and falsehood when they were forced to rely on the reported experience of others?[90] In the rest of this chapter, we will consider several aspects of trust and reporting in early modern natural history. Each of these cases reveals a different approach to trust and facts; the issue of trust was solved piecemeal by sixteenth- and early seventeenth-century naturalists.

The natural history of the Arctic demonstrates how naturalists—in particular, those interested in animals—combined humanist textual analysis with artifacts to establish matters of fact about the walrus, elk, and other large fauna of the North. The natural history of the Indies, both East and West, reveals a different dynamic, closer to that of Europe-based naturalists, in which a reporter's knowledge and skills played a greater role in assessing the accuracy of his claims. I end by considering the relation between practices of observation and reporting experience on the one hand, and the increasing skepticism of seventeenth-century natural history on the other hand. Renaissance naturalists sought to extend their knowledge as far as possible, even if it was imperfect; their successors in the late seventeenth and eighteenth centuries, armed with increasingly fine-grained systems, rejected whatever they did not know well enough to classify.

The Wonders of the North

Though the Scandinavian lands and northern Muscovy were part of Europe and had been Christianized for centuries, they were culturally and geographically marginal.[91] At the edge of the Catholic world or in the grasp of schismatics, both still home to pagan cults, these lands were only vaguely known to the learned world of the sixteenth century.[92] Even the introduction of the Reformation in Sweden provoked little response from Rome, concerned with more important matters in the Holy Roman Empire and heresy in Italy.[93] Naturalists' reports on such fauna as the reindeer, the elk, the walrus, and the wolverine were thus based not on immediate, personal experience, but on travelers' tales and the remains of dead animals, often of uncertain provenance, that made their way south.

Before 1555, humanist-educated naturalists may have drawn most of their meager knowledge of Scandinavia from classical sources or their Renaissance mouthpieces. Most classical authors were silent about the region. But a few passages in Caesar, Pliny, and Pausanias provided grist for Renaissance naturalists' mills. Julius Caesar described two beasts inhabiting the "Hercynian wood," in modern Thuringia and Bohemia. One was a "bos cervi figura" with one branching horn in the forehead of both male and female animals. The other, called elk (*alces*), was goat-like but larger, with mutilated horns and unarticulated haunches.[94] Caesar wrote from hearsay, but his terse, impersonal prose disguised the fact. In this matter, as in others, Renaissance scholars attributed great weight to his *auctoritas*.[95]

Caesar's description, perhaps garbled in translation, may have served as the basis for Pliny the Elder's description of two Northern animals. According to this indefatigable encyclopedist, the North produced troops of wild horses; the elk (*alcen*), similar to a mare but with taller ears and neck; and the *achlin*, a creature similar to the elk but which could not bend its knees. Because of the size of its upper lip, it had to walk backwards to graze.[96] Pliny seems to have combined elements of Caesar's report with travelers' tales; in the process, he turned one animal into two.

The Greek traveler Pausanias had actually seen an elk (alive or dead, we do not know) in Rome. Male elk, he wrote, have horns on their brows, but not the females. In form, the creature was intermediate between the deer and camel, and it breeds in Celtic lands. Due to its sagacious avoidance of humans, it was but rarely seen and hence little known.[97] Pausanias's commentator Frazer has remarked that this description is fairly accurate;[98] but it was less detailed than Caesar's and Pliny's, and Renaissance writers gave it less attention.

To these three classical writers, early sixteenth-century naturalists could add a handful of more recent texts. Albertus Magnus, in his commentary on Aristotle's natural history, had mentioned the elk and the reindeer (confusing their names mightily, as Conrad Gessner later complained) and described the walrus as a kind of whale.[99] Descriptions of natural history could also be gleaned from a few early sixteenth-century travel accounts. Maciej z Miechowa, a Polish diplomat sent to Moscow, and Damião da Goes, a Portuguese legate to Sweden, described the walrus, elk, reindeer, and "gulo" (our wolverine).

The authors of these descriptions, be they soldier, diplomat, or scholar, were all fascinated by the wondrous fauna of the North. The elk, reindeer, walrus and the like, creatures they had never before seen nor heard of, demanded their attention. Their remarks, made in passing, consisted of brief physical descriptions and one or two reports on the creatures' strange behavior. The Renaissance writers were particularly interested in the economic uses of the walrus and the reindeer, a preoccupation they shared with later explorers of the Barents and White Seas.

Olaus Magnus's illustrated map of the North (1539) and history of the northern peoples (1555) provided a much broader perspective on Arctic natural history.[100] Though his map stressed wondrous animals and their sagacious behavior, it did remind its readers that bear, squirrels, rabbits, and other prosaic creatures were as much denizens of the North as its unique productions. And the sheer bulk of Olaus's history, whose last six books were devoted to natural history, ensured him a prominent place in the works of later Renaissance naturalists. Olaus was born in Uppsala and educated in Rostock; he gained much firsthand knowledge of northern Scandinavia during a visitation of the churches of Norway and Sweden in 1518–19. Though Olaus did not correspond with naturalists, and natural history was not his main focus, his status as a firsthand observer gave him great authority in the eyes of contemporary naturalists.[101]

But classical texts, travelers' tales, and humanist histories were not the only sources for Northern natural history. Even before the Arctic expeditions of the later sixteenth century, *naturalia* made their way from north to south. In Albertus's day, walrus-hide thongs were sold in the Cologne marketplace.[102] This trade seems to have slacked off, perhaps due to the demographic calamities of the fourteenth century, but Renaissance collections of curiosities boasted the occasional antler or hide.[103] Sometimes of uncertain provenance, these *naturalia* nonetheless served to bolster accounts based primarily on literary sources.

Even the memory of such objects could serve natural history. The walrus head sent by the archbishop of Trondheim to Pope Leo X (1513–21)[104] may well have vanished in the Sack of Rome, or perhaps it was improperly pickled,[105] but during its voyage south a Strasbourg artist painted it on a wall in the city's town hall. Next to the picture, doggerel set out the beast's origin and lamented its fate:

> Solt ich mein zeyt auß mögen läben/
> Ich hett nichts umm all wallfisch gäben.
> Von Nidrosia der Bischoff batt
> Mich stechen lassen an dem gstad.
> Bapst Leo meinen kopff geschickt
> Gen Rom/da mich manch mensch anblickt.[106]
> [Had I lived my life to its end, I would have cared nothing for any whale.
> The Bishop of Trondheim had me stabbed on the shore and sent my head
> to Pope Leo in Rome, so that many people might see me.]

Albrecht Dürer's drawings from his journey through the Netherlands preserve another walrus, captured by Dutch fishers in the North Sea and sketched by the German master in 1521 (figure 5.1).[107]

These texts and objects had a great impact on the learned naturalists of Europe. But they seem not to have penetrated more broadly into Renaissance culture. The explorers and traders of the sixteenth century who encountered the walrus described it as something completely new; they had not read of it in Albertus or Maciej. Their descriptions, in turn, were not taken up by naturalists until the seventeenth century. Jacques Cartier had encountered the beast during his voyage through the St. Lawrence Strait in 1534, and described it as something completely new: "Around this isle [Brion] there are several great beasts like great cattle. They have two teeth in their mouths, like elephants, and they go into the water."[108] Cartier's relation was unpublished and little known until it was published in Ramusio's 1556 collection of voyages; a 1544 manuscript *Cosmographie* mentions Cartier's animals, but only as "other fishes that also resemble elephants. They have horns like elephants and come onto the land."[109]

Likewise, the English explorers and traders who penetrated the Barents Sea in search of the Northeast Passage to Cathay described the walrus as a new marvel. Richard Chancellor, who sailed to Archangel in 1553,

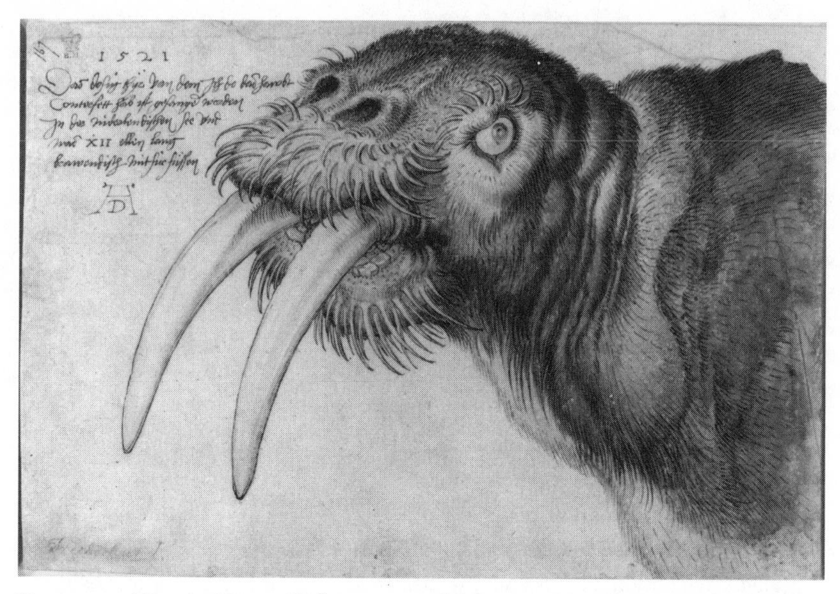

Figure 5.1. Albrecht Dürer, *Walrus*, pen and ink, 1521. British Museum, inventory no. SL, 5261–167. © Copyright The Trustees of The British Museum.

remarked in a letter home that in the White Sea "there are also a fishes teeth, which fish is called a Morsse."[110] In this initial report commercial considerations were foremost, but Chancellor's longer debriefing, taken in Latin by an English schoolteacher on the navigator's return, provided a somewhat lengthier description of the beast:

> The sea adjoyning breedes a certaine beast, which they call the Mors, which seeketh his foode upon the rockes, climing up with the helpe of his teeth. The Russes use to take them, for the great vertue that is in their teeth, whereof they make as great accompt, as we doe of the Elephants tooth.[111]

These relations were not published until the 1590s, in Richard Hakluyt's *Principal Navigations,* and like Cartier's account, they had little impact on the learned world before then. British sailors and merchants in the 1550s were increasingly familiar with the beast but had little interest in it. Steven Burrough, navigating toward the river Ob in 1556, encountered Russians going morse-hunting; at Vaigats Island he saw three recently killed morses, but he remarked only that the hunters were demanding a ruble for a rather small tooth.[112] The Muscovy Company was interested only in goods that would

Figure 5.2. A sixteenth-century walrus hunt, woodcut from Gerrit de Veer, *Reizen naar het noorden* (1598).

sell in England: primarily train-oil, flax, hemp, and cheap furs. They had trouble selling seal-skins, and walrus hide (not to mention the entire beast) would have been too bulky to be worth the cost of shipping.[113] Except for the handful of sailors who navigated to Archangel and back, the British encountered walruses only after they had been rendered into train oil. Continental naturalists were certainly not going to learn about them that way.

Naturalists in Europe encountered these creatures at firsthand only at the end of the century. The English Muscovy Company had brought back walruses only as train oil, but Dutch explorers returned with more substantial portions of the beasts. In August 1594 Willem Barents's crew brought "a 'Wal-rusch' to Amsterdam, a fish of marvelous appearance, which they had captured on an ice floe and killed."[114] In the voyage account, first published in 1598, readers found not only brief descriptions of this beast but also an illustration of the walrus hunt (figure 5.2). Conrad Gessner might have quibbled about some of the details, but it was the first description published with a picture drawn from life. However, its impact on the learned world seems to have been minimal. Even Clusius, who translated Gerrit de Veer's account of the voyage, did not mention walruses in his *Exoticorum libri.*[115]

Other Northern fauna were also slow to penetrate further south. The grand duke of Tuscany had, in Aldrovandi's day, a live elk at his court, and he made drawings available to the Bolognese scholar.[116] After it died, the beast was stuffed, and it still existed in Pisa when John Ray and Francis Willughby toured the ducal collection in 1664.[117] But less powerful and wealthy collectors had to content themselves with antlers and skins,[118] and for the most part, naturalists continued to draw on the same sources that had been available in the 1550s. The way this information was evaluated and presented by naturalists reveals the limits of the Renaissance science of describing when naturalists tried to apply its methods secondhand.

Conrad Gessner's encyclopedic *Historia animalium,* like many of Gessner's other works, testifies to his breadth and depth of learning, as well as his immense stamina. Because of its scope, there was no way Gessner could personally observe everything that he described; this was especially true of zoology, since it was much more difficult to keep and observe animals in menageries than plants in gardens.[119] Nonetheless, Gessner was aware of the problems of relying on others' reports, and he explicitly raised the question in his preface to the readers of his work.

In tone and substance, his defense was humanistic. He had read and excerpted books, just as Pliny had done, but far more than the ancient Roman had used. "No one without experience in the matter can easily know how difficult and dull it is to compare the works of different authorities and compile them all in one body, as it were, so that nothing is omitted, and nothing is unnecessarily repeated." In presenting his compilation, Gessner chose to use a simple style, "for in those texts that are concerned with the knowledge of things (as Massarius says), one must seek to express the uncorrupted truth, not the beauty of sparkling words." To the humanist Gessner and his audience, simple and direct speech, rather than rhetorical tricks, implied truth. But the consensus of authorities was the most important guarantor of secondhand truths:

> As far as things themselves, and their verity and certitude, in many cases I don't commit myself but am content to give the names of the authorities who claim something. Claims that are supported by the consensus of many erudite authors, for many centuries, are for the most part worthy of faith. For this reason I have named several authors and repeated myself even when it otherwise seemed of little import. For if a thing is asserted with the same words by many witnesses, it is more worthy of faith.[120]

This position seems eminently reasonable—that is, assuming that the testimony of each author is independent. Such was not necessarily the case.

Fifteenth- and sixteenth-century humanists often piled Pelion on Ossa in listing sources for a claim without realizing that many classical writers were merely citing their predecessors, and that their testimony was thus of no additional weight.[121] In order to determine the virtues of Gessner's method, and to see whether he followed it, it is necessary to turn to concrete examples.

Gessner was certainly thorough in gathering material. His descriptions of northern fauna drew not only on Olaus and Damião but also on Albertus Magnus, Julius Caesar, Pausanias, Pliny, Maciej, and even Aelian—not to mention Johannes Agricola Ammonius, Paulus Venetus, Sebastian Münster, and others who gave secondhand accounts of northern animals.[122] Gessner had earlier published a *Bibliotheca universalis* (1545), a guide to all literature, extant or not, written in Latin, Greek, or Hebrew.[123] He was well versed in ancient and modern natural history and topography, and he could provide his readers with a long list of sources on which his work was based.[124]

After collecting and arranging his sources, Gessner criticized them.[125] His description of the reindeer began with Damião da Goes's account and Sebastian Münster's additions or corrections: "The Lapps use, in the place of horses, animals that they call 'raingi' in their own language. These creatures have the size and color of an ass (Münster says hair almost like that of an ass, and that they are hairy), and hooves cloven in two," adding that according to Olaus they were bigger than deer.[126] Gessner thus realized that Münster's account was derived from Damião, and that Olaus was his source for the additions and corrections to Damião's account. Gessner then summarized descriptions from Albertus, Joannes Agricola Ammonius, Paulus Venetus, Julius Caesar, and even Aelian. But the chapter concluded with a compilation of Olaus's remarks, as the last word (literally) on the creature and its uses.

Gessner's careful collation of texts ensured that dependent sources (like Münster) were not accorded more importance than they were worth. This methodological achievement was, however, vitiated by his excessive trust of classical texts, such as Caesar's, that purported to be firsthand accounts. After determining the relations between texts, Gessner proceeded to evaluate them as sources. In this, he valued both learning and personal experience. Hence Olaus, "a man who is worthy of faith on account of his erudition and his experience of the region, attests that elk are very swift." Gessner's preference for allegedly firsthand sources is both clear and comprehensible. His esteem for erudition, however, was not just a humanist prejudice. He valued learning because the unlearned could write confusedly even about things they knew. This was true of Albertus Magnus, whom Gessner called "a man who knew a lot about things but who clumsily abused their names."

Albertus had discussed the same animal in different places under different names, for Gessner an unforgivable crime.[127] To Gessner, Olaus's account of the elk was preferable to Albertus's on both grounds.

Gessner applied the same criteria to his illustrations. He pointed out if they were drawn from life or agreed with it, as with the elk, whose "picture I received from a certain painter; eyewitnesses (*oculati testes*) certify that it is accurate." In this case, he treated the picture like a verbal description, guaranteed by witnesses to the thing itself. Gessner had to take some images on trust, but when he did, he informed his reader: his picture of the reindeer was taken from Olaus, as was that of the gulo. Of the latter, Gessner noted wryly, "I won't claim is drawn from life, but I reproduce it as I found it." Finally, there were *naturalia* themselves. They could play a decisive role in certain instances, but they did not always speak unambiguously. In some cases, it was difficult even to identify them. In his article on the reindeer, Gessner gave a drawing of an antler "from certain old horns that a citizen of our city keeps—whether of from a reindeer or some other animal, I don't know."[128] Uncertain provenance made the artifact's meaning doubtful.

When Gessner turned from land quadrupeds to sea creatures, he was faced with similar problems and adopted the same approach. Living on a small salary in landlocked Switzerland, unable to travel to the ocean or have sea creatures shipped to him at great expense, he had to rely on the reports and the books of others.[129] He turned to the monstrous fishes of the North, to which Olaus had devoted an entire book, in a section "De cetis quibusdam ex Olai Magni Septentrionalis oceani europaei descriptione" (On certain whales from Olaus Magnus's description of the Northern European ocean). At the beginning of this section, he criticized Olaus's style in his drawings:

> Let the author guarantee the accuracy of these images. I have taken them all from his map of the North. It seems that he depicted many according to seafarers' tales rather than from life. I scarcely approve of certain heads which are too similar to terrestrial animals, nor the feet armed with claws. . . . I have added the author's descriptions, in part from the map itself, in part from a German pamphlet that is slightly more detailed.[130]

After this general disclaimer, Gessner enumerated the strange sea creatures found in Olaus's table. He knew most of them only through Olaus's map, and he contented himself in those cases with listing them and providing the prelate's short descriptions. But in the section "De Rosmaro" (On the walrus) he had more to say.

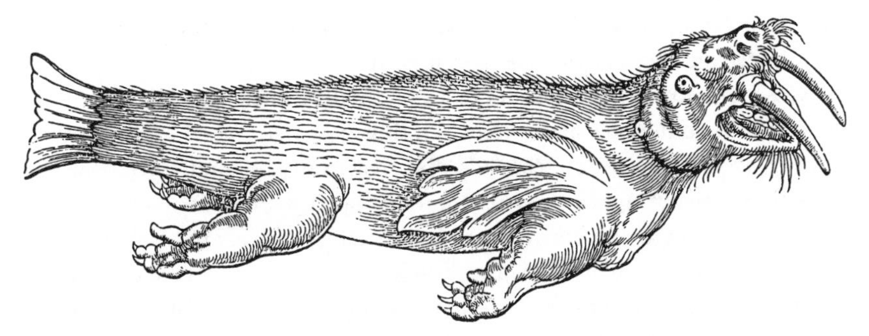

Figure 5.3. Walrus picture from Strasbourg town hall, woodcut from Gessner, *Historiae animalium liber IV* (1558).

Characteristically, he had more to say because he had independent sources: a "chorographica Moschoviae tabula" (map of Muscovy) and a copy of the illustration from Strasbourg, which he reproduced in the text along with its verse description (quoted in part above). Based on these sources, he criticized both of the illustrations. Olaus was scolded for showing the teeth pointing up—because the Strasbourg picture and the *tabula chorographica* depict them the other way, but also because Olaus's description says that the beasts hang from rocks with their teeth, and Gessner was hard put to imagine how they could do that if their teeth pointed upwards.[131]

Gessner's comments on the Strasbourg picture (figure 5.3) are particularly interesting. He considered this picture more accurate than the other he reproduced. But it was not perfect: fish don't have feet. He admitted that in the skeletons of large fish, fins can sometimes look like feet, but the Strasbourg artist had taken license a bit too far.[132] The head, though, had really existed—the one sent four decades earlier from Trondheim to Rome—and so Gessner judged that in this respect the picture was trustworthy. In this case, even the report of an artifact—clearly identified in the picture, more so than in the antler itself, of uncertain provenance, that Gessner saw in Zürich—served to confirm some knowledge and put other claims into question.

Gessner did not have Olaus's *Historia* to hand. If he had, he would have learned little more than he already knew, other than the way they were hunted. From scattered textual clues, Gessner had assembled a reasonably coherent natural history and granted it his authority. At least one Renaissance naturalist, Ferrante Imperato of Naples, was impressed enough to include Gessner's rosmarus from the Strasbourg illustration among the sea creatures painted on the ceiling of his museum.[133] Out of two descriptions, two illustrations, and the reported existence of an artifact, the Swiss scholar had created an animal from the frozen North.

Gessner's rosmarus, reproduced in Imperato's museum and, fifty years later, in John Johnston's natural history, thus possessed more authority than Olaus's account, despite Gessner's distance from the Arctic regions. His critical exposition tipped the balance in his favor. Unlike Olaus, whose description rested only on his own word, Gessner made clear to his readers not only what he believed but why. Doubtless much of his readers' trust derived not from his method per se but from his reputation. But that reputation itself was due to his position within the community of naturalists and his mastery of their tools and techniques.

The subsequent history of the walrus underscores the importance of erudition to early modern natural history and stresses the dangers of a little learning. For despite Olaus's popularity and Gessner's authority, Ulisse Aldrovandi's natural history of fishes mentions neither of these sources. Instead, Aldrovandi cited the philosopher of nature Girolamo Cardano, who in turn drew on Maciej z Miechowa. Indeed, Aldrovandi barely mentioned the creature at all. He discussed it only in his general description of marine monsters, where he wrote that some have teeth, some not. Of the former, Cardano calls the teeth similar to ivory but stronger and more resistant to wear; according to Maciej, the Muscovites use them to make handles for weapons and tools. Aldrovandi also included a picture, not of the walrus, but of a tusk, with the caption, "Tooth of a great marine fish called a Morse."[134]

Maciej was also the source, apparently the only source, for the description of the "morss piscis" by Juan Eusebio Nieremberg, whose *Historia naturae* (1635) was a collection of exotic beasts, mostly but not entirely from the New World. Nieremberg's description goes back to the Polish writer; his illustration may have been drawn from a skin (figure 5.4).[135] Imperato, who must have had access to the same source for the illustration, hedged his bets and included this painting next to Gessner's on the ceiling of his Naples museum (figure 5.5).[136]

Gessner had deduced that the "rosmarus" and "morsz" were the same creature, presumably due to similarity between Olaus's and Maciej's descriptions. But he discussed them, in the article on "Cete," and they were not indexed. Hence the Renaissance reader would have needed to note this particular fact among all of Gessner's volumes and remember it later. John Johnston did so; Ferrante Imperato too; but Aldrovandi and Nieremberg missed it. Even Gessner's knowledge had its limits. He had heard of a fifty-foot beast called the Ruswal but decided it was not the same creature. Only in 1693 could John Ray state confidently the animal's synonyms.[137]

MORSS PISCIS.

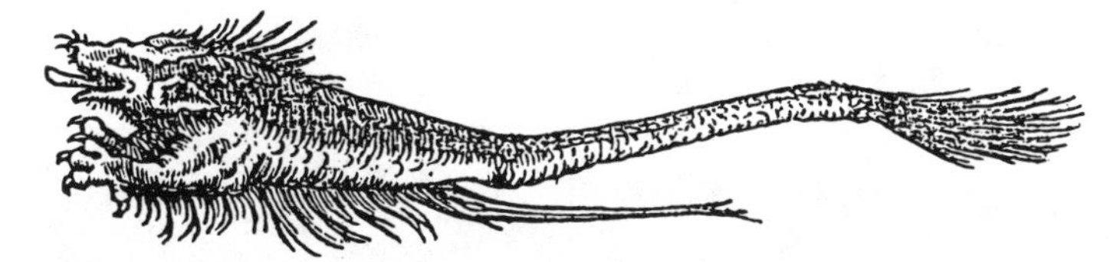

Figure 5.4. Morss piscis, woodcut from Nieremberg, *Historia naturae* (1635).

Figure 5.5. Ferrante Imperato's museum (detail), engraving from Imperato, *Dell'historia naturale* (1599).

Though Aldrovandi's learning slipped in this instance, with other animals he was on firmer ground. In his account of the reindeer, he proceeded much like Gessner. After its names, he considered its form. Albertus describes the creature, he wrote, as having three horns. Other writers claim it has three *pairs* of horns, but they must be mistaken—perhaps its horns are twisted, so that it appears to have more sets than it does. Olaus, whom Aldrovandi glossed as "a man who is worthy of faith on account of his erudition, his way of living, and his experience of the region," agreed with Albertus in this point, which settled it for Aldrovandi.[138] Unfortunately, Aldrovandi did not compare Olaus's account very carefully with Albertus's description, for the Swedish archbishop's words were lifted, without acknowledgement, from the Great Doctor's text.[139]

Like Gessner, Aldrovandi confronted contradictory sources and used a similar method to resolve them. First he set them out clearly. According to Damião da Goes, the reindeer's horns were smaller than those of deer, but Olaus said they were larger. Aldrovandi, convinced that Olaus was more reliable—indeed, the bulk of his article was extracted from Olaus's

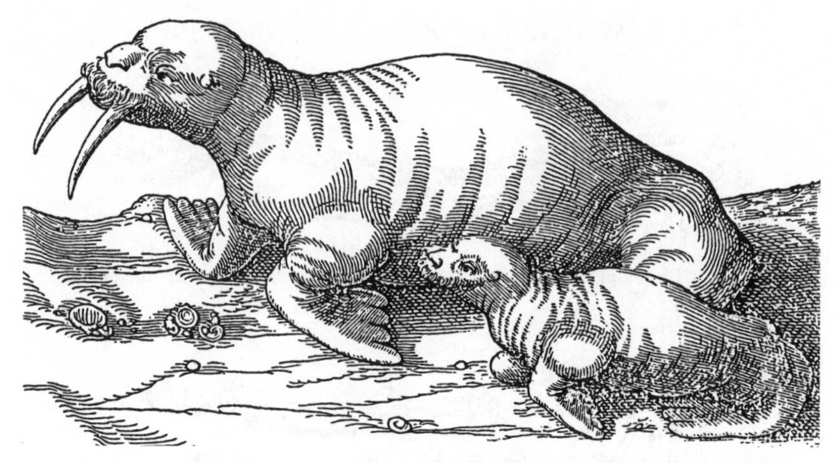

Figure 5.6. Walrus with cub, woodcut from De Laet, *Novus orbis* (1633).

Historia—concluded that Damião had confused the reindeer with the elk. Ferdinando de' Medici, grand duke of Tuscany, had sent Aldrovandi a picture of an elk, part of his menagerie, with quite small antlers. Aldrovandi must have concluded that this beast, also used as a beast of burden (per Olaus), was the one Damião had seen. Though Aldrovandi knew that the horns of the creature depicted by Ferdinando's artist were not fully developed, his measurements of another elk antler convinced him that they were not so large as Gessner and other authors had claimed.[140]

By the time Aldrovandi was writing, at the end of his life in the 1590s, *naturalia* from the north were more common than before; he could confidently identify an elk's antler as such, unlike Gessner, who had confused it with a reindeer's. But specimens were often unique, and as a consequence, learned naturalists' representations of exotic nature were not always representative. However, unlike Olaus and Münster, Gessner and Aldrovandi provided the information necessary for the reader to assess the accuracy of their conclusions. This conscientious scholarship meant that they continued to be authorities into later ages, even as fewer of their stories were believed and as natural history took a distinctly different turn.

Still, naturalists preferred firsthand observations. In 1612, a baby walrus, its teeth not yet erupted, was taken to the Netherlands, where it was observed and described by Aelius Everard Vorst, professor of medicine at the University of Leiden. Vorst's description of the baby, another adult, and a "picture ... accurately drawn from life" (figure 5.6) were first published by Jean de Laet in his 1633 description of the New World, *Novus orbis,* as a gloss to Cartier's cryptic remarks on the sea-horses of the St. Lawrence Strait.[141] Natural histories throughout the seventeenth and eighteenth centuries referred to Vorst's description, not to Gessner's learned discourse.[142] Even John Johnston, an uncritical compiler from the mid-seventeenth century, ignored Gessner's text in this instance.[143]

Renaissance naturalists' treatment of the natural history of the North provides an important counterpoint to the history of observation and experience presented in the preceding three chapters. When looking at what surrounded them, European naturalists could make hundreds or thousands of observations, gather and exchange material between different localities, and thus come to a rich understanding of the flora and fauna of France, Italy, and the Empire. Beyond these confines, even within the traditional bounds of Europe, they were forced to rely on techniques of comparison and collation of texts and isolated objects, techniques that were part of a broader humanist literary culture. Granted, a naturalist such as Gessner applied these techniques at far greater length, and more thoroughly, than a polymath like Sebastian Münster, who devoted a few lines rather than several pages to the animals described in his *Cosmographiae.* Nonetheless, the descriptions they produced were radically different from those based on personal observation and experience. The acuity of the science of describing depended on a dense network of scholars making repeated observations and comparisons. In its absence, Renaissance naturalists were myopic if not completely blind.

Beyond Europe: The Natural History of the Indies

The Scandinavian peninsula did not attract any traveling naturalists in the sixteenth or early seventeenth centuries. Olaus, Damião, Maciej, and others who wrote at first hand on its wonders were not trained as naturalists or even physicians, and they emphasized the strange and marvelous. Wonders and marvels were also a prime concern of sixteenth-century travelers to the East and West Indies, but many of these travelers were physicians in service of the Spanish or Portuguese crown, and beginning in the 1590s, of the Dutch East India Company. Some of these physicians were familiar with the techniques and traditions of sixteenth-century natural history. The few ancient testimonies on the natural history of India were thus complemented by recent descriptions, written by competent observers on the basis of personal investigation. The history of their reception by naturalists in Europe differs significantly from that of Arctic natural history.

Nonetheless, the European community of naturalists did not immediately accept everything that physicians like Garcia da Orta, Cristobal Acosta, and Nicolas Monardes, and observant travelers like Jan Huygen van Linschoten, claimed to have seen in their travels. Their works went through a careful process of evaluation before being assimilated into the body of knowledge, or rejected as untrustworthy. In this case too the limits of judgment when personal experience was lacking were made painfully clear.

Rather than attempt to survey the vast field of the natural history of both Indies, I have chosen three examples that illuminate the methods European naturalists—in particular, Clusius—used when evaluating their sources, and the problems they faced: the tract by Cristobal Acosta on the medicines of India; birds of paradise, a subject of much debate in the sixteenth century; and the history of the banyan tree. These examples highlight the interrelation of observation, interrogation, and corroborative testimony in Renaissance accounts of exotica. The section concludes by looking at how stay-at-home naturalists tried, and generally failed, to make ship's surgeons and physicians into reliable observers.

Acosta on the Medicines of India

Cristobal Acosta was neither the first nor the most famous writer on the simples of the East Indies. That honor belongs to Garcia da Orta, a Portuguese physician who settled in Goa in 1534 and quickly established a flourishing practice. Garcia's *Coloquios dos simples, e drogas he cousas mediçinais da India,* published in Goa in 1563, described in dialogue form several vegetable products of the East and their medicinal uses. The following year, Carolus Clusius obtained a copy in Portugal. On his return to the Low Countries in 1565, Clusius began an abridged translation of Garcia's text, abandoning the dialogue format and adding illustrations and notes. The work appeared from Plantin's Antwerp press in 1567.[144] Although Clusius had published a brief treatise on gums and resins and a Latin translation of a Florentine apothecary's manual, his annotated epitome of Garcia made his reputation. After obtaining a copy, Ulisse Aldrovandi wrote to Clusius to initiate a correspondence: from the epitome he had learned that "you delight greatly in the same studies of the natural philosophy of sensible things as I do," and hoped that Clusius would favor him with a response.[145] By the time his *Exoticorum libri* appeared in 1605, Clusius had prepared five editions of his translation of Garcia.[146]

Clusius was thus in an ideal position to judge another work on Indian natural history, Cristobal Acosta's *Tractado delas Drogas, y medicinas de las Indias Orientales* (1578). He obtained this book on a trip to England in 1581 and, as he had done with Garcia, prepared an annotated epitome that Plantin published in 1582.[147] In the dedicatory letter to Wilhelm, Landgraf of Hessen, Clusius set out his reasons and methods in preparing his version:

> When a friend gave me this book during my journey in Britain, not only did I read it avidly, but I thought it would be useful for those who do not read Spanish if I were to translate it into Latin. I was led to this by my nature, which has always delighted in the knowledge of such things.
>
> However, when I compared this work more carefully with what Garcia da Orta wrote on this matter many years earlier, I found that much had been taken from his work. For this reason, I thought it would be useless work, and tedious to the reader, if I were to repeat what Garcia said in his history of aromatics.
>
> Hence, by altering the order of the chapters, which was pretty confused, and reducing the entire work to an epitome, I have rendered in Latin only those things that are not in Garcia or that are described in a somewhat different fashion.
>
> The illustrations, which the author boasts are drawn from life, have been eliminated, since they were clearly inept and do not depict real plants; this can be seen from the effigy of the clove tree, which I left in that it might be compared with the true image in my epitome of Garcia.
>
> Moreover, I have added notes to some chapters, as I am accustomed to do in this type of work, especially when I have read of something similar or diverse in other authors, or have learned of it through my own experience.[148]

This letter traces the stages of Clusius's experience with the text: first excitement at new reports of the plants of India, then disappointment at the large amount of text copied from Garcia and the ludicrously inadequate illustrations. Nonetheless, Clusius thought that some of Acosta's claims were worth crediting.

Acosta's own self-presentation in his introduction certainly fit the late Renaissance ideal of the experienced naturalist. He was, he wrote, so impressed by Aristotle's dictum in the *Metaphysics* that all men desire knowledge that "I left my country and decided to seek out, in different regions and provinces, wise and curious men from whom I could learn something new every day." In order that he profit from the travel itself, "I took care to observe the variety of plants that God created for human salvation."[149]

In Goa, he met Garcia, who was very learned, but due to the low quality of printing in the city, there were some errors in his text. Moreover, Garcia had not had time to add pictures to his book.

Acosta decided that it would profit his countrymen "if they were led to knowledge of the good things in Garcia's book, and their eyes presented with examples and pictures. But no one can easily do that, except for someone who has seen them with his own eyes and has experienced them." He expected to be criticized by those who carp at what they don't understand, especially since he criticizes many Greek and Arab writers who set down hearsay as fact: "but I should not be judged harshly, if I try to set down certain and true things that I have seen with my own eyes."[150]

One can see why Clusius was excited after reading this preface. Acosta presented himself in the proper fashion of a sixteenth-century naturalist: someone who had traveled widely, conversed with many scholars, and carefully observed plants during his travels. Indeed, someone very much like Clusius himself—yet fortunate enough to have traveled not only within Europe but with the Portuguese fleet to India. He thus had firsthand knowledge of what Clusius knew only from reports and artifacts. However, as Clusius read further he was less and less enchanted: for Acosta's claims often contradicted both other reports and Clusius's own experience, as he put it, of the objects described.

The *Musa* or *Ficus indica*—the banana—for example, was better described by Oviedo, whose description Clusius had translated and included in his second edition of Garcia. Likewise, Garcia's description of the tamarind was more accurate than Acosta's, and De l'Obel's illustration was the most reliable. And neither Acosta's nor Garcia's description of the *Galanga* satisfied Clusius, at least not if the imported medicine called *Galanga maior* in Europe was the true plant. Its root, he argued, seemed more closely related to the iris than to asphodel or ginger, with which Garcia and Acosta compared it.[151]

Clusius's dissatisfaction with Acosta comes out most clearly, however, in the chapter on the clove tree. Acosta described the tree and its fruit briefly. The latter, in Acosta's words, "are born in the fashion of figs, scattered among the branches at the base of leaves, two, three, or four joined together, or occasionally one."[152] This passage, and the simple illustration that Acosta included (figure 5.7), immediately attracted Clusius's scorn. "I had not thought to translate the chapter on the clove tree," he sneered,

> for most of its description is taken from Garcia's history of aromatics (like most of the rest of this book). Contrary to the opinion of the other

Figure 5.7. Clove tree, woodcut from Acosta, *Aromatum liber* (1582).

> authors who have written on cloves, he asserts that they grow from the branches among the leaves. I thought it worthwhile to present this claim to the reader, in order to show how little faith this writer deserves at times: although he boasts that he tells the truth, and has had plants drawn from life, none of his illustrations imitate the living image of any plant, especially those that I have been able to see. In my version of Garcia's history of aromatics, I certainly gave a legitimate image of cloves, drawn by a diligent, skilled artist from a pickled branch of the tree (many of which are brought to Antwerp). Furthermore, this year I saw nine-inch and twelve-inch long dried branches of cloves, brought back from the Moluccas last September by Francis Drake, an English sailor who circumnavigated the world. All of these had the fruit on the ends, on their own little stalks, just as shown in my illustration.[153]

Clusius's note dwarfs Acosta's chapter, at least in Clusius's epitome. It reveals clearly his disappointment in the dissonance between Acosta's claims to firsthand observation and the accuracy of his actual claims.

Clusius's methods for uncovering this dissonance are broadly similar to those used a century earlier by Leoniceno in his criticism of Pliny. He compared authoritative sources and the objects themselves with the text in question and found it wanting. But there were important differences between the two. For Clusius, the object itself, and accurate illustrations of it, were of paramount importance. The accuracy of Garcia's text was vouchsafed by the correspondence between his text and the branches from clove trees brought to Antwerp by merchants and to London by Drake. Of course, the objects themselves could not verify the accuracy of an entire description: they were not entire trees, merely short branches. But they had a talismanic effect: the agreement between Garcia's description and the branches implied that his account was accurate in other regards. Conversely, Acosta's clear blunder rendered the rest of his text suspect.

Compared with Gessner's account of Northern animals, Clusius's commentary on Acosta placed more emphasis on material objects and less on travelers' claims. The Portuguese bases in the Indies and the proximity of Clusius to the entrepot at Antwerp certainly made it easier for him to assess the natural history of the Indies than it had been for Gessner to control his information. But Clusius's focus on the *naturalia* themselves reveal a subtle shift in emphasis from Gessner's humanist submersion in the world of texts. Texts were still necessary, for no one could see all; but they were increasingly subject to control and censure based on comparison with objects rather than other texts.

Birds of Paradise

If Clusius's writing on the clove tree demonstrated the power of comparing words and things in early modern natural history, the case of the bird of paradise exhibits its pitfalls. The story of these creatures—that they allegedly spend their entire lives in the air, never alighting on the ground, and hence having no legs—was well known in the Renaissance, and is often offered as evidence of the credulity of early modern naturalists.[154] Contemporary emblem books rehearsed, and moralized, the story of this bird, which was utterly unconcerned with earthly matters (figure 5.8). All this despite the fact that the first person to describe it, Antonio Pigafetta, who sailed with Magellan, stated clearly that it had legs!

But was it, in fact, the bird of paradise? Pigafetta's account is unclear. He wrote that the king of Bachian (or Badjan) sent several gifts to the Spanish court, including

> two extremely beautiful dead birds. Those birds are as large as thrushes, and have a small head and a long beak. Their legs are a palm in length and as thin as a reed, and they have no wings, but in their stead long

Figure 5.8. Bird of paradise, woodcut from Camerarius, *Symbolorum & emblematum centuria tertia* (1596).

> feathers of various colors, like large plumes. Their tail resembles that of the thrush. All the rest of the feathers except the wings are of a tawny color. They never fly except when there is wind. The people told us that those birds came from the terrestrial paradise, and they call them *bolon diuata,* that is to say, "birds of God."[155]

Pigafetta's text does not mention the name "bird of paradise," though that is close enough to birds of God, and he does not assert that they are legless. It would seem that the birds brought back by the crew of the *Victoria* should have settled the question.

However, another contemporary narrative muddles the matter. Maximilianus Transylvanus, a secretary at the Imperial court, interviewed the surviving crew of the *Victoria* and wrote a short tract "De Moluccis insulis," as a letter to the cardinal of Salzburg, describing the voyage of circumnavigation.[156] This tract gives the bird a different name and provides the first account of its wholly aerial life:

> A few years ago the kings [of] Marmin began to believe that the soul is immortal. They were induced to believe this solely from the following reason, that they observed that a certain very beautiful small bird never settled on the earth, or on anything that was on the earth; but that these birds sometimes fell dead from the sky to the earth. And when the

> Mohammedans, who visited them for trading purposes, declared that these birds came from Paradise, the place of abode of departed souls, these princes adopted the Mohammedan faith, which makes wonderful promises respecting this same paradise. They call this bird Mamuco Diata.

Transylvanus also mentioned the birds given by the king of Bachian to the Spaniards:

> The most remarkable curiosities were some of the birds, called Mamuco Diata, that is the Bird of God, with which they think themselves safe and invincible in battle. Five of these were sent, one of which I procured from the captain of the ship, and now send it to your lordship [the cardinal of Salzburg]—not that you will think it a defence against treachery and violence, but because you will be pleased with its rarity and beauty.[157]

The name is slightly different from that given by Pigafetta, but the circumstances similar enough to conclude that they are describing the same bird. Transylvanus omitted the description, but then he was sending the bird itself along with his letter. Aside from the name, the most important circumstantial detail on which the accounts vary regards the bird's habitat: Transylvanus's narrative mentions the fact that the creatures spend their whole lives in the air, but not that they lack legs.

This element entered the history of the animal not from a text but from the objects themselves. Beginning in the 1540s, the skins of birds of paradise were sent from the Aru Islands, via Amboina or Banda, to Lisbon, whence they reached the cabinets of curiosity of the rest of Europe. The hunters of the Aru Islands had developed a method for preserving the skins that involved cutting off the legs, and often the wings. Hence, those naturalists fortunate enough to possess such a skin, or to know someone who did, could verify themselves that the birds possessed no legs.[158]

This observation rendered plausible Transylvanus's claim that the birds spent their entire lives in the air. Without legs, they could scarcely do otherwise. Girolamo Cardano asserted that the birds lived off of heavenly dew, and that they mated and hatched their eggs in flight. Various forms of this hypothesis were adopted by other naturalists, such as Julius Caesar Scaliger and Ulisse Aldrovandi, though Aldrovandi insisted that dew was hardly nourishing and that the birds' beaks indicated they took solid nourishment, probably flying insects.[159]

Though initially unwilling to get involved in disputes over the bird of paradise, Carolus Clusius eventually took up the matter in an appendix to his *Exoticorum libri.* He remarked that Aldrovandi's treatment was most precise, due to his accurate and subtle disputations and his careful description of the five species he had observed. But Aldrovandi, and all others, were wrong when they said that the bird had no feet and impugned Pigafetta's honor. The error was introduced by the birds themselves brought back by Spanish and Portuguese sailors: "All the birds that these scholars saw had had their feet removed."[160] He himself had held the same, erroneous belief, for all of the specimens he had seen, both in Spain and in the Low Countries, had lacked feet: in fact, they were gutted skins. This despite Aristotle's assertion that all birds have feet! Clusius ruefully admitted that he should have taken the Stagirite more seriously in this point.

But that was a judgment made in hindsight. What convinced Clusius of his and his contemporaries' error was a new set of specimens brought back from the East by Dutch sailors. He had not seen them himself, but had heard reports from witnesses who had seen them in Amsterdam. "Nonetheless, I desired to see them, and had I been able to get my hands on even one, I would have had it drawn so that I could present it to the reader's eyes, and uphold Pigafetta's faith."[161] But Rudolf II's desire to possess them all outstripped Clusius's wish to make them known to all: by the time Clusius wrote to Amsterdam, the birds had been sent to Frankfurt, the first stage on their journey to Rudolf's Kunstkammer in the Hradschin, where they joined earlier, footless specimens.[162] Clusius's informants reported that the sailors who brought back the earlier, footless birds knew full well from their native suppliers that the birds had feet, and landed on the ground like other birds, but that their guts and feet were removed by native taxidermists.

This case raises the question of trust directly. Clusius had not trusted Aristotle, and thus was led astray. As it seems, he put too much trust in the skins he had seen (which, he later admitted, he had not observed too closely) and not enough in Pigafetta's eyewitness report. And then, he credited another set of eyewitnesses, ones who had seen the legs themselves. What made them more reliable than Pigafetta? Clusius did not name them, but presumably they were known to him, fellow professors at Leiden, or perhaps merchants or one of his noble friends. These witnesses, who could be cross-examined, were more credible than a half-educated merchant who had sailed some eighty years earlier—and who reported twelve-foot-tall giants in southern South America.

Erwin Stresemann's account of the history of the bird of paradise places a caesura between the "period of speculation," from 1522 to 1599, and the "period of 'objectivity without experience' [*unerfahrener Sachlichkeit*]." Yet in both periods, the sources of information about the bird and its habits were the same: native reports and dead birds' skins. Only in 1824 did a European naturalist succeed in seeing a live bird of paradise. Moreover, Clusius's report was not immediately accepted; the question of whether birds of paradise had feet was debated at length in the seventeenth century.[163] Stresemann's division thus seems determined more by presentist concerns—for the birds do, indeed, have feet—than by historical issues of method. Both sixteenth- and seventeenth-century accounts of the bird of paradise show the limits of Renaissance naturalists' notion of observation and experience: once more, in the absence of firsthand testimony from learned men, things were allowed to speak for themselves. This method worked in Europe, where the naturalist could see hundreds of plants of a given kind, or dozens of animals. For rare, singular, exotic imports, it posed grave problems.

The Ficus Indica: *Reliable Witnessing*

If European naturalists were distrustful of Pigafetta's claim that birds of paradise did, indeed, have feet, they were less so of other travelers' reports. A case in point is the *Ficus indica.* Clusius provided his contemporaries with a description and illustration of this plant at the beginning of his *Exoticorum libri.* He had never seen this tree, but he had heard about it from the "distinguished man, the well-traveled Fabrizio Mordente of Salerno (who lived for some years in Goa, and explored the areas around that city)."[164] Mordente visited Vienna during the last years of Maximilian's reign, where he met Clusius and described the tree.[165]

The woodcut illustration (figure 5.9) and detailed description make it clear to the modern reader that this "Indian fig" is the banyan tree. According to Fabrizio, the tree grows out of the ground and sends out branches from which fibers dangle toward the ground. When they reach it, they root and thicken, eventually sending out their own branches with fibers that repeat the process. They also grow upwards, so that "one tree makes a dense forest." The foliage is so thick that the sun cannot penetrate it, and within the tree voices sometimes echo. Fabrizio claimed that he had camped with a force of eight hundred men under one of these trees, and that it could have sheltered three thousand. He added descriptions of the leaves and the fruit, of which elephants were especially fond.[166]

The woodcut itself is a strange amalgam of Fabricius's observations, Clusius's botanical knowledge, and an unnamed illustrator's skill. "I had this tree sketched," Clusius reported "as elegantly as it could be, from

Figure 5.9. *Ficus indica* (Banyan tree), woodcut from Clusius, *Exoticorum libri* (1605).

Fabricius's narration, and he approved it."[167] This is one of the few instances in which Clusius published an illustration that was not drawn from life. Elsewhere in the *Exoticorum libri,* and occasionally in his other works, he included illustrations taken over from other books or from his correspondents, but these were themselves the result of firsthand observation.[168] Compared with European trees, the banyan was odd enough that even an inaccurate illustration would stress its peculiarities; perhaps this is why Clusius included it.

But why did Clusius believe Fabrizio? In the first instance, he was a well-traveled man who had lived on the spot and claimed to have seen the tree firsthand. It would have been impolite to challenge him. There was another reason, however. Clusius had independent testimony, provided by Theophrastus. Indeed, he saw himself as being in the same position as the ancient Greek. Just as Clusius gathered information from Fabrizio, so too did Theophrastus "carefully collect the narrations of those who were soldiers with Alexander in the expedition to India."[169] Aside from a few details, Fabrizio's account agreed with that of Theophrastus in his *Historia plantarum.*[170] The conjuncture of their two accounts, in turn, allowed Clusius to reject a third description: that provided by Jan Huygen van Linschoten in his immensely popular *Itinerario* of a voyage to the East. Linschoten not only described the "Arbor de rayz," the root-tree as the Portuguese called it, but provided an illustration. This illustration, quite different from Clusius's, depicted so many roots that he could scarcely believe any tree or bush could produce so many. "Hence I am highly suspicious of it, and I think that the one I had drawn from Fabrizio's description is truer, and closer to the ancients' description."[171]

The situation seems paradoxical: Clusius thought that an illustration based on a description from memory was more accurate than one purportedly drawn on the spot. His justification, that the latter was botanically improbable, is one that he advanced elsewhere: for instance, sailors' probably mistaken claim that they found a palm branch on a rock jutting out of the ocean.[172] But this is a slippery criterion: had not Clusius himself accepted the zoologically suspicious claim that the bird of paradise had no feet and lived perpetually in the air, until observation proved otherwise?

What tilted the balance was the independent testimony of reliable witnesses, especially Theophrastus. Mordente was not trained as a naturalist, but he was available for questioning, and he vouchsafed the drawing made from his description. The description and drawing, in turn, agreed with ancient sources describing what Clusius judged to be the same plant. Linschoten's inexpert testimony could not withstand this combined assault. As Clusius knew only too well, illustrations made by the uneducated or unguided artist could mislead. Only an experienced naturalist, or a witness who could be examined by one, could produce reliable knowledge.

Training Witnesses

One solution to the problem of finding reliable reporters was to train them. Clusius and his colleague at Leiden, Pieter Paaw, were quick to grasp the potential usefulness of Dutch foreign trade companies as a source of information and *naturalia* from East Asia. In the summer or fall of 1599, the two Leiden professors formed a plan to send a trained physician on one of the Dutch voyages to the Indies. They asked the curators of the university, with support from the States of Holland and West Friesland, to petition the Old East India Company in this matter.[173] On November 29, the States formally requested that the company collect Indian herbs, seeds, flowers, and the like, according to more specific instructions to be drawn up by Paaw.[174]

The physician who sailed from the Netherlands late in 1599, Dr. Nicolaas Coolmans, shared the fate of many later East India Company employees: he died on the journey out. But before perishing, he collected a number of plants and described at least one more. His effects were returned to Amsterdam in 1601, and the company's officers sent his collection of plants to Paaw in Leiden. Clusius obtained permission to describe them in his *Exoticorum libri,* where they occupy five folio pages.[175] Unfortunately for Clusius, Coolmans was not an organized collector. Most of the plants were well-preserved between sheets of paper, but they were not identified. Possibly they had become separated from descriptive notes, since at least one description, written in Dutch and pertaining to a plant no longer among the material, was found along with the plants. Clusius chose to include this description, of *Sedum Madagascaricum,* in his book: if Coolmans was disorganized, at least Clusius considered him reliable. But his death prevented further contributions to European natural history.

The number of physicians who were both willing to travel to the East and interested in natural history was necessarily small, especially given the high mortality of early voyages. In their absence, Clusius decided to resort to another method of gathering reliable information. Sometime early in 1602, he sent a memorandum to the Heren XVII of the new East India Company (Verenigde Oost-Indische Compagnie, VOC) with instructions for the "Apothecaries and Surgeons who are to sail with the fleet to the East Indies in 1602."[176] If university-educated collectors could not be sent, at least the surgeons and apothecaries—who had some interest in plants due to their trades—could be told what to look for.[177]

Since the document is so short, I will reproduce it in its entirety:

> They should bring back branches, placed between sheets of paper, with their leaves, fruit, and flowers when possible, from the following plants: both sorts of muscat nuts, male and female; black pepper; white pepper; long pepper betel; cubebs; mangoes; mangosteens; and similar beans from a kind of cotton that grows near Bantam, with branches and flowers; they should also ask how the locals name that plant.
>
> Also, branches from all other kinds of tree that are strange [*vreemd*] and that grow there, with flowers, leaves, and fruit; when possible, to make a drawing of the tree's form; whether it is large or small, whether it remains green in the winter or not. Their names in the local tongue, and to what uses they are put.
>
> Because one must know all these things, in order to make a good description [*om wel te connen beschrijven*].
>
> One finds, on such voyages,[178] also various plants, and small shrubs [*boomkens*] of diverse sorts and colors, which appear to be strange: they should also be brought back.
>
> Also diverse kinds of strange fish, when they aren't too big.
>
> In sum, those who are diligent should find enough to bring back.

> Many other trees and fruits that grow there should also be brought back, if their names and uses are known.

These instructions emphasized collecting and preserving two kinds of information: the objects themselves, as complete as possible and suitably pressed in paper, and their local names and uses. Clusius requested further information, such as drawings, only when the object itself could not be brought back. Although his memorandum implied that this information would suffice to "make a good description," such a description would omit much of the precise detail that characterized Clusius's own descriptions in the Hispanic and Austro-Hungarian floras.

Clusius's instructions reveal a deep mistrust of collectors who were not themselves *studiosi*. In this respect, they instantiate his judgments in the case of birds of paradise and the *ficus indica*. He knew how much work was involved in careful observation, and how prone to error even dedicated naturalists were. The simple mistakes, or deliberate lies, told by sailors about the bird of paradise did not predispose him to accept other accounts. When Clusius decided to interview someone who had been on the Dutch attempt to circumnavigate the globe, he singled out Sebald de Wirt, the captain of a ship that returned rather than complete the dangerous passage of the Strait of Magellan.[179] (Unfortunately, De Wirt was not at home, but he later answered Clusius's questions by letter.) Clusius probably thought that De Wirt, an officer, would be more reliable than a common member of the crew.

In any case, Clusius was disappointed with the results of his memorandum. It is uncertain whether it was even issued to the fleet. He complained in his *Exoticorum libri* that he had not received a single item from the VOC's voyages.[180] Even had he received the objects and information he wanted, it would not have been satisfying. Indeed, the entire *Exoticorum libri* show the weakness of attempting to write the natural history of non-European regions at the end of the sixteenth century. It could not truly be known, according to contemporary standards, until naturalists spent time exploring indigenous plants and animals *in situ*. Only a dense series of repeated observations could produce descriptions that met the best sixteenth-century standards.

❧

The history of exotic natural history in the sixteenth century, from the Arctic wastes to tropic heats, reveals that naturalists unable to observe exotica closely and repeatedly, in person, fell back on humanist methods of collation and comparison. They recognized these latter methods were distinctly inferior. Still, naturalists strove to make the best of these markedly inferior sources. They were willing to use any method that could help them catalogue nature as exhaustively as possible.

Nonetheless, personal observation remained the primary means of judging the accuracy of claims. Clusius began his *Exoticorum libri* by contrasting the work with his earlier histories:

> Some years ago I took the care to write the history of various plants that I had observed, as a youth and as an adult, in my travels, and to present this history for public judgment. Later I augmented it with further observations. Now, in my old age, when due to my bodily weakness I can scarcely walk, so as not to pass my life completely in leisure, I have applied my mind to the observation of those exotic plants and other things that are brought from foreign parts. Now I have taken on the task of writing the history of all the exotic things that I have acquired in recent years, and that through great efforts I have been able to obtain. I hope that this history, which I have written with great faith and the greatest diligence, will stimulate young men to take up this study in the same way that my earlier observations led them to the study of other plants.[181]

Clusius's repetition of the word *observationes* emphasizes his preference for personal visual inspection of objects. The ideal natural history would contain more—names, uses, place and time of growth, and the other items on his questionnaire for the East India Company doctors and surgeons. But the essential element of his catalogue of nature was the precise description of an object. Clusius knew that his *Exoticorum libri* were imperfect, but he also knew that natural history was a cumulative process. A glance at his own career sufficed to show him that; had he not discovered and described hundreds of previously unknown plants? The imperfections of his history of exotica should spur coming generations, young men who were still capable of enduring hardship for the sake of natural history, to do the work better.

In the early seventeenth century, when Clusius finished his life's work, naturalists were still relatively fact-poor. They knew of several thousand species of plants—almost all of them vascular plants—and several hundred species of vertebrate animals. They had just begun to explore the vast world of insects and other invertebrates. And their knowledge of nature beyond the bounds of western and central Europe was meager. In such circumstances, imperfect descriptions were better than none. Over the course of the seventeenth century, the circumstances would change. The Dutch and English commercial empires would bring vast quantities of *naturalia* to Europe, much of it for the growing market in curiosities, and some trading officials and colonists devoted themselves to the study of nature. Natural history became fact-rich; by the end of the century travelers' tales and strange objects from afar were commonplace. In these circumstances, naturalists no longer needed to describe any strange object that caught their eye, no matter what its provenance. Instead, they could travel or send their students or servants to observe and collect material in its natural habitat. Surfeited with new, firsthand information gathered by trained observers, European naturalists would turn a skeptical eye to what had fascinated their Renaissance predecessors.

Epilogue: A New Sensibility?

When Francis Bacon sat down to survey the whole of learning at the turn of the seventeenth century, he saw little amiss with natural history as it was practiced. Certainly, its scope needed to be expanded:

> History of Nature is of three sorts; of nature in course, of nature erring or varying, and of nature altered or wrought; that is, history of Creatures, history of Marvels, and history of Arts. The first of these no doubt is extant, and that in good perfection; the two latter are handled so weakly and unprofitably, as I am moved to note them as deficient.[182]

But the history of creatures, by which Bacon no doubt meant the natural histories of Aldrovandi, Gessner, Dodoens, Clusius, and others, was more or less perfected. Little remained to do with it.

Years later, in the *New Organon,* Bacon rethought his position. As with much else, he reflected the contemporary positions and presaged later developments, though—again, as with much else—his programmatic ideas expressed an extreme that was not attained by anyone else in practice. The third part of his Great Instauration was to be a natural history, a collection of facts, out of which philosophical axioms could be drawn. Bacon saw this work as surpassing the powers of any one individual, as necessarily a collective enterprise;[183] his own *Sylva sylvarum,* published posthumously, was only a step in the right direction. And given the magnitude of the work, Bacon wanted it to be free of certain superfluities.

In striking at the first of these, Bacon aimed at the heart of Renaissance natural history, which relied on careful identification and quotation of sources as a guarantee of its authority.

> First then, away with antiquities, and citations or testimonies of authors, and also with disputes and controversies and differing opinions—everything, in short, which is philological. Never cite an author except in a matter of doubtful credit; never introduce a controversy unless in a matter of great moment.[184]

In characterizing all of these as "philological," Bacon effaced a distinction that was very clearly made by Renaissance authorities. Gessner distinguished in his own work philology from natural history proper. By philology he meant the historical, emblematic, symbolic, and other literary use of animals; he apologized for including more of this material than was perhaps proper in a work on natural history. (Aldrovandi had no such qualms.) But for Gessner, a careful account of names, citation, and comparison of sources were not philology but natural history proper. No one naturalist could describe everything. Bauhin's *Phytopinax* (1596) and *Pinax* (1623) both began with a list of abbreviated references that looks familiar to any modern reader. Yet no one would accuse Bauhin of filling his works with irrelevant philology; his citations were essential.

Not so to Bacon, who thought it antithetical to the succinctness and order he desired in his natural history. In matters doubtful it was sufficient to add a qualification such as "it is reported," "I have heard from a person of credit," and so forth. More detail was a laborious interruption. "Nor is it of much consequence to the business at hand because (as I have said in the 118th aphorism of the first book) mistakes in experimenting, unless they abound everywhere, will be presently detected and corrected by the truth of axioms." Only when the matter is a singularity of utmost importance need the author and circumstances of a doubtful fact be mentioned.[185] Since natural history is not an end in itself, merely a propaedeutic to a natural philosophy, the occasional error does not cause any great difficulties.

Bacon's practice in the *Sylva sylvarum* allows one to doubt the efficacy of this criterion—he included, inter alia, the sympathetic action of gemstones on the human spirit—and, in order to eliminate as many false beliefs as possible, he did call for "a calendar of popular errors: I mean chiefly, in natural history such as pass in speech and conceit, and are nevertheless apparently detected and convicted of untruth; that man's knowledge not be weakened nor imbased by such dross and vanity." It is not enough for individual skeptics to pass over falsehoods that they encounter: errors "must be in express words proscribed, that the sciences may be no more troubled with them."[186]

Sir Thomas Browne conceived his *Pseudodoxia epidemica: or, Enquiries into very many received tenents, and commonly presumed truths* (first edition 1646), as this housecleaning of received but false beliefs. Browne located the disposition to error in original sin and the confusion of common people, ultimately due to Satan's machinations. But the most immediate causes of error were credulity and adherence to authority.[187] Authority, he wrote, has no power in mathematics and only little in philosophy; and though we must admit its use in history and related disciplines, it must be treated with care. Just as legal testimony must be corroborated, so must the testimony of authorities.[188] "Lastly," he wrote, "the strange relations made by Authors, may sufficiently discourage our adherence unto Authority, and which if we belieeve we must be apt to swallow any thing." Such remarks, supported only by the author's words, "being neither consonant unto reason, nor correspondent unto experiment," are to be treated as "things unsaid."[189] To Browne, once said is not said at all.

Unlike Gessner and Aldrovandi, Browne thought that Olaus Magnus made enough suspicious claims to rate careful analysis of the rest. Some of the archbishop's remarks were simply not worth consideration—such as his ludicrous assertion that Biarmia, not more than 70 degrees of latitude, verged on the North Pole, or that a magnetic island could be found there. Browne included Olaus among a list of writers, ancient and modern, whose works contained a great deal of fabulous material and should be read with great caution. Browne's treatment of pygmies provides a clear example of his method at work. Many authors affirm their existence, he wrote, "yet that there is, or ever was such a race or Nation, upon exact and confirmed testimonies, our strictest enquiry receaves no satisfaction."[190]

Note Browne's emphasis on *exact* and *confirmed* testimony. Reports of pygmies are not exact, Browne argued, because they gave different descriptions of their size, location, and customs. Furthermore, the testimony is not confirmed: "for though *Paulus Iovius* delivers there are Pygmies beyond Japan, Pigafeta, about the Molucca's, and Olaus Magnus placeth them in Greenland; yet wanting frequent confirmation in a matter so confirmable, their affirmation carrieth but slow perswasion."[191] Hence, Browne concludes, we are unsure whether pygmies exist. In keeping with his method, Browne treated Olaus as more reliable in two matters where other authors, and ocular testimony, confirm his reports: the horns of certain large fishes, and the white coats and feathers of northern animals and birds. This is not because Browne rejected all tales of wonder—he believed in the existence of unicorns.[192] Rather, he demanded that these tales be corroborated by independent sources.

Browne's notions of proper method in natural history do not represent a radical break with the past; at most, they demand a shift of emphasis. Conrad Gessner, in the middle of the sixteenth century, and Carolus Clusius, at its end, had carefully compared the claims of different writers, and placed them side-by-side with artifacts, in order to corroborate or reject them. Furthermore, they tried to determine when sources were independent and when one writer merely repeated what he had read in another book. They were perhaps more willing than Browne to accept singular relations from people they knew and trusted, but they emphasized above all the reliability of observations—their own observations of *naturalia* that they had before them—and used such observations, for good or ill, as the ultimate judge of other observers' claims.

The difference found in some seventeenth-century writers is, perhaps, an insistence on the particulars of an observation as a way of guaranteeing its authority to the skeptical reader. In Bacon and Browne we find explicit the attitude that reports such as Vorst's description of the walrus—with a clear statement of where a phenomenon was observed, when and by whom—were by their nature more reliable than other reports that did not give such particulars, even when their authors could be presumed knowledgeable about the events they described. This attitude was certainly not unique to them. Their contemporary Benedetto Ceruto, in his description of Francesco Calceolari's museum in Verona, applied the same criteria to reports of the rhinoceros and the unicorn. That these creatures differ from one another is supported not only by Pliny, Aelian, Gessner, and others—"who knew of them by hearsay but were not eyewitnesses [*oculati testes*]"—but, more significantly, by Garcia da Orta and his translator and annotator Carolus Clusius, "who, although they had not seen those animals, could nevertheless acquire more certain testimony about them."[193] Garcia's sources, according to Ceruto, demonstrate clearly that not only are these two creatures distinct but also that their horns are efficacious against toxins; he reports an experiment to that effect. Browne might have questioned Ceruto's judgment of his sources, but he would have found nothing wrong with his method.

This critical attitude was common in the later seventeenth century. In a letter written in Copenhagen, Thomas Henshaw told Henry Oldenburg that he had been reading Ole Worm's book on the Norwegian Mouse, which the locals called lemmings. Henshaw then inquired "of Monsr Guldenlew ye viceroy, and Monsr Cruys Tolmaster, who both affirmed to me by a voluntary Oath," that the animals plague the country in certain years. "I had met with some accounts of them long agoe in Olaus Magnus and Heylin, but gave little Credit to it, but here it is a thing so well testified that Nobody doubts of it."[194] Here the testimony of credible witnesses vouchsafed a report from Olaus, no longer a credible witness himself. This is not simply because Olaus reported wonders. The next letter in Oldenburg's correspondence, from Oldenburg to the zoologist Francis Willughby, asked Willughby and John Ray to investigate reports—ludicrous to us—of a maggot as big as a man's thigh.[195] Rather, Olaus reported them with neither the circumstantial detail required to make his account believable nor the skeptical distance required of a dubious fact. After all, Oldenburg asked Willughby and Ray to see whether the reports, of uncertain provenance, were in fact true.

John Ray's zoological epitome, the *Synopsis methodica quadrupedum* (1693), demonstrates the effects of Bacon's and Browne's programmatic remarks on late seventeenth-century natural history. In many respects, the epitome rejected the method of Gessner, Aldrovandi, and Johnston. Faced with those writers' tomes, Ray asked his readers, why write anything more? Had they not said it all? His first response was that the mass of Gessner's and Aldrovandi's volumes terrifies the reader. Second, their books are hard to come by, and expensive. Third, many new species have been discovered. But Ray's chief reason for writing was probably his last: "that I might dispose all animals that are known with certainty in a more accurate order [*methodo accuratiore*], one more suited to their natures than the order currently in use."[196]

Natural order was important to Ray, who had contributed to John Wilkins's search for a universal, natural language.[197] But equally important was his emphasis on excluding not only the fabulous, but also the uncertain, from his account of nature. "I have excluded not only fabulous and feigned animals, but also doubtful, uncertain, and suspect ones." His readers would find in his book no Hippocentaurs or Chimeras, but also no Indian asses, no Suhak, no American Taxus, no Gulo, nor

> others of that sort, which are found in Nieremberg. I will not resolutely deny that there are such animals in nature. But I have not been able to determine whether they do in fact exist, or whether—if they do exist—they are indeed distinct from those better known animals that I have included in this synopsis. I have given no place to any form that I have not seen myself or that does not have eyewitness [*autoptas*] and trustworthy authorities.[198]

Hence at least in theory, Ray's text relied only on descriptions provided by either him, such as his observations on the dissection of a porpoise, or by his immediate sources. Any further sources were to be ignored, as tales rather than reliable observations.

How did the walrus fare? Ray accepted its existence, though he was uncertain in what class to place it. His description, in which it is placed along with the seal under "quadrupeda vivipara unguiculata, multifida, carnivora maiora capite longiore, seu Caninum genus," followed Jean de Laet's and Aelius Everard Vorst's account, adding only that the creature is called *equus marinus* or, falsely, *hippopotamus*, Morse (from the Russian name) by the English, Walrus by the Dutch, and Rosmarus by the Scandinavians, and that the English call its tusks "sea-horses teeth."[199] There is no mention of Johnston's recent compilation, nor of Gessner nor Aldrovandi, to whom Ray refers his readers on other animals. At least in this instance he stuck to his stated purpose of giving only his own descriptions or those from eyewitnesses. The learned structure of Renaissance erudition, of textual collection and comparison, had vanished. Ray chose the best authority and stuck with it—a practice that would be followed by later writers, whether in Linnaeus's terse descriptions or Buffon's flowing prose.[200]

The same is true of Ray's accounts of other northern animals. By the 1690s, the reindeer and the elk were well known, and Ray could give brief descriptions with references to the best accounts—Gessner and Ole Worm, the seventeenth-century Danish collector. The "Norwegian mouse (*mus norwagicus*)" was more problematic. "Olaus Magnus was the first to mention this animal," according to Ray; "Conrad Gessner, Julius Scaliger, Jacob Ziegler, and Johnston took everything they said about it from the same source, that is, Olaus's *Historia*." But Ray did not bother to quote or even summarize Olaus's chapter. Rather, he noted, "after investigating its nature more accurately, Ole Worm wrote a detailed history of it," which Ray reproduced at length.[201]

By Ray's time, naturalists could afford to be deeply skeptical of hearsay. Renaissance naturalists—like Renaissance antiquarians and historians—were skilled at untangling the conflicting reports of their sources and reaching at least a provisional conclusion. And, as we have seen, they preferred eyewitness reports of things they had not themselves observed. But in the absence of conflicting accounts, they were content to accept, summarize, and hence validate the reports they could get.[202] Gessner expressed some skepticism about Olaus's pictures, and about creatures such as the *marticora*—there he had Aristotle backing him—but his work did not systematically sift out wonders and exclude them from its pages. As Sebastian Münster had remarked, such marvels were expected at the edges of the world.[203] Ray, on the other hand, excluded such *dubia*. His natural history had no room for Catoblepas and Centaurs—nor for wolverines and armadillos.

When Ray did include an animal, he gave only the most accurate description he had available: either his own observations or reliable firsthand accounts. He omitted Olaus Magnus's description of the lemming, and its analysis by Gessner, Scaliger, Ziegler, and Johnston, in favor of Ole Worm's detailed account. In his appendix to Francis Willughby's *History of Fishes,* Ray announced that he was preparing a comprehensive account of fishes. His preparations included visiting fish markets in order to inspect catches and compare published species with what he could find—especially as he suspected his predecessors of unnecessarily multiplying species based on unusual specimens. Given the amount that had been written, especially by Rondelet and Gessner, Ray doubted that he would add many new species. Rather, he claimed that he would give more accurate descriptions and organize his material according to a better method.[204]

Accuracy, for Ray, required firsthand observation and detailed description. He saw no need to accumulate citations, and he excluded hearsay and secondhand accounts. His history of the whale quotes from the descriptions of Polydore Vergil and Guillaume Rondelet, but only those parts they guaranteed that they themselves had seen.[205] The same is true of his accounts of the reindeer, the elk, and the walrus. The amount of detail Ray provided in these descriptions varied widely according to his sources and his judgment as to whether an animal was well known or rare.[206] But in all cases, he refused to include anything that "eyewitness and trustworthy" writers had not corroborated. He broke the chain of citation and commentary that had characterized Gessner's and Aldrovandi's natural histories of animals.

In the end, then, Ray's practice in the *Synopsis methodica quadrupedum* and other works represents not a radical reconfiguration of method in natural history, but rather its limitation to the best sixteenth-century methods. Certainly much in Ray was not there a century earlier in Clusius: above all his concern for taxonomic method. But the need for taxonomy had been produced by a century of discovery and description. And Ray's practices of firsthand observation and long experience, and the techniques that allowed for the gathering and preservation of such experience, were those of his Renaissance predecessors. They had been elaborated and passed down by a community that has existed continuously from the sixteenth century to today, even as its contours have shifted and its concerns have changed. In this sense, all modern natural history is a child of the Renaissance science of describing.

CHAPTER SIX

Conclusion: What Was "Renaissance Natural History"?

The plant that Linnaeus christened *Drosera rotundifolia* has a peculiar habit. Its small, fleshy leaves glisten with tiny drops of liquid, caught at the end of little hairs (figure 6.1). Attracted by the liquid, a small insect occasionally lands to rest or feed. But the liquid is not water or nectar; it is sticky, and the insect finds itself trapped. Its struggle serves only to cover it in more of the viscid liquid. And as it struggles, the leaf on which it has landed slowly curls to surround it. The movement is slow, but the insect is trapped and the plant has time to spare. The liquid is not only sticky but acidic, and with its prey in hand the plant secretes another fluid containing digestive enzymes. Trapped, soon enclosed by the plant, the insect dies and is gradually digested. The leaf uncurls, and the plant prepares itself for another meal.[1]

I find the sundew, the Venus fly-trap, the pitcher plant, and other carnivorous plants both fascinating and mildly unsettling. We vertebrates tend to be upset when our place at the pinnacle of the evolutionary hierarchy and the food chain is called into question. When invertebrates eat vertebrates, something seems unnatural: it's just not right that giant desert centipedes will attack, kill, and devour lizards and small rodents.[2] And it is even more unnatural for a plant to eat an animal, even if it is only an insect. The thriving contemporary community of carnivorous plant aficionados takes an interest in these insectivores from a complex amalgam of delight at their beauty, wonder at their unusual ability, and disgust at their inversion of the natural order.[3]

Renaissance naturalists should have been even more unnerved by carnivorous plants, for they upset not only the hierarchy of nature but also its laws. Aristotle divided the potentialities of living beings into several types: nutritive, sensitive, motive, and rational.[4] Plants had only the first;

Figure 6.1. Sundew (*Drosera* sp.) digesting an insect. Photo by the author.

sensation, to say nothing of motion, was reserved for animals. The *Mimosa pudica*, or "sensitive plant," native to the tropics, got its name precisely because it acted as plants should not; it even inspired a long poem by Shelley. But the sensitive plant was unknown in the Renaissance, while Renaissance naturalists simply did not notice that some plants caught and ate insects. It makes no sense to blame them for not noticing something that they would have considered unnatural; they did not go so far as Linnaeus, who in the 1760s openly rejected the idea that the Venus fly-trap nourished itself with the insects it caught.[5] Only in the 1870s did Charles Darwin definitively demonstrate that many plant species nourished themselves on insects, a demonstration that required painstaking experimentation.[6]

Naturalists in the sixteenth century had, of course, noticed that plants grew, and some had even remarked on the apparent motion of a few species. In an emblem book drawn from natural history, Joachim Camerarius the Younger noted that the "Peruvian Chrysanthemum," or sunflower, turned its head to follow the rays of the sun. But Camerarius treated this behavior as an effect of the sun, not a natural motion of the plant.[7] Plants could grow—an aspect of the soul's nutritive faculty, according to Aristotle—and they could move in response to external forces. But they could not initiate motion themselves, especially not motion directed to a specific end. Had Renaissance naturalists noticed how the sundew caught and then enveloped insects, they might have been led to reconsider the Aristotelian psychology that made plants almost entirely passive creatures.

Figure 6.2. Sundew, woodcut from Conrad Gessner, *De raris herbis* (1556).

They did notice the sundew (figure 6.2). Though it was a small plant, Renaissance naturalists considered it a marvel. Valerius Cordus called it the rorella or salsirora and noted of its leaves that "even in the driest weather they have little drops like dew."[8] A decade later Dodoens stated what Cordus implied: "This herb has a very strange and marvelous nature, for no matter how long and strongly the sun shines on it, you will find it always damp and watered, and the hairs on it always have little drops of water. And the stronger the sun shines on this plant, the more it is damp and watered."[9] From Dodoens's entry we know that already in the middle of the sixteenth century the plant was called "sundew": "rosée de soleil" in French and "Sundauw" in Flemish. Conrad Gessner attests that the English used the same name, though "some learned men" called it "Lunaria."[10] A humble, small plant, grouped with the mosses by Dodoens, the sundew was nevertheless a marvel because its dew was especially resistant to the sun's evaporating rays.

But of course it was not dew that glittered on the sundew's leaves. John Ray noted that the "dew" was "adhesive, tenacious and viscid, and it can be drawn into filaments if touched with a finger."[11] The earliest naturalist who made this observation and bothered to record it was Caspar Bauhin, who wrote:

> It is a worth marveling that that damp substance contained in the vessels [*sic: acetabulis*] of the leaves, if lightly touched under sunlight while the plant is still in the ground or has recently been dug out, can be drawn out into long silky or bright filaments that immediately harden and remain so indefinitely.[12]

If earlier naturalists had noted this property of the "dew," they had not deemed it worth publishing in their descriptions. They saw the liquid, noted it, and described it. They must have seen the insects too—but they did not observe them closely. Dew that survived the blazing sun was worth their attention; insects on a plant were not.

The sundew reveals, in exaggerated form to be sure, the limits of observation in Renaissance natural history. Naturalists must have seen the bugs on the sundew, but they did not observe closely enough for the bugs to register as interesting. They were interested in form, in morphology—in describing a plant as accurately as possible and differentiating it from other plants. Animal behavior was worth noting; plants had no behavior to note. Even the heliotropism of the sunflower was not considered anomalous enough to call into question Aristotle's division between plants' and animals' souls. Becoming a naturalist in the Renaissance meant training oneself to observe certain aspects of nature very carefully and to report those aspects precisely. Naturalists might notice other aspects, but those aspects were less salient. As morphology and its variations came to consume naturalists, other aspects of the natural world faded into the background.

There is a deep irony in this development. Renaissance natural history developed in scholarly settings, in the attempt to restore the medical, agricultural, and philosophical knowledge of the ancient world. Yet the humanists' approach to reconciling ancient descriptions and modern observations led quickly to the formation of a new disciplinary community whose motives were far more complicated than their pragmatic origins. The Renaissance science of describing was *sui generis*. It contributed to resolving problems in medicine, agriculture, and—for some thinkers like Girolamo Cardano, Andrea Cesalpino, and Francis Bacon—natural philosophy, but it was not a part of any of those disciplines.

Renaissance naturalists were unabashedly concerned with particulars and their description. Depending on the context, these particulars could be individual plants and animals, or they could be species—a term that was not precisely defined but taken over from commonplace usage.[13] Classification, these species' place in a larger scheme of nature, was of little import for most naturalists. If there was a prime motivation for the pursuit of natural history in the sixteenth century, it was aesthetic. Description was a way to grasp the multiplex beauty and diversity of the natural world. In an age that placed an aesthetic value on curiosity and rarity as well as beauty, the communication of uncommon or unusual specimens, or their descriptions, between widely separated locales was itself a means of satisfying aesthetic urges. The process of finding new plants in the wild, or cultivating them in the garden, served the same end.

The collecting impulse, the desire to possess as many and as varied things as possible, contributed to the pursuit of natural history, in particular to its pursuit in gardens, herbaria, and cabinets. This impulse itself drove natural history along a path toward less immediate experience of nature *in situ*. There is something that strikes the field naturalist as bizarre and sad at the thought of Caspar Bauhin sitting in his study, turning pages of his herbarium, and looking at specimens that his students had gathered on the sunny slopes of the Pyrenees or in Prussian meadows. Yet this practice, too, had its aesthetic element: in this case the more formal delight at perceiving, and putting in order, the similarities and differences between hundreds and thousands of individual species. The search for a formally elaborated system in nature, one that would go beyond tacit perceptions of relationship, has its origins in this pursuit.

Moreover, these two moments of natural history, the careful vicarious descriptions of Clusius and the herbarium-based work of Bauhin, were products of the same impulse to describe and love the particular. Renaissance natural history was indelibly marked by its origins in the debate over classical *materia medica* and the humanist epistemology of the particular. Leoniceno and Euricius Cordus compared individual plants to the descriptions found in the ancients; Hieronymus Bock and Valerius Cordus turned instead to careful description of the individual species they observed, while modeling their descriptions on those of Dioscorides. The botanical garden and the herbarium were products of this concern for the particular, at the same time reinforcing it. If the vicarious descriptions of Clusius, or the alpine effusions of Gessner, were a product of the sixteenth-century science of describing, they were possible only for a brief moment, a generation or two, when phytographers believed it was still possible to catalogue the world while lingering on each of its individuals. By Bauhin's time, that hope had passed; brevity, not vicariousness, had become the hallmark of a good description, and the rise of the herbarium, the collection of stiff, dry plants, was accompanied by the stiff, dry language of Bauhin's *Prodromos*, a fitting precursor to the dessicated prose of Linnaeus's *Systema naturae*.

But the lonely impulse of delight that so clearly shines out of Clusius's descriptions did not vanish from natural history. It simply found another expression. Systematic natural history treatises, from the seventeenth century to the present, do not satisfy the urge to know what it is like to experience nature firsthand, but other books do. From Izaak Walton to Gerald Durrell, in passing by Buffon, Gilbert White, Charles Darwin, John Muir, and countless others, naturalists have written engaging, lively descriptions of their experiences of nature. Direct descendants of Euricius Cordus's *Botanologicon*, these books have inspired amateur naturalists not only to explore nature but also to master, at least to a certain degree, the dry, technical systematic natural history that is essential for the amateur to completely enjoy nature. As Anne Secord has shown, the artisans who created and animated amateur botanical societies in early nineteenth-century England demanded of their peers a high level of mastery of the Linnaean system—not because they hoped to become professionals but because their pleasure in the field depended on being able to identify the plants they observed and collected.[14] For those artisans, as for me in my attempts at amateur botanizing in the New England hills, being able to put a name to a plant, and thus to locate it precisely within the global vegetable kingdom, brought an intellectual satisfaction to the immediate pleasure of the outdoors that enriched and completed their experience—just as it did for the elder and younger Cordus, Gessner, Clusius, and their contemporaries. Nonetheless, in their scholarly publications naturalists took a scalpel to nature, cutting plants and, to a large degree, even animals out of their contexts. Naturalists were aware of some of the associations of plants and animals that comprised the ecology of Europe, but a science of ecology would have to wait until the nineteenth century.

The aesthetic impulses that drove early modern naturalists led, in a certain sense, to a less anthropocentric study of nature. *Materia medica*, with its focus on the practical employment of plants and animals to restore and preserve human health, was unabashedly anthropocentric. Unlike the natural history of the Enlightenment, Renaissance naturalists had little interest in the economic aspects of their pursuit, aside from the occasional gain to be made by selling rare plants. Pierre Belon's call for an economic natural history, for employing naturalists' knowledge to acclimatize tropical plants and make the kingdom of France more productive, had to wait for Linnaeus to find a true successor. In this sense, it was also less anthropocentric than what was to follow.

On the other hand, the reasons for pursuing natural history were clearly anthropocentric. As we have seen, sixteenth-century classifications placed the naturalist's aesthetic sense at the center, with such groups as "coronariae," "flowers that please," "plants that smell good," and the like. The high-level division of plants into trees, shrubs, sub-shrubs, and herbs, which was definitively challenged only in the seventeenth century, reveals a deeper anthropocentrism, one grounded not in human wants and desires but in the human cognitive apparatus.[15] Nonetheless, the close attention to morphology promoted by studies in gardens and herbaria, along with the exchange of plant material across long distances, fostered comparisons between species that eventually led to the rejection of both aesthetic criteria and common sense in classification and their replacement by a more 'natural' system. The aesthetic motivations of Renaissance naturalists may well have been an important precondition for that later development.

Be that as it may, the descriptive natural history of the sixteenth century was up close and personal. Renaissance naturalists were in a bad position to judge in matters where they had no personal experience. They preferred, at least, to have corroborating evidence for others' claims, in the form of plants they could grow themselves or *naturalia* they could display in their cabinets. The strange history of the bird of paradise reveals the pitfalls inherent in that approach. Naturalists in the later seventeenth century solved this problem not by developing new techniques but by avoiding it: by going to places to experience nature firsthand; by sending trained students or collectors to bring it back; or by suspending their judgment. Renaissance naturalists, despite their belief in the inexhaustibility of nature, wanted to know it all. Their successors, surfeited with facts, were faced with a different problem: how to make sense out of what they already knew. Theirs was the age, not of description, but of systems.

NOTES

CONVENTIONS

1. Caspar Bauhin, *Pinax* (1623).

2. Rembert Dodoens, *Histoire des plantes* (1557); Frederick G. Meyer, Emily Emmart Trueblood, and John L. Heller, *Great Herbal of Fuchs* (1999); Brigitte Hoppe, *Kräuterbuch des Hieronymus Bock* (1969).

CHAPTER ONE

1. The following discussion is condensed from Brian W. Ogilvie, "Natural history" (2005).

2. See the brief biobibliography in Charles B. Schmitt et al., eds., *Cambridge history of Renaissance philosophy* (1988), 838, and the review essay by Riccardo Fubini, "Umanesimo ed enciclopedismo" (1983).

3. Giorgio Valla, *De expetendis* (1501), title page.

4. Valla, *De expetendis*, books 20–23 (physiologia), 21.1–101 (soul); 24.24–105, 25.1–107, 26.24–26, 29.60–103 (medicine); books 42–44 (economy). Domestic plants and animals are discussed, in alphabetical order, 42.25–131; book 26 takes up common barnyard animals from poultry to cattle.

5. Valla, *De expetendis*, 1.1; Gianpietro's dedication to Giovanni Jacopo Trivulzio, sig. vii^V.

6. On Petrarch and later humanists, see chapter 3.

7. Polydore Vergil, *On discovery* (2002), 1.21, pp. 162–69, 2.8.1, pp. 246–47, 2.19.7, pp. 316–17, 3.2–4, pp. 354–81. All quotations from Vergil are from Brian Copenhaver's translation.

8. Vergil, *On discovery*, 1.21.6–7, pp. 166–69, 2.12, pp. 263–69, 2.13.12, pp. 278–79, 3.5, pp. 380–92, 3.6.7, pp. 396–97, 3.13.11, pp. 454–55. I have omitted some similar references; see the index in *On discovery* for a fuller account.

9. Vergil, *On discovery*, 1.12, pp. 106–13, 1.4.1, pp. 54–55.

10. Juan Luis Vives, "De tradendis disciplinis" (1785), 262, 264–65.

11. Juan Luis Vives, "De causis corruptarum artium" (1785), 190.

12. Vives, "De causis corruptarum artium," 186.

13. Francis Bacon, *Advancement of learning* (1863), 184.

14. Francis Bacon, *De dignitate et augmentis* (1864), 2: 189–93, 196–97.

15. My notions about culture are drawn chiefly from Clifford Geertz, *Interpretation of culture* (1973); Pierre Bourdieu, *Outline* (1977); Anthony Giddens, *Central problems* (1979); William H. Sewell, Jr., "Theory of structure" (1992); and, for the importance of practices and implicit meaning, Peter L. Berger and Thomas Luckmann, *Social construction* (1966).

16. On the role of socialization in the sciences, in particular with regard to tacit knowledge of practices, see Michael Polanyi, *Personal knowledge* (1962).

17. Cf. the remarks of W. Scott Blanchard, "*O miseri philologi*" (1990), 101, on fifteenth-century philology. Blanchard recognizes that "amateur" and "professional" are problematic categories but he offers no alternative.

18. Blanchard, "*O miseri philologi*," analyzes analogous problems in Renaissance philology.

19. In thinking about communities and their concerns, I have been helped greatly by David L. Hull, *Science as a process* (1988).

20. This fact is given a negative connotation by historians of biology for whom, like Bacon, observation is only a preliminary to theory: for example, Frederick Simon Bodenheimer, "Towards the history of zoology and botany" (1960). Needless to say, I differ from the implication.

21. Bacon, *De dignitate et augmentis*, 197–98; Francis Bacon, *Novum organum* (1861), 308 (book 1, aphorism 98).

22. Svetlana Alpers, *Art of describing* (1983).

23. For instance: Antonello Gerbi, *Nature in the New World* (1985); Wolfgang Harms, "Natural history and emblematics" (1985); F. David Hoeniger, "Plants and animals" (1985); Oliver Impey and Arthur MacGregor, eds., *Origins of museums* (1985); Krzysztof Pomian, *Collectionneurs* (1987); Antoine Schnapper, *Le géant* (1988); Christa Riedl-Dorn, *Wissenschaft und Fabelwesen* (1989); William B. Ashworth, Jr., "Emblematic world view" (1990); Maria Suutala, *Tier und Mensch* (1990); Allen J. Grieco, "Social politics" (1991); Ilse Jahn, "Descrizione nella storia naturale" (1991); Karen M. Reeds, *Botany* (1991); Giuseppe Olmi, *Inventario del mondo* (1992); *Botany in the Low Countries* (1993); Paula Findlen, *Possessing nature* (1994); Laurent Pinon, *Livres de zoologie* (1995); N. Jardine, J. A. Secord, and E. C. Spary, eds., *Cultures of natural history* (1996); Sachiko Kusukawa, "Fuchs on pictures" (1997); Claudia Swan, "Lectura-imago-ostensio" (2000); and Pamela H. Smith and Paula Findlen, eds., *Merchants and marvels* (2001). See the bibliography for further titles.

24. E.g., Ilse Jahn, *Grundzüge der Biologiegeschichte* (1990); Ernst Mayr, *Growth of biological thought* (1982); and Scott Atran, *Cognitive foundations* (1990), which is nonetheless crucial for understanding Renaissance natural history (see chapter 5, below).

25. Leonhart Fuchs, *De historia stirpium* (1542), sig. α2v–6r, and Thomas Johnson's prefatory letter "To the Reader" in John Gerard, *Herball* (1633). The details of the story vary but its outline is the same.

26. Gerard, *Herball*. See chapter 3, below, for a fuller discussion of such emblematic title pages.

27. I follow the King James version, which would have been familiar to Johnson and his contemporaries.

28. Polydore Vergil, following in the tradition of Euhemerus, explained that these and other goddesses were mortals whose discoveries had been so beneficial that grateful men deified them: Vergil, *On discovery*, 6–7. On Euhemerus, see Jean Seznec, *Survival of the pagan gods* (1972).

29. Andreas Vesalius, "Preface to *De fabrica*" (1932).

30. See Wallace K. Ferguson, *Renaissance in historical thought* (1948), 59–67.

31. On Latin scholarship, see Ernst Robert Curtius, *European literature* (1953); L. D. Reynolds and N. G. Wilson, *Scribes and scholars* (1991); and Peter Godman, *Silent masters* (2000). Any good survey of the history of art will discuss both the continuity of artistic practice in the Latin Middle Ages and the stylistic discontinuities, as well as frequent returns to classical models. The subject is treated more systematically in the brilliant study by Erwin Panofsky, *Renaissance and renascences* (1972).

32. Nancy G. Siraisi, *Medieval and early Renaissance medicine* (1990), 86–97; Bernard Schultz, *Art and anatomy* (1985).

33. Vivian Nutton, "Prisci dissectionum professores" (1988).

34. Katharine Park, "Criminal and saintly body" (1994).

35. Reynolds and Wilson, *Scribes and scholars;* L. D. Reynolds, ed., *Texts and transmission* (1983); R. H. Rouse, "The transmission of the texts" (1992).

36. Cf. Peter Dear, *Discipline & experience* (1995), 116–19, who discusses the creation of a legitimating tradition by late sixteenth- and early seventeenth-century experimental natural philosophers.

37. Marshall G. S. Hodgson, *Venture of Islam* (1974), 1: 80. I encountered Hodgson's definition as this book was almost complete, but it expresses so clearly and succinctly a distinction that I had been struggling to articulate that I have incorporated it here.

38. Hodgson drew on anthropology, but his concern with the high culture of the Islamic world led him to formulate a definition better suited to literate cultural traditions than to the broader sense of "culture" in anthropology. His primary focus on religion led him to emphasize above all the centrality of the initial creative act for participants in the tradition: that is, Muhammad's recitation of the Qur'an for followers of the religion of Islam. My concern with natural history as a cultural tradition has led me to a different sense of the initial creative act: it is not, in the case of natural history, the production of a text and a way of life oriented toward it but, rather, a problem and a method for solving that problem. As we will see, the problem and method of Renaissance natural history did not remain static, but they evolved continuously within a community that was committed to applying and refining them.

39. For example, Mayr, *Growth of biological thought.*

40. An early critique of the presentist, "judgmental" history of science, with particular attention to Julius Sachs's influential *Geschichte der Botanik*, was offered in 1936 in Wilhelm Jan Lütjeharms, *Geschichte der Mykologie* (1936), chapter 2; Lütjeharms draws on Dilthey, Troeltsch, and especially on Hélène Metzger, "L'historien des sciences" (1933).

41. Humanism, a nineteenth-century reification, has been the subject of a vast scholarly literature: as a starting-place for investigation, see Albert Rabil, Jr., ed., *Renaissance humanism* (1988).

42. Discussed at greater length in chapter 3, below; see Karen M. Reeds, "Humanism and botany" (1976), and Charles G. Nauert, Jr, "Humanists, scientists, and Pliny" (1979).

43. On education, see Paul F. Grendler, *Schooling* (1989); Paul F. Grendler, ed., "Education in the Renaissance" (1990); George Huppert, *Public schools* (1984); T. W. Baldwin, *Small Latine and less Greeke* (1944); Anthony Grafton and Lisa Jardine, *Humanism* (1986); and, most recently, Robert Black, *Humanism and education* (2001).

44. Grafton and Jardine, *Humanism*.

45. On humanist alphabets, the classic study is B. L. Ullman, *Humanistic script* (1960).

46. The history of science, in particular, suffered in past decades from a tendency to label sixteenth-century science as humanistic, and thus based on book-learning and ancient authority rather than empirical observation. (See, for example, Marie Boas, *Scientific Renaissance* [1962].) More recent investigation has dispelled this myth, but many historians still seem to assume that book-learning is opposed to empirical observation. Only in the last decade has this assumption been explicitly challenged (e.g., Steven Shapin, *Social history of truth* [1994]), and several writers have begun to address how books and observation were interrelated (e.g., Anthony Grafton, April Shelford, and Nancy Siraisi, *New worlds* [1992]).

47. See (inter alia) Allen G. Debus, *Chemical philosophy* (1977) (for Paracelsus); Donald R. Kelley, *Foundations* (1970) (for Alciati); Lorraine Daston and Katharine Park, *Wonders* (1998) (for Palissy and Paré). Recent studies of early modern empiricism include Mary Poovey, *Modern fact* (1998), Barbara J. Shapiro, *A culture of fact* (2000), and the essays in Anthony Grafton and Nancy Siraisi, eds., *Natural particulars* (1999). Further references are found in chapter 4, below.

48. Daston and Park, *Wonders*; Dear, *Discipline & experience*.

49. Daniel 4.2–3; I Kings 4.33 (authorized version). Solomon was often described and depicted in Renaissance natural histories as an illustrious forebear; see chapter 3, below.

50. A delightful introduction is provided by Joy Kenseth, ed., *Age of the marvelous* (1991), the catalogue of an exhibit at the Hood Museum of Art at Dartmouth College in Hanover, NH.

51. Jay Tribby, "Body/building" (1992). See however the cautionary remarks of Schnapper, *Le géant*, 14, regarding princely collections. The use of collections for a more direct control over nature, in an occult mode, is illustrated by Thomas DaCosta Kaufmann, *Mastery of nature* (1993), chapter 7.

52. Paula Findlen, "Economy of scientific exchange" (1991), and Findlen, *Possessing nature*, are prime examples.

53. Bourdieu, *Outline*, esp. 171–83.

54. See chapter 4 on exchange.

55. Ernst Heinrich Krelage, *Bloembollenexport* (1946), provides a systematic overview of Dutch commercial flower gardening.

56. See chapters 2 and 4 below.

57. The same was true much later: see Anne Secord, "Artisan botany" (1996), and, more generally, D. E. Allen, *Naturalist in Britain* (1994).

58. These remarks are expanded on in chapter 2, below.

59. See Peter Dear, *Mersenne* (1988), 1.

60. Arthur O. Lovejoy, *Great chain of being* (1964 [1936]); E. M. W. Tillyard, *Elizabethan world picture* (1943).

61. For a concise critique of Foucault's approach, see Nelly Bruyère, *Méthode et dialectique* (1984), 406–10; a fuller response is given by Ian Maclean, "Foucault's Renaissance episteme" (1998). A more historically specific and generally convincing analysis of late Renaissance intellectual life in terms of a sense of universal harmony and the interrelationships of macrocosm and microcosm, restricted to the court of Rudolf II and its environs, is offered by R. J. W. Evans, *Rudolf II* (1973), chapters 6–8 (especially chapter 7).

62. Cf. Ashworth, "Emblematic world view," and more recently, William B. Ashworth, Jr., "Emblematic natural history" (1996).

63. E.g., Andrea Cesalpino, Adriaan van de Spiegel.

64. Gessner included a large amount of what he called *philologia* in most entries; this section, which he designated with the letter H, he considered to be superfluous to natural history properly speaking: Conrad Gessner, *Historiae animalium lib. I* (1551), sig. α6r (publisher's advertisement), β1v, γ2v–3r. Ashworth has repeatedly insisted that Gessner did not recognize this distinction, a claim that I find puzzling in light of Gessner's clear words to that effect: "Philologiam autem appello, quicquid ad grammaticam, & linguas diversas, proverbia, similia, apologos, poetarum dicta, deniq; ad verba magis quam res ipsas pertinet" (sig. β1v). It is worth noting that Gessner's publisher omitted the *philologia* from the German epitome of the *Historia animalium*.

65. For an extended consideration, see Margery Corbett, "Emblematic title page" (1981).

66. Agnes Arber, *Herbals* (1986), 262–63.

67. Important work includes Barbara Shapiro, "History and natural history" (1979); Mary M. Slaughter, *Universal languages* (1982); Joseph M. Levine, "Natural history" (1983); Olmi, *Inventario del mondo*; Ashworth, "Emblematic world view." For a comprehensive bibliography, though heavily weighted toward later periods, see Gavin D. R. Bridson, *History of natural history* (1994).

68. *Metaphysics*, 980b24–981a30; *Posterior Analytics*, 100a4.

69. David Stewart and Algis Mickunas, *Exploring phenomenology* (1990), 63–89.

70. The echoes are largely unconscious, and progressive historians of experience would probably be shocked and dismayed to realize their similarity with the crusty, conservative Hegelian. Despite Arthur Danto's cogent argument against the possibility of thinking the thoughts of past actors, insofar as the past really is different from the present, I am wholly in sympathy with the attempt. See Arthur C. Danto, *Narration and knowledge* (1985), 285–97.

71. Peter Dear, "Miracles" (1990); Peter Dear, "Narratives, anecdotes, and experiments" (1991); Dear, *Discipline & experience*. The notion of "fact," similarly, has a history: see Lorraine Daston's "Marvelous facts" (1991) and "Baconian facts" (1991). Different seventeenth-century notions of experiment and fact are analyzed by Daniel Garber, "Experiment" (1995). A brief, not entirely objective summary is provided by Steven Shapin, *Scientific revolution* (1996), 80–96.

72. Aristotle, *Physics* 196b10, *Historia animalium* 571a25, and especially *De generatione animalium* 770b10.

73. Dear, "Narratives, anecdotes, and experiments," 146–48.

74. Dear, "Narratives, anecdotes, and experiments," 161–63. Cf. the discussion of "virtual witnessing" in Steven Shapin and Simon Schaffer, *Leviathan* (1985), 60–65.

75. Bacon, *Novum organum,* 308 (book 1, aphorism 98).

76. My thanks to Lorraine Daston for bringing this to my attention.

77. Dear, "Narratives, anecdotes, and experiments," 150.

78. Olaus Magnus, *Historia de gentibus septentrionalibus* (1555), 59: "aliorum assertione"; "propria inspectione, vel experientia."

79. Niccolò Leoniceno, *De Plinii erroribus* (1529), 15 ("sensu ipso"), 27–28, 48, 61, 95.

80. Clusius to Camerarius, Sept. 7, 1580, in Friedrich Wilhelm Tobias Hunger, *Clusius* (1927–43), 2: 379.

81. Dominicus Baudius, in Carolus Clusius, *Exoticorum libri* (1605), sig. †4r: "multum experientis Ulyssei."

82. See e.g., Ernst H. F. Meyer, *Geschichte der Botanik* (1854–57); Arber, *Herbals;* even Michel Foucault, *Les mots et les choses* (1966), revolutionary in many regards, focuses almost exclusively on texts. An important exception is Charles E. Raven, *English naturalists* (1947); Raven includes naturalists such as Thomas Penny who published little or nothing.

83. My thinking in this regard has been particularly influenced by Giddens, *Central problems,* and Bourdieu, *Outline;* in science studies, Thomas S. Kuhn, *Structure* (1970) emphasizes the importance of paradigmatic practices (problems to solve, experiments) for scientists' understanding of their fields.

84. Hull, *Science as a process,* makes a similar argument, though in the context of an evolutionary model for intellectual history.

85. The question of truth is equated with that of trust in the provoking, if ultimately unconvincing, book by Shapin, *Social history of truth.*

86. In the Baconian context, see e.g., Sir Thomas Browne's skeptical analysis of "popular errors"; a critical edition with commentary is provided by Thomas Browne, *Pseudodoxia epidemica* (1981). Chapter 5, below, treats Browne at more length.

87. Aristotle's criticisms of Ctesias and Herodotus are paradigmatic: on the former, see J. M. Bigwood, "Elephant" (1993); on the latter, see Aristotle, *Generation of animals* (1942), 2.2, 736a10. Mary B. Campbell, *Witness* (1988), provides a marvelous account of travelers' tales and their reception during the Middle Ages and Renaissance.

88. For the headaches that minerals gave to Renaissance naturalists, see Martin J. S. Rudwick, *Meaning of fossils* (1985); Alix Cooper, "Museum and the book" (1995); and David Freedberg, *Eye of the lynx* (2002).

89. F. S. Bodenheimer, *Geschichte der Entomologie* (1928–29).

90. Readers interested in Renaissance zoology more broadly will soon be able to consult Laurent Pinon's *these d'état* in published form.

91. Giuseppe Olmi, *Ulisse Aldrovandi* (1978); Margherita Azzi Visentini, "Il giardino di semplici di Padova" (1980); Olmi, *Inventario del mondo;* Findlen, *Possessing nature;* Paula Findlen, "Formation of a community" (1999); and Freedberg, *Eye of the lynx.*

92. Henry Lowood, "New World" (1995).

93. José Pardo Tomás, *Ciencia y censura* (1991); José Pardo Tomás and María Luz López Terrada, *Las primeras noticias* (1993); José M. López Piñero, "Pomar Codex" (1992); José Maria López Piñero, "Las 'nuevas medicinas' americanas" (1990); and Antonio Barrera, "Local herbs" (2001).

94. E. C. Spary, *Utopia's garden* (2000); Antonio Barrera, *Experiencing nature* (forthcoming).

CHAPTER TWO

1. Clusius's biography is given succinctly but romantically by Johannes Theunisz, *Carolus Clusius* (1939), and more soberly but unimaginatively by Friedrich Wilhelm Tobias Hunger, *Clusius* (1927–43). No adequate biography exists in any language but Dutch, but French readers can consult, *faute de mieux*, E. Roze, *Charles de l'Ecluse* (1899).

2. On Mattioli's place in sixteenth-century Italian natural history, see Paula Findlen, "Formation of a community" (1999).

3. Caspar Bauhin, *Animadversiones* (1601).

4. Caspar Bauhin, *Phytopinax* (1596), sig. α3v–4v. Bauhin never completed this project. He contributed to it with the *Prodromos* (1620) and the *Pinax* (1623), an updated and expanded version of the *Phytopinax*, but he died in 1624 without finishing the *Theatrum botanicum*. One volume, edited by his son, was finally published in 1658.

5. Busti to Bauhin, Nov. 20, 1609 (Basel UB, MS. Fr.Gr. II.1, 105–106).

6. Busti mentions Bauhin's letter in a further letter to Bauhin, Sept. 22, 1610 (Basel UB, MS. Fr.Gr. II.1, 106–107). Bauhin's relations with his contemporaries and students will be discussed below.

7. Busti to Bauhin, Nov. 20, 1609.

8. On Pona and Mount Baldo, see Paula Findlen, *Possessing nature* (1994), 180. Pona's description was first published in Latin in Carolus Clusius, *Rariorum plantarum historia* (1601); Clusius instructed his publisher to send twenty offprints of the description to Pona for him to distribute to his friends (Clusius to Giovanni Vincenzo Pinelli, April 4, 1601, Leiden UB, MS. FACS UB A 139 [facsimile of the original in the Biblioteca Ambrosiana, Milan, Cod. S 107]). Bauhin mentioned Pona's description and his correspondence with Pona in two letters to Leonhard Dold of Nuremberg from 1603 and 1606 (Erlangen UB, Trew-Briefsammlung, Bauhin, Caspar, 28 and 38).

9. Charles G. Nauert, Jr, "Humanists, scientists, and Pliny" (1979).

10. Carlo Maccagni, "Le raccolte e i musei" (1981), 284–85.

11. F. David Hoeniger, "Plants and animals" (1985), 131–32.

12. The generational approach to Renaissance intellectual history has been defended with verve by Margaret L. King, *Venetian humanism* (1986), 261–63, 269–72, especially 263 n. 9.

13. For Leoniceno's biography, see Daniela Mugnai Carrara, "Profilo di Nicolò Leoniceno" (1978).

14. See chapter 3, p. 121, below, for a discussion of the polemic over Pliny in terms of its methodological presuppositions and results.

15. Niccolò Leoniceno, *De Plinii erroribus* (1529), fol. 1v–3r, p. 1–4; Daniela Mugnai Carrara, *Biblioteca di Leoniceno* (1991), 25–31; Peter Dilg, "Botanische Kommentarliteratur" (1975), 235–40.

16. See Nauert, "Humanists, scientists, and Pliny," and my discussion in chapter 3, p. 121ff.

17. See the inventory in Mugnai Carrara, *Biblioteca di Leoniceno*.

18. Mugnai Carrara, *Biblioteca di Leoniceno*, 44. The books mentioned in the rest of this paragraph can be found using the author index in Mugnai Carrara's catalogue.

19. The manuscript of Theophrastus gives the title as *peri phyton*; this could refer to either the *History of plants* or the *Causes of plants*, or more likely to both.

20. On the history of this text, often attributed (as in Leoniceno's manuscript) to the late antique bishop Epiphanios, see the introduction to *Physiologus* (1979).

21. Mugnai Carrara, *Biblioteca di Leoniceno*, 147, suggests that the book by "Joannis de Busta" is probably the *Difficiliorum herbarum explanatio* by Jacopo Manlio de Bosco, an excerpt from De Bosco's *Luminare maius*.

22. Agnes Arber, *Herbals* (1986), 13–37; Harold J. Abrahams and Marion B. Savin, "The *Herbarius latinus*" (1975).

23. The importance of up-to-date editions of ancient medical texts in the late fifteenth and early sixteenth century is emphasized by Vivian Nutton, "Prisci dissectionum professores" (1988). That some ancients were novel to Renaissance readers is a central theme of Anthony Grafton, April Shelford, and Nancy Siraisi, *New worlds* (1992).

24. See the discussion in chapter 3, below.

25. Conrad Gessner, *Bibliotheca universalis* (1545), fol. 492r, complained that Virgilio's commentary was useless for learning about plants but admitted that his philology was sound.

26. Gessner, *Bibliotheca universalis*, fol. 317r–18r, 451v–52r, and 92r, notes reprints of Barbaro's Plinian castigations in 1534; his corollaries to Dioscorides with Virgilio's translation in 1529; and Ruel's 1516 Dioscorides translation in a revised version in 1539.

27. Karen M. Reeds, *Botany* (1991), 24–26, 73–75; Findlen, *Possessing nature*, 251, 261–72.

28. Karen M. Reeds, "Humanism and botany" (1976), 534–35.

29. On Helmstedt, see Reeds, "Humanism and botany," 535; on Leiden, see Pieter Smit, "Clusius and the beginning of botany" (1973), and P. C. Molhuysen, ed., *Bronnen* (1913–24), 1: 42*, 192*.

30. Maccagni, "Le raccolte e i musei."

31. On the relationships between Leoniceno and the German naturalists of the early sixteenth century, see Dilg, "Botanische Kommentarliteratur."

32. Richard J. Durling, "Gesner's *Liber amicorum*" (1965).

33. Conrad Gessner, *Historiae animalium lib. I* (1551), sig. γ1vr–v; Conrad Gessner, *Historiae animalium liber III* (1555), sig. a6r; Conrad Gessner, *Historiae animalium liber IIII* (1558), sig. b5r.

34. Conrad Gessner, *Horti Germaniae*, (1561), fol. 237v–39v.

35. On Ghini and herbaria, see A. Louis, *Mathieu de l'Obel* (1980), 292–93. Louis argues that Ghini *invented* the technique. Given that drying flowers in the pages of

books is an age-old practice, one practiced among others by medieval pilgrims to the Holy Land (see chapter 3), it is more likely that Ghini systematized the practice and taught it to his many pupils at Pisa.

36. Euricius Cordus, *Botanologicon* (1534), 20; Peter Dilg, *Botanologicon des Euricius Cordus* (1969), 75–81; Gessner, *Horti Germaniae*, fol. 239r.

37. Carolus Clusius, *Exoticorum libri* (1605), 68, 87.

38. See the frequent references in Rembert Dodoens, *Histoire des plantes* (1557) to plants growing "es jardins d'apothequaires."

39. Gessner, *Horti Germaniae*, fol. 239r. On Coudenbergh's garden, see chapter 4, below.

40. Cordus, *Botanologicon*, 65–66, though one of the participants in the dialogue, a former student of Cordus's, Johannes Ralla, was himself an apothecary in Leipzig. Paracelsus, "Herbarius" (1930), 492, though Paracelsus's own knowledge of plants was meager.

41. Findlen, *Possessing nature*, 145, 157, 251. Findlen's conclusion that apothecaries were marginal in later sixteenth-century Italian natural history, based on entries in the visitor's register in Ulisse Aldrovandi's museum, seems to me untenable. Aldrovandi's register is filled largely with the sorts of learned people who traveled widely in the early modern period: clerics and academics. Apothecaries, usually artisans, did not have the same reasons to travel, though many of them were famous collectors, such as Ferrante Imperato and Giovanni Pona.

42. And in many other directions, including collecting more generally. See my remarks in chapter 1, p. 13, about natural history and collecting.

43. Julius Caesar Scaliger, *In libros de plantis* (1556); *In libros De causis plantarum* (1566).

44. Fuchs to Camerarius, undated (Erlangen UB, Trew-Sammlung; Fuchs, Leonhard, 1).

45. Much of sixteenth-century scholarship was pursued by "amateurs," in the sense that their professional activities were not directly related to the subject of their investigation. For example, Antonio Agustín, archbishop of Tarragona, made important contributions to philology and antiquarianism: M. H. Crawford, ed., *Antonio Agustín* (1993).

46. Gessner, *Historiae animalium lib. I*, sig. γ1vr–v; *Historiae animalium liber III*, sig. a6r; *Historiae animalium liber IIII*, sig. b5r.

47. Sebastian Münster, *Cosmographia* (1550) (and many later editions in several languages); Olaus Magnus, *Historia de gentibus septentrionalibus* (1555) (many other editions and epitomes in several languages). See chapter 5, below, for a discussion of these works and their reception.

48. The history of Renaissance herbals is set out admirably in the classic text by Arber, *Herbals*; a more popular, anecdotal, and impressionistic account can be found in Frank J. Anderson, *Illustrated history of herbals* (1977). See also Laurent Pinon, *Livres de zoologie* (1995).

49. On medieval herbals and plant lists, see Jerry Stannard, "Medieval herbals" (1974) and chapter 3, below.

50. Otto Brunfels, *Herbarum vivae eicones* (1532).

51. Leonhart Fuchs, *De historia stirpium* (1542), reprinted as vol. 2 of Frederick G. Meyer, Emily Emmart Trueblood, and John L. Heller, *Great Herbal of Fuchs* (1999).

52. A fact recognized by Gessner, *Bibliotheca universalis*, fol. 329r.

53. Hieronymus Bock, *Kreütterbuch* (1577) (originally published 1539); *De stirpium commentaria* (1552); the 1577 edition includes a "Teutsche Speisekammer" with an account of the medicinal properties of food and the four elements.

54. Rembert Dodoens, *Cruydeboeck* (1554); *Histoire des plantes*.

55. John Gerard, *The herball* (1597); *Herball* (1633).

56. Such indexes are found, for example, in Dodoens, *Histoire des plantes*, and the 1577 edition of Bock's *Kreütterbuch*.

57. Dioscorides, in Mattioli's commentary or another edition, was prescribed by the Leiden University curators in the late sixteenth century (the university was founded in 1575) as the textbook in *materia medica:* Molhuysen, ed., *Bronnen*, 1: 113. Dioscorides was also taught, along with other texts such as Valerius Cordus's *Historia plantarum*, in Montpellier: Reeds, *Botany*, 57, 76–79.

58. Gessner, *Bibliotheca universalis*, fol. 317r, 492r, 520v.

59. Brunfels, *Herbarum vivae eicones*, vol. 2; see also Gessner, *Bibliotheca universalis*, fol. 532r.

60. Conrad Gessner, *De raris herbis* (1556).

61. Reeds, *Botany*, 110.

62. See Hunger, *Clusius*, 1: 186. In the end, Cluyt became prefect under the supervision of Clusius; the degree to which Clusius supervised Cluyt is, however, open to question. I am not as convinced as Hunger was that Clusius commanded and Cluyt obeyed. See Claudia Swan, "Lectura-imago-ostensio" (2000), 144–45, and Claudia Swan, "Jacques de Gheyn II" (1997), 179ff.

63. The notion that natural history was still "subject" to medicine and became "emancipated" only slowly has been remarkably persistent in the secondary literature—most recently, D. E. Allen, "Walking the swards" (2000), 335–36.

64. Gessner, *Horti Germaniae*, 236v.

65. Adriaan van de Spiegel, *Isagoge* (1606), 128.

66. Caspar Bauhin to Leonhard Dold, Jan. 14, 1609 (Erlangen UB, Trew-Sammlung. Bauhinus, Caspar, 44).

67. On Clusius's relations with Coudenbergh and Plantin in the late 1550s, see *Antidotarium* (1561), 4–5.

68. Hunger, *Clusius*, 1: 101ff., on Brancion and Clusius.

69. See Leo J. Vandewiele, "Garden of Coudenberghe" (1993).

70. The early modern "culture of curiosity" has been the subject of many fine scholarly studies; the most important for my understanding are Krzysztof Pomian, *Collectionneurs* (1987); Oliver Impey and Arthur MacGregor, eds., *Origins of museums* (1985); Antoine Schnapper, *Le géant* (1988); Findlen, *Possessing nature*; Justin Stagl, *History of curiosity* (1995); Lorraine Daston and Katharine Park, *Wonders* (1998); and the essays in the *Journal of the History of Collections*. Joy Kenseth, ed., *Age of the marvelous* (1991), the catalogue of an exhibition at the Hood Museum of Art in Hanover, N.H., offers a beautifully illustrated introduction.

71. Daston and Park, *Wonders*, 77–86, with illustrations.

72. Gessner, *Historiae animalium liber IIII*, 250. Two more famous papal animals, the elephant and rhinoceros sent to Leo by Manoel I of Portugal, are discussed in the charming book by Silvio A. Bedini, *The pope's elephant* (1998).

73. Gessner, *Historiae animalium lib. I*, 950.

74. Findlen, *Possessing nature*, 37ff., discusses the museum catalogues in which Italian collectors explained their collections to the public.

75. Raphael Pelecius to Caspar Bauhin, Aug. 20, 1594 (Basel UB, MS. Fr.Gr. II.1, 42, IV).

76. Rotraud Bauer and Herbert Haupt, "Die Kunstkammer Rudolfs II." (1976), 19 (italics in original).

77. François Rabelais, *Gargantua and Pantagruel* (1990), 158; Baldesar Castiglione, *The book of the courtier* (1959), book 1.

78. Jay Tribby, "Body/building" (1992), 147.

79. Hunger, *Clusius*, 1: 125–27; 1: 138, on Clusius's dismissal from court, after a period in which his status was ambiguous. Hunger attributes the dismissal to Rudolf's strict Catholicism, but cf. R. J. W. Evans, *Rudolf II* (1973), chapter 3, which dispels the myth that Rudolf was a Catholic bigot and pawn of the Jesuits. More likely, Clusius was dismissed because of his contretemps with the Hofmarschalk, Adam von Dietrichstein, as Clusius himself thought: Gyula Istvánffi, *A Clusius-codex mykologiaia méltatása* (1900), 181, quoting letters from Clusius to his friends; Hunger, *Clusius*, 1: 139–41.

80. Clusius, *Exoticorum libri*, 97.

81. Clusius, *Exoticorum libri*, 359. These specimens were important because most naturalists, including Clusius, had thought that the birds were footless and spent their entire lives in the air. See the discussion in chapter 5.

82. Bauer and Haupt, "Die Kunstkammer Rudolfs II."

83. Thomas DaCosta Kaufmann, *Mastery of nature* (1993), chapters 4 and 7.

84. Walther Rytz, "Herbarium Felix Platters" (1932–33), 37–38, citing an inventory made at the end of the sixteenth century.

85. In the years 1602–12 Platter received 179 pounds "Mein cabinet und garten zezeigen," including a chain from the Landgraf of Hessen: Felix Platter, "Tagebuch" (1878), 341.

86. Rytz, "Herbarium Felix Platters," 38. Platter pointed out to visitors, for example de Thou, the material he had from Gessner.

87. Findlen, *Possessing nature*, 17–31, 122–24, 136–46, provides a detailed account of Aldrovandi's museum, its contents, and its visitors.

88. Raphael Pelecius to Caspar Bauhin, Aug. 20, 1594 (Basel UB, MS. Fr.Gr. II.1, 42, IV).

89. Inventoried by Hendrik Engel, *List of Dutch cabinets* (1986); see Pomian, *Collectionneurs*.

90. Mary Alexandra (Alix) Cooper, "Inventing the indigenous" (1998).

91. Clusius to Curatores of Leiden University, Aug. 27, 1592 (o.s.) (Leiden UB, MS. BPL 885, s.v. "Clusius").

92. Clusius to J. van Hoghelande, Oct. 15/25, 1592 and Jan. 9, 1592 (o.s.) (Leiden UB, MS. BPL 885, s.v. "Clusius").

93. Gessner, *Horti Germaniae*, 237v–38r.

94. Schnapper, *Le géant*, 36, 38.

95. Spiegel, *Isagoge*, 78.

96. Clusius to Camerarius, Dec. 26, 1584 (n.s.) (Hunger, *Clusius*, 2: 403); Clusius, *Exoticorum libri*, 87–88.

97. Rauwolf to Clusius, Sept. 7, 1584 (Leiden UB, MS. Vul. 101, s.v. "Rauwolff"). Rauwolf's herbarium traveled almost as much as Rauwolf himself: he sold it to the Herzog von Bayern for 200 gulden, who put it in the Munich Kunstkammer. During the Thirty Years' War it was taken by Swedish soldiers as booty to Stockholm; in the middle of the seventeenth century Queen Christina presented it to Isaac Vossius, who took it to Holland and then to London. It reached its current home in 1688, when the University of Leiden bought Vossius's library from his heirs: Karl H. Dannenfeldt, *Leonhard Rauwolf* (1968), 229–30.

98. Some small slips in the Basel Universitätsbibliothek testify to this: Basel UB, MS. Fr.Gr. I, 12, nos. 15, 18, 19.

99. Rytz, "Herbarium Felix Platters," 22–28.

100. As with Stephen Pighius's use of Cardinal Granvelle's collection: e.g., Stephanus Vinandus Pighius, *Annales Romanorum* (1615), 1: 306–07.

101. Jacob Spon, *Voyage d'Italie* (1724), 1: 185–187, quoted by Schnapper, *Le géant*, 267.

102. Caspar Bauhin, *Pinax* (1623), sig. *4v.

103. The 1610 engraving of the Leiden garden shows several people, only a few of whom are actually looking at the plants. More generally, see Erik de Jong, *Natuur en kunst* (1993), chapter 6, esp. 190–95.

104. See the engravings of collections in Ferrante Imperato, *Historia naturale* (1599), Ole Worm, *Museum Wormianum* (1655); and Benedetto Ceruto and Andrea Chiocco, *Musaeum Calceolari* (1622). The collection of the Franckesche Stiftungen in Halle an der Saale (Germany), assembled early in the eighteenth century, still has a small reference collection now used by the museum's curators to identify objects in the collection.

105. Gessner, *Historiae animalium lib. I*, sig. γ1v–3r, justified this arrangement.

106. His collected works, the *Rariorum plantarum historia* of 1601 and the *Exoticorum libri decem* of 1605, are exceptions, though even they do not match Gessner in bulk.

107. Carolus Clusius, *Stirpium per Pannoniam historia* (1583), sig. *3r.

108. Clusius studied medicine at Montpellier, but probably because it was the only way to receive formal training in natural history. He was encouraged to go to Montpellier by Melanchthon after showing his interest in plants while a student at Wittenberg; he seems never to have practiced medicine. Hunger, *Clusius*, 1: 15–16, 21–34, 325; 2: 214–15. Clusius's correspondent François Vertunien Le Vau apologized for addressing him as a physician: Vertunien to Clusius, April 7, 1599 (Leiden UB, MS. VUL 101, s.v. Vertunien, 1); on a letter from Lorenz Scholz Clusius noted, apparently as a memorandum for his reply, "me non esse Medicum, nec Medicinam unquam fecisse" (Scholz to Clusius, January 27, 1593, Leiden UB, MS. VUL 101, s.v. Scholzius, 2). Scott Atran, *Cognitive foundations* (1990), 161, claims that John Ray was "the first eminent naturalist who was not a physician," but Clusius is arguably Ray's predecessor by a century in this regard.

109. Rembert Dodoens, *Florum et coronariarum historia* (1568).

110. Mathieu de L'Obel and Pierre Pena, *Stirpium adversaria nova* (1570); De L'Obel, *Plantarum seu stirpium historia* (1576).

111. On Plantin's business see Leon Voet, *Golden Compasses* (1969–72).

112. On Plantin and natural history publishing, see Louis, *Mathieu de l'Obel*, 167–76, 486, 490.

113. Writing to the antiquarian Jacopo Strada, Plantin emphasized that he was unwilling to take on books that had no market. Plantin to Strada Oct. 1578, in Christophe Plantin, *Correspondance* (1883–1918), 6: 32.

114. Arber, *Herbals*, 228–33.

115. The title page does not indicate a publisher, and the colophon notes that the book was produced by "the press of Thomas Purfoot": Louis, *Mathieu de l'Obel*, 130 n. 20.

116. Louis, *Mathieu de l'Obel*, 131 n. 22, 157.

117. Clusius mentions one of De l'Obel and Pena's descriptions in a letter to Camerarius, Oct. 24, 1575 (Hunger, *Clusius*, 2: 314).

118. Findlen, "Formation of a community."

119. Findlen, "Formation of a community," 374.

120. Pietro Andrea Mattioli, *De plantis epitome* (1586); Pietro Andrea Mattioli, *Opera omnia* (1598).

121. Pál Gulyás, *A Zsámboky-könyvtár katalógusa* (1992). I identified 100 printed natural history books and 3 manuscripts in this catalogue. Any reader who would like a complete list may write me care of my publisher. For Sambucus's biography, see Endre Bach, *Humaniste hungrois* (1932).

122. Bauhin, *Phytopinax*, fol. β2r. I have broken down the list by Bauhin's categories, as the individuals were identified in the list. One of them was listed both as Magister and as Med. Cand.; I have included him among the latter. Ulisse Aldrovandi is identified as a physician, though he was also a Bolognese patrician.

123. Some of the letters are in Leiden UB, MS. VUL 101, s.v. "Peiresc," and MS. BPL 2724a (Clusius to Caccini).

124. Basilius Besler, *Hortus Eystettensis* (1613); Crispijn de Passe, *Hortus floridus* (1614).

125. Elisabeth B. MacDougall, "Paradise of plants" (1991), 152, 155; Kenseth, ed., *Age of the marvelous*, 371–72 (catalogue no. 148).

126. On flower collectors in the seventeenth century, see Schnapper, *Le géant*, and Elizabeth Hyde, "Cultivated power" (1998).

127. Arber, *Herbals*, 245–46.

128. John Parkinson, *Paradisus terrestris faithfully reprinted* (1904).

129. Ludwig Jungermann, *Catalogus plantarum circa Altorfium* (1615); Caspar Bauhin, *Catalogus* (1622); a similar, though not as extensive, list of plants to be found in the region around Paris is provided by Jacques Cornut, "Enchiridium botanicum Parisiense," in Jacques Cornut, *Canadensium plantarum historia* (1635).

130. By the seventeenth century, the two groups were largely distinct, though there was some overlap. See Schnapper, *Le géant*, 39, on the distinction between collectors of species and flower breeders.

131. Clusius to Camerarius, April 22, 1579 (Hunger, *Clusius*, 2: 370).

132. Brian W. Ogilvie, "Encyclopædism" (1997).

133. Bauhin, *Phytopinax; Pinax*. The latter was intended to be merely the index to a universal history of plants, the *Theatrum botanicum*, but in the event only

one volume was ever published, several decades after Bauhin's death: Caspar Bauhin, *Theatri botanici liber primus* (1658).

134. See chapter 4, below, for an extended comparison of Clusius's and Bauhin's descriptive style in its relation to their techniques.

135. Brian W. Ogilvie, "Books of nature" (2003).

136. D. E. Allen, *Naturalist in Britain* (1994), 7.

137. Harold J. Cook, "Physicians and natural history" (1996).

138. And later; see Wolf Lepenies, *Ende der Naturgeschichte* (1976), 72–77.

139. Pinon, *Livres de zoologie*, 13–18.

140. Pinon, *Livres de zoologie*, 16.

141. Penny to Carolus Clusius, March 31, 1583 (o.s.), Leiden. UB. MS. VUL 101, s.v. "Penneius"; Charles E. Raven, *John Ray* (1950), 390–91, on the eventual publication of Moffett's work. Leonhard Dold referred to "creeping things and worms" (*repentia et vermes*) as "insects" in a letter to Jacob Zwinger, December 18, 1595 (probably o.s.), Basel UB, MS. Fr.Gr. I.12, 123. Moffett's name is also spelled Moufet and Muffet or Muffett; his daughter is supposedly the "Little Miss Muffett" of the nursery rhyme.

142. Moffett's correspondence reveals some of his troubles, especially in securing privileges for the book (to provide a legal defense against pirated editions): Moffett to Joachim Camerarius the Younger, March 27, 1590 and October 1,1590 (both probably o.s.), Erlangen UB, Trew-Briefsammlung, Moffett, Thomas, 3–4; and Moffett to Carolus Clusius, April 1, 1590 (again, probably o.s.), Leiden UB, MS. VUL 101, s.v. Moffetus.

143. See Leonhard Dold to Jacob Zwinger, December 18, 1595 (probably o.s.), Basel UB, MS. Fr.Gr. I.12, 123, and various letters to Jacob Zwinger, Basel UB, MS. Fr.Gr. I.13, especially Pascal Le Coq to Zwinger, March 15, 1596 (no. 82).

144. Raven, *John Ray*, 390–91. On Swammerdam, see A. Schierbeek, *Jan Swammerdam* (1967).

145. On physiology, see Thomas S. Hall, *History of physiology* (1969).

146. Quoted by Edward Lee Greene, *Landmarks* (1983), 1: 367.

147. Dodoens, *Histoire des plantes*, sig. *iiiv.

148. Cf. Regiomontanus's laudation of astronomy: Noel Swerdlow, "Science and humanism" (1993).

149. Carolus Clusius, *Stirpium per Hispanias historia* (1576), 4.

150. Hunger, *Clusius*, 1: 135. This figure is taken by Hunger from a letter from Clusius to Johann Crato von Krafftheim.

151. Andrea Cesalpino, *De plantis* (1583), sig. a2v.

152. Clusius, *Stirpium per Pannoniam historia.*

153. André Cailleux, "Nombre d'espèces" (1953).

154. Spiegel, *Isagoge*, 131.

155. See Peter Novick, *That noble dream* (1988), chapter 1; Lorraine Daston and Peter Galison, "Image of objectivity" (1992); Lorraine Daston, "Moralized objectivities" (1999).

156. Daston and Galison, "Image of objectivity."

157. Daston and Park, *Wonders*, use the term "naturalist" for "those who engage in the systematic study of nature," in order to avoid "scientist" and other anachronisms (p. 373 n. 4). My usage is more precise for the period that is the main focus of

this book. However, I will also use the term loosely to refer to ancient and medieval investigators, in the sense that Daston and Park use it. The context should be clear enough to avoid confusing the reader.

158. "Naturalist" is attested in English from 1587 in the sense, now obsolete, of "natural philosopher." It is also attested from 1600 in the sense of a scholar of botany or zoology. Neither of these uses is common before the middle of the seventeenth century. *Oxford English Dictionary*, 2nd ed., s.v.

159. Cf. Reeds, *Botany*, 7.

160. Clusius, *Stirpium per Pannoniam historia*, sig. *3r; Clusius, *Exoticorum libri*, sig. †2r.

161. Paludanus to Caspar Bauhin, Mar. 8, 1611 (Basel UB, MS. Fr.Gr. II.1, 43–44, VII); Cargill to Caspar Bauhin, Aug. 7, 1598 (Basel UB, MS. Fr.Gr. II.1, 71–72, XLII).

162. Jacob Zwinger to Caspar Bauhin, June 14, 1592 (Basel UB, MS. G^2 I 13b, fol. 199–200).

163. Ulisse Aldrovandi to Carolus Clusius, Feb. 8, 1569 (Leiden UB, MS. Vulcanius 101, s.v. "Aldrovandi," no. 1); Spiegel, *Isagoge*, 124.

164. Passim in the archives I have consulted.

165. Ole Worm to Caspar Bauhin, Mar. 8, 1617 (Basel UB, MS. Fr.Gr. II.1, 143–144): Worm includes himself among those "quibus studia botanica cordi sunt."

166. My models for the notion of "profession" in the sixteenth century are medicine, law, and theology. Each of these fields comprised a higher faculty in the universities, which granted degrees entitling their holders (often with additional licensing from local authorities) to act as physicians, jurisconsults, or theologians.

167. On imagined traditions, the *locus classicus* is Eric Hobsbawm and Terence Ranger, eds., *Invention of tradition* (1983). See also, for a contemporary example in the history of mathematical physics, Peter Dear, *Discipline & experience* (1995), 93–97, 115–21.

168. Naturalists' family members may be a significant exception to this generalization. Alix Cooper is currently studying their contributions to early modern natural history.

169. This fact was noted by contemporaries such as Clusius, *Stirpium per Hispanias historia*, 3; Pinon, *Livres de zoologie*, 15.

170. See Henry Lowood, "New World" (1995).

171. Spiegel, *Isagoge*, 128; Gessner, *Historiae animalium liber IIII*, 520.

172. Clusius, *Exoticorum libri*, passim: references both to oral reports, transmitted by Clusius's correspondents, and to published "diaria" of voyages; Clusius, *Stirpium per Pannoniam historia*, 10.

173. Spiegel, *Isagoge*, 128.

174. On Latin's place in Renaissance culture, see Walter J. Ong, *Ramus* (1958), 10–14. Latin's persistence since the end of Middle Ages as scholarly language and, more significantly, as cultural marker has been perceptively examined by Françoise Waquet, *Le latin* (1998).

175. Bauhin, *Phytopinax*, fol. β1r–2r.

176. Leiden UB, MS. VUL 101, and a few other signatures; cf. the *Brievencatalogus* that is available in the reading room of the Dousa (the Western manuscript and rare book library). Clusius's command of Italian is all the more remarkable as he had never

been to Italy, though in his youth he longed to go there: Clusius to Matteo Caccini, 29. September 1606 (Leiden UB, MS. BPL 2724a, no. 1); Hunger, *Clusius*, 1: 30, 37, 76.

177. Clusius's range of languages was exceptional, though then as now scholars could read more languages than they could actively use.

178. Platter, "Tagebuch," 183–85, 188.

179. Clusius's side of the correspondence is in the Biblioteca Ambrosiana in Milan; I have consulted the facsimiles in the Leiden Universiteitsbibliotheek (MS. Facs UB A 139) and the printed version in G. B. De Toni, "Carteggio degli italiani" (1912).

180. *Antidotarium*, 9, presents the physicians' point of view that apothecaries needed to know Latin to read Dioscorides, Serapion, Galen, and other writers on *materia medica*; R. Schmitz, "Apotheke" (1980), col. 797.

181. On Renaissance education, see Paul F. Grendler, *Schooling* (1989), George Huppert, *Public schools* (1984), and most recently Robert Black, *Humanism and education* (2001). On German humanists, see Joël Lefebvre, "Le latin et l'allemand" (1980).

182. Clusius's Latin translations of Garcia da Orta, Nicolas Monardes, Christobal Acosta, Pierre Belon, and Guillaume Rondelet, as well as Thomas Hariot's and Gerrit de Veer's travel narratives and the Florentine *Antidotarium*, are exemplary in this regard.

183. *Honestissimus:* e.g., Clusius, *Exoticorum libri*, 87, for Christianus Porretus, Petrus Garretus, and "honestissimus vir Ioannes Pona, pharmacopoeus Veronensis" on the same page where Prosper Alpino is referred to as "Cl. V."; Garretus is referred to as "honestissimus" passim. On p. 97 a councilor to the count of Solms's widow, Antonius de Fleury, is also referred to as "honestissimo." The use of the superlative is normal and honorific in humanist Latin, though on p. 49 William Parduyn is called "honestus," as is Dirk Cluyt (Theodoricus Clutius) on p. 73. *Ornatissimus:* p. 1 of a Salernitan (possibly a military officer) who had passed time in Goa; Walichius Syvertz, a ship's captain, p. 47; p. 68 for Porretus; p. 96 for Jacques Plateau.

184. Clusius to Matteo Caccini, Feb. 27, 1609 (Leiden UB, MS. BPL 2724 a, no. 22).

185. Two of her letters to Clusius are quoted in Hunger, *Clusius*, 1: 169, 177.

186. William J. Bouwsma, *Waning of the Renaissance* (2000), 146–50.

187. Other social conventions could be relaxed as well: thus Ulisse Aldrovandi wrote to Clusius without first having been introduced through an intermediary: Aldrovandi to Clusius, Feb. 8, 1569 (Leiden UB, MS. VUL 101, s.v. "Aldrovandi").

188. Hunger, *Clusius*, 1: 5, 123, 126.

189. *Antidotarium*, 4–5.

190. Anthony Grafton, "Scaliger's table talk" (1988), 583.

191. Zwinger to Bauhin, 18.5.1592 (Basel UB, MS. G2 I 13b, fol. 197r–198r).

192. Findlen, *Possessing nature, passim* on Imperato (see the index), esp. 31–33 in this context.

193. Onorio Bello, a Venetian physician resident in Cydonia, Crete, corresponded with naturalists in the European mainland, including Jacob Zwinger in Basel: Basel UB, MS. Fr.Gr. I, 12, no. 28.

194. On Rauwolf, see Dannenfeldt, *Leonhard Rauwolf*; Prosper Alpino, *De plantis Aegypti* (1592).

195. My discussion in this paragraph is based chiefly on Findlen, "Formation of a community."

196. See the discussion of gardens in chapter 4, p. 151, for further details.

197. Findlen, "Formation of a community," 380–81.

198. Fuchs to Camerarius, August 10, 1565, Erlangen UB, Trew-Briefsammlung, Fuchs, Leonhard, 25; November 24, 1565, ibid., 26.

199. Gessner to Jean Bauhin, November 14, 1563, in Johann Bauhin, *De plantis a divis nomen habentibus* (1591), 126–27 and Conrad Gessner, *Vingt lettres* (1976), 28–29.

200. Petrus Monau to Joachim Camerarius the Younger, July 25, 1582, Erlangen UB, Trew-Briefsammlung, Monau, Petrus, 23; Joachim Jungerman to Joachim Camerarius the younger, November 5, 1585, Erlangen UB, Trew-Briefsammlung, Jungerman, Joachim, 9.

201. Mattioli, *De plantis epitome*; Mattioli, *Opera omnia*.

202. Eric Cochrane, *Florence in the forgotten centuries* (1973), chapter 2, on Scipione Ammirato and his contemporaries. Cf. a similar phenomenon in material wealth, as discussed by Richard A. Goldthwaite, *Wealth and the demand for art* (1993).

203. See Françoise Waquet, *Modèle français* (1989).

204. Peter Burke, *European Renaissance* (1998), discusses the domestication of the Renaissance, and the complex interactions in northern Europe between the reception of antiquity and the reception of Italian culture.

205. Mercati's letters to Camerarius, written between 1580 and 1583, are in Erlangen UB, Trew-Briefsammlung, Mercatus, Michael, 1–8. A quaestor was a Roman magistrate whose duties included charge of the public treasury; "Camerarius," an equivalent of the German Kammermeister, referred to a similar office. Hence the pseudonym was apt. It may have been inspired, initially, by the same humanist urge that led Philipp Schwarzerd to adopt the Greek translation of his name, "Melanchthon"; in a letter written c. 1521 from Ferrara to the elder Camerarius, Euricius Cordus also used the name Quaestor for his correspondent: Erlangen UB, Trew-Briefsammlung, Cordus, Euricius, 1.

206. On Clusius's reputation and his translation of Garcia da Orta, see Hunger, *Clusius*, 1: 97–98. Ulisse Aldrovandi began his correspondence with Clusius after being impressed by the latter's translation of Garcia: Aldrovandi to Clusius, 8. February 1569 (Leiden UB, MS. VUL 101, s.v. "Aldrovandi"). See also Lowood, "New World," though Lowood exaggerates the importance of printing as a factor in communicating natural history to the European community.

207. I. H. Burkill's analysis of the areas in which modern botany originated (1953), cited with map in William T. Stearn, "Botanical exploration" (1956/57), is based on the locations where botanical books were printed between 1500 and 1623. Though the area sketched by Burkhill—northern Italy, eastern France, western Germany, and the Low Countries—is, roughly speaking, the area my own researches have specified, there are important differences, for example Austria and Hungary, explored by Clusius and his companions in the 1570s and 1580s. There are two sources of error in Burkhill's method. First, books were not always printed where they are written: for example, Clusius published with Plantin in Antwerp even while working in Vienna. Second, as I have discussed at more length in my introduction, natural history is more than the publication of natural history books.

208. Gessner probably met most of his English correspondents while they were on the continent during the Marian Exile; see Charles E. Raven, *English naturalists*

(1947), 82, 153, 155; this was not true, however, of Gessner's friend John Caius (139).

209. On natural history activity in sixteenth-century Italy, see above all Findlen, *Possessing nature.*

210. Gessner, *Horti Germaniae,* fol. 237v–39r. However, Carolus Clusius visited the gardens of Paris in 1571 and found them rotten and without interesting plants: Jean de Brancion to Clusius, July 26, 1571 (Leiden UB, MS. VUL 101, s.v. "Brancion"); Brancion to Clusius, Sept. 6, 1571 (quoted in Hunger, *Clusius,* 1: 117).

211. Gessner, *Horti Germaniae,* fol. 239v.

212. Hunger, *Clusius,* 1: 51–52; Clusius, dedicatory letter in *Antidotarium,* 4.

213. Clusius to Joachim Camerarius II, 7 March 1574 (Hunger, *Clusius,* 2: 296).

214. Kristof Glamann, "Der europäische Handel" (1979), 276–80, on the location of these areas along major trade routes.

215. Clusius, *Stirpium per Hispanias historia,* 3.

216. Spanish natural history in the Americas has been largely neglected by the historiography in English; this is about to change with the publication of Antonio Barrera, *Experiencing nature* (forthcoming).

217. Bauhin, *Phytopinax,* fol. 2r.

218. E.g., Clusius passed on to Camerarius some bulbs that he had received from members of the imperial delegation to Constantinople: Clusius to Camerarius, Feb. 13, 1575, in Hunger, *Clusius,* 2: 306. See my discussion below, in chapter 4, p. 162.

219. The Hansestädte and the cities of the northern Netherlands had been involved in the Baltic trade since the Middle Ages, but this high-volume trade in commodities did not produce the high profits and subsidiary trades that the luxury trade, increasingly important in the North, did: Jonathan Israel, *Dutch Republic* (1995), 24–25, 315–18.

220. I stress local and international as the real binding ties: "national" contexts seem particularly unsuited to analyzing the interactions between naturalists in different locations during this period. See Roy Porter and Mikulás Teich, eds., *Scientific Revolution* (1992), and Lorraine Daston, "Several contexts" (1994).

221. Gessner, *Horti Germaniae,* fol. 238v.

222. Gessner, *Historiae animalium lib. I,* sig. γ1r–v.

223. Rytz, "Herbarium Felix Platters," 9; also Elisabeth Landolt, "Felix Platter als Sammler" (1972), 245–60.

224. Landolt, "Felix Platter als Sammler," 246, 250–60.

225. Some of the notes sent back and forth are preserved in the Basel Universitätsbibliothek.

226. Ulisse Aldrovandi to Jacob Zwinger, Oct. 4, 1596 (Basel UB, MS. Fr.Gr. I.13, no. 44); Heinrich Cherler to Jacob Zwinger, Aug. 21, 1599 (*ibid.,* no. 64); Pascal Le Coq to Jacob Zwinger, Mar. 15, 1596 (*ibid.,* no. 82); Leonhard Dold to Jacob Zwinger (Basel UB, MS. Fr.Gr. I.12, no. 123).

227. Noel L. Brann, "Humanism in Germany" (1988), 126–27; Karl Weiß, *Geschichte der Stadt Wien* (1883), 2: 463–64.

228. On the university, see Kurt Mühlberger, "Bildung und Wissenschaft" (1992).

229. Christiane Thomas, "Wien als Residenz" (1993), 107–11, 116–17.

230. Susanne Claudine Pils, "Stadt als Lebensraum" (1993), 125–26.

231. Hunger, *Clusius*, 1: 125–26.

232. See Mühlberger, "Bildung und Wissenschaft," 212–17.

233. Dirk Jacob Jansen, "Instruments of patronage" (1992), 198–201.

234. See in general Clusius's letters to Joachim II Camerarius, which form the most detailed source of his naturalizing in the 1570s and 1580s; reprinted in Hunger, *Clusius*, vol. 2.

235. Hunger, *Clusius*, 2: 40 n. 2.

236. Hunger, *Clusius*, 1: 131.

237. So Clusius thought: Clusius to Balthasar Batthyány, October 21, 1577, in Istvánffi, *A Clusius-codex mykologiaia méltatása*, 205; Hunger, *Clusius*, 1: 139.

238. Clusius mentioned some purchases in letters to Camerarius, May 22, 1582 and July 8, 1582 (Hunger, *Clusius*, 2: 388, 390). Clusius refers to a "Domina de Thaw patritia Viennensis" and to "Matronae nobiles" who cultivated exotic bulbous plants: Clusius, *Stirpium per Pannoniam historia*, 135–36.

239. Evans, *Rudolf II*, 120–21.

240. The rest of this paragraph follows Kaufmann, *Mastery of nature*, chapters 4 and 7.

241. This discussion is based on eighteen letters from Jacques Levenier to Carolus Clusius, written between December 1597 and June 1606: Leiden UB, MS. VUL 101, s.v. Venerius, 1–18.

242. Levenier to Clusius, December 1, 1597 (Leiden UB, MS. VUL 101, s.v. Venerius, 1).

243. Levenier to Clusius, April 18, 1598 (Leiden UB, MS. VUL 101, s.v. Venerius, 2).

244. On *amateurs de fleurs*, see Schnapper, *Le géant*.

245. Levenier to Clusius, November 26, 1599, July 8, 1601, August 28, 1601, June 25, 1602, August 20, 1602, October 4, 1603, July 6, 1605, and June 23, 1606 (Leiden UB, MS. VUL 101, s.v. Venerius, 4, 10, 11, 13, 14, 16, 17, 18).

246. Levenier to Clusius, February 18, 1600, November 7, 1600, and February 10, 1601 (Leiden UB, MS. VUL 101, s.v. Venerius, 6, 8, 9).

247. Levenier to Clusius, July 8, 1601.

248. Levenier to Clusius, August 28, 1601.

249. Chmielecius to Jacob Zwinger, July 22, 1592 (Basel UB, MS. Fr.Gr. I.12, no. 98); see below, p. 78.

250. Paaw to Clusius, August 10, 1593 (Leiden UB, MS. VUL 101, s.v. Pauw); Swan, "Jacques de Gheyn II," 186, and Hunger, *Clusius*, 1: 215–16, on the later relationship between Paaw and Clusius.

251. Conrad Gessner, "Descriptio Montis Fracti," in Gessner, *De raris herbis*, 44–47, 50–51.

252. Joannes Rhellicanus, "Stockhornias," in Gessner, *De raris herbis*, 77–82. On humanist rhetorical exercises, see Michael Baxandall, *Giotto and the orators* (1971), especially 32ff.

253. Thomas Platter, d. J., *Beschreibung der Reisen* (1968), 1: 90–91, 154–55.

254. Carolus Clusius, *Stirpium nomenclator Pannonicus* (1584), sig. †1v–†2r.

255. Basel UB, MS. Fr. Gr. I, 12.

256. Martin Chmielecius longed to be with Zwinger and Le Coq on their trip to Padua: Chmielecius to Zwinger, Nov. 15, 1591 (Basel UB, MS. Fr. Gr. I, 12, no. 95).

257. Clusius, *Stirpium nomenclator Pannonicus,* is a dictionary of Hungarian plant names and their Latin equivalents, compiled by Clusius with information from Beythe. Clusius, *Stirpium per Pannoniam historia,* often refers to the "local" name for a plant, and it is likely that Beythe was a major source for this information as well, in the case of plants found in Hungary.

258. Platter, *Beschreibung der Reisen,* 1: 159–60.

259. Spiegel, *Isagoge,* 92–93, 128.

260. Reeds, "Humanism and botany."

261. Spiegel, *Isagoge.*

262. Reeds, *Botany,* 55–72, 88–89.

263. See for example the documents regarding the appointment of Bernardus Paludanus (who rejected the post), Carolus Clusius, and Pieter Pauw and Gerard Bondt as praefecti: Molhuysen, ed., *Bronnen,* 1: 71–72, 83 (appointment of Dirk Cluyt as Clusius's assistant), 112–14, 180*, 381* (Rekest van studenten aan Curatoren over de Praefectura Horti, 18 June 1598).

264. Reeds, *Botany,* 110.

265. Martin Chmielecius to Jacob Zwinger, July 22, 1592 (Basel UB, MS. Fr.Gr. I, 12, 98).

266. Platter, *Beschreibung der Reisen,* 1: 85–88.

267. Reeds, *Botany,* 89.

268. Platter, *Beschreibung der Reisen,* 1: 159–60.

269. Joannes Posthius, "De obitu D. Guillelmi Rondeletii," quoted by Reeds, *Botany,* 66.

270. Molhuysen, ed., *Bronnen,* 113, 154.

271. Molhuysen, ed., *Bronnen,* 401*.

272. Mathieu de L'Obel, *Plantarum seu stirpium icones* (1581).

273. Bauhin to Vorst, Dec. 11, 1611 (Erlangen UB, Trew-Briefsammlung. Bauhinus, Caspar, 51).

274. Vorst to Bauhin, Mar. 27, 1612 and Mar. 19, 1615 (Basel UB, MS. G2 I 1, fol. 221–22, 224).

275. Bauhin, *Catalogus* (1622). Cf. the brief description in Reeds, *Botany,* 124–25.

276. A copy of the third edition of this book (1671) in the Basel Universitätsbibliothek is copiously annotated on the blank leaves (Basel UB, MS. K III 45).

277. Caspar Bauhin, *Catalogus* (1671), t.p.

278. Spiegel, *Isagoge,* 77.

279. See the lists of contributors to Gessner, *Historiae animalium lib. I,* sig. γ1r–v; Gessner, *Historia animalium liber III,* sig. a6r; Gessner, *Historiae animalium liber IIII,* sig. b5r and his list of German gardeners: Gessner, *Horti Germaniae,* fol. 237v–39v.

280. Almost all the contributors to Bauhin's *Phytopinax* were physicians, medical students, or apothecaries: Bauhin, *Phytopinax,* sig. β2r.

281. On late medieval and Renaissance travel see John Hale, *Civilization of Europe* (1995), 144–49, 180–84.

282. J. C. Russell, "Bevölkerung Europas" (1978), 16–18; Dietrich Denecke, "Straßen, Reiserouten und Routenbücher" (1992), 238.

283. Erasmus to Beatus Rhenanus, c. Oct. 15, 1518, in Desiderius Erasmus, *Praise of folly and other writings* (1989), 251–59.

284. Justus Lipsius, *Iusti Lipsi Epistolae* (1978–), 3: 307–9; 5: 9.

285. Erasmus, *Praise of folly and other writings*, 256.

286. Platter, "Tagebuch," 181.

287. Levenier to Carolus Clusius, August [20], 1602 (Leiden UB, MS. VUL 101, s.v. Venerius, 14).

288. Platter, "Tagebuch," 179–80, 182.

289. Justin Stagl, "Ars apodemica" (1992), 141–42.

290. Stagl, "Ars apodemica"; *History of curiosity*, 47–49.

291. Michel de Montaigne, *Complete essays* (1958), "Of the education of children," 112, 116.

292. Platter, "Tagebuch"; Platter, *Beschreibung der Reisen*; Hunger, *Clusius*, 1: 21–34; Basel UB, MSS. G2 1 I, Fr.Gr. I 12, and Fr.Gr. I 13 (correspondence of and referring to Jacob Zwinger).

293. For example, Carolus Clusius and Mathieu de l'Obel, both of whom studied with Guillaume Rondelet in Montpellier. Hunger, *Clusius*, 1: 22–24; Louis, *Mathieu de l'Obel*, 94–122.

294. Platter to Camerarius, July 8, 1593 (Erlangen UB, Trew-Sammlung: Plater, Felix, 6).

295. Zwinger correspondence, Basel UB, MS. Fr.Gr. I.12, I.13, I.22.

296. Platter, *Beschreibung der Reisen*, e.g., 154–65.

297. Clusius to Camerarius, Feb. 5, 1583 (Hunger, *Clusius*, 2: 395).

298. Dold to Zwinger, Oct. 15, 1594 (Basel UB, MS. Fr.Gr. I, 12, 122).

299. Papius to Zwinger, July 29, 1609 (Basel UB, MS. Fr.Gr. I, 12, 215).

300. Bauhin to Camerarius, July 8, 1593 (Erlangen UB, Trew-Sammlung. Bauhinus, Caspar, 11).

301. Bauhin to Weyer, Dec. 9, 1587 (Erlangen UB, Trew-Sammlung. Bauhinus, Caspar, 54).

302. Dold to Zwinger, Jan. 25, 1598 (Basel UB, MS. Fr.Gr. I, 12, 131).

303. On the *peregrinatio academica*, see Heinz Schneppen, *Niederländische Universitäten* (1960), 64–67, and Stagl, "Ars apodemica."

304. Ludovic Legré, *Les Bauhin, Cherler, Dourez* (1904).

305. Bock, *Kreütterbuch* (1577), sig. b6r.

306. Felix Platter to Rennward Cysat, July 2, 1596, in Theodor von Liebenau, "Plater und Cysat" (1900), 106.

307. Landolt, "Felix Platter als Sammler," 261ff.

308. Hunger, *Clusius*, 1: 142, 145. Hunger surmises that Clusius went to England to meet Drake.

309. Clusius, *Exoticorum libri*, 55, 83–84.

310. Bruce T. Moran, "Princes, machines and precision" (1977), 214–15. My thanks to Zeno Swijtink for this reference.

311. Clusius, *Exoticorum libri*, 97–99.

312. Giuseppe Olmi, "Molti amici" (1991).

313. Gessner, *Historiae animalium liber IIII*.

314. Clusius to Camerarius, March 13, 1574 (Hunger, *Clusius*, 2: 297).

315. Gessner, in Valerius Cordus, *Annotationes* (1561).

316. Hunger, *Clusius*, 1: 3; Florence Hopper, "Clusius' world" (1991), 17, 29 n.17.

317. Busbecq to Clusius, July 18, 1585 and April 28, 1586 (Leiden UB, MS. VUL 101, s.v. "Busbecq," nos. 3, 4).

318. Felix Platter to Jacob Zwinger(?), s.d. (Basel UB, MS. Fr.Gr. I.12, no. 272).

319. Boodt to Clusius, Oct. 12, 1602 (Leiden UB, MS. VUL 101, s.v. "Boodt," no. 2).

320. Chmielecius to Zwinger, July 22, 1592 (Basel UB, MS. Fr.Gr. I.12, no. 98).

321. Chmielecius to Zwinger, April 27, 1593 (Basel UB, MS. Fr.Gr. I.12, no. 101). Chmielecius did not explicitly name Platter, but given the earlier letter, it seems clear whom he meant.

322. Levenier to Clusius, December 1, 1597 and April 18, 1598 (Leiden UB, MS. VUL 101, s.v. Venerius, 1, 2).

323. Ernst Heinrich Krelage, *Bloembollenexport* (1946), 8–11.

324. Clusius to Matteo Caccini, Dec. 8, 1608 (Leiden UB, MS. BPL 2724a, no. 19). He expressed a similar opinion in letters to J. van Hoghelande, Jan. 9, 1592 (Leiden UB, MS. BPL 885, s.v. "Clusius"), in which he wrote that he had never raised plants to sell them but considered their cultivation to be a liberal art, and to Justus Lipsius (Istvánffi, *A Clusius-codex mykologiaia méltatása*, 192–93); Istvánffi does not mention the date of the latter letter, but it was written not long after Clusius arrived in Leiden in 1593. Some of Clusius's purchases are documented in letters to Joachim Camerarius of May 22, 1582 and July 8, 1582 (Hunger, *Clusius*, 2: 388, 390).

325. Clusius to Camerarius, Aug. 19, 1581, May 15, 1582, and May 22, 1582 (Hunger, *Clusius*, 2: 384, 387–88).

326. Clusius to Bernard Paludanus, April 6/16, 1598 (Leiden UB, MS. Facs UB A 138).

327. Hunger, *Clusius*, 1: 237.

328. Levenier to Clusius, December 18, 1599 and February 10, 1601 (Leiden UB, MS. VUL 101, s.v. Venerius, 5, 9).

329. Johannes van Meurs, *Athenae Batavae* (1625), 33. This *Lex horti* was set up c. 1600.

330. For Gessner, see above; for Dalechamps, see Baudouin Van den Abeele, "Albums ornithologiques de Dalechamps" (2002).

331. Clusius to Camerarius, Oct. 7, 1573 (Hunger, *Clusius*, 2: 295).

332. Letters from Aldrovandi to Clusius, Leiden Universiteitsbibliotheek, MS. VUL 101, s.v. "Aldrovandi" (seven letters dated from Feb. 8, 1569 to April 17, 1596).

333. Clusius to Rehdiger, March 1, 1567, in Carolus Clusius and Conrad Gessner, *Epistolae ineditae* (1830), 9.

334. Hunger, *Clusius*, 1: 101–2, 122. But Clusius did not want to presume too much on his friend's largesse: to disguise his penury, he borrowed money from his former charge Thomas Rehdiger: Clusius to Rehdiger, s.d. and March 14, 1571, in Clusius and Gessner, *Epistolae ineditae*, 11–13. Hunger, *Clusius*, 1: 112 dates the first letter in mid-1570.

335. Clusius's surviving correspondence with Lipsius is catalogued by Alois Gerlo, *Inventaire* (1968), and is being published in Lipsius, *Iusti Lipsi Epistolae*. On Lipsius and Clusius, see Mark Mark Morford, "Stoic garden" (1987), and Swan, "Jacques de Gheyn II," 144–45.

336. "Ad Clusii nomen lusus.
Omnia Naturae dum, Clusi, arcana recludis:

CLUSIUS haud ultra sis sed APERTA mihi."
[Pun on Clusius's name: When you open, Clusius, all the secrets of nature, / you are, to me, no longer CLUSIUS (closed) but OPEN.] In Clusius, *Stirpium per Pannoniam historia*, sig. *3v, along with a eulogy of the book by Lipsius.

337. Lipsius was, at least, instrumental in persuading Clusius to accept the position: Lipsius to Clusius, Feb. 16, 1592 and Mar. 21, 1593; Clusius to Lipsius, April 18, 1593 (o.s.) (Lipsius, *Iusti Lipsi Epistolae*, 5: 112–13; 6: 120–21, 135–37).

338. Clusius to Crato, Aug. 1, 1561, in Clusius and Gessner, *Epistolae ineditae*, 14–15.

339. Platter to Cysat, July 2, 1596, in Liebenau, "Plater und Cysat," 106; see also Platter to Cysat, Nov. 12, 1592 (p. 100) and March 5, 1593 (pp. 103–4).

340. See John Monfasani, "Humanism and rhetoric" (1988), esp. 192–95, on epistolography. See also the list of two hundred ways to say "I enjoyed reading your letter" in Desiderius Erasmus, *De copia verborum et rerum*, (1988), 76–82. I am unaware of any treatments of humanist letter writing from a social perspective, other than particular studies such as George W. McClure, *Sorrow and consolation* (1991).

341. Leonhard Dold to Jacob Zwinger, Oct. 15, 1594 (Basel UB, MS. Fr.Gr. I, 12, 122).

342. Gessner to Johann (Jean) Bauhin, Oct. 28, 1563, in Gessner, *Vingt lettres*, 31.

343. This analysis, of course, draws on Benedict Anderson, *Imagined communities* (1983).

344. E.g., by Marc Fumaroli, *Âge de l'éloquence* (1980), 155, 430–32.

345. Dena Goodman, *Republic of letters* (1994), 12–23. The most important overview of the issue is Françoise Waquet, "République des lettres" (1989). The notion of the bourgeois public sphere emerges from the work of Reinhart Koselleck, *Kritik und Krise* (1973), and Jürgen Habermas, *Strukturwandel der Öffentlichkeit* (1990).

346. Goodman, *Republic of letters*, 3–5; Daniel Gordon, *Citizens without sovereignty* (1994). See Lorraine Daston, "Ideal and reality" (1991).

347. Elizabeth L. Eisenstein, *Printing press* (1979), 137 n. 287; Waquet, "République des lettres."

348. Paul Dibon, "Communication" (1978).

349. Gessner, *Historiae animalium lib. I*, sig. 3r.

350. Bachmeister to Bauhin, July 11, 1603 (Basel UB, MS. Fr.Gr. II.1, pp. 29–30).

351. Aldrovandi to Clusius, April 3, 1590 (Leiden UB, MS. VUL 101, s.v. "Aldrovandi," 5).

352. Worm to Bauhin, March 8, 1617 (Basel UB, MS. Fr.Gr. II.1, pp. 143–44).

353. Bello to Zwinger, Aug. 15, 1596 (o.s.) (Basel UB, MS. Fr.Gr. I.13, no. 52).

354. The range of Gessner's interests is apparent in most studies on him. See especially Lucien Braun, *Conrad Gessner* (1990); H. Fischer, ed., *Gessner Universalgelehrter* (1967); Urs B. Leu, *Gesner als Theologe* (1990); and Hans H. Wellisch, *Conrad Gessner* (1984).

355. Hunger, *Clusius*, 1: 78, 82, 90–91, 108.

356. Cesalpino to Zwinger, Dec. 22, 1593 (Basel UB, MS. Fr.Gr. I.12, 68).

357. On Scaliger, see the letters of Carolus Clusius to Bernardus Paludanus, March 26, 1601 (Leiden UB, MS. PAP 2, s.v. "Clusius"), and Raphael Pelecius to Caspar

Bauhin, Aug. 20, 1594 (Basel UB, MS. Fr.Gr. II.1, p. 42, IV). On Lipsius, see Morford, "Stoic garden."

358. Dold to Zwinger, Dec. 10, 1598 (Basel UB, MS. Fr.Gr. I.12, 132).

359. Daston, "Ideal and reality."

360. Raphael Pelecius to Caspar Bauhin, Aug. 20, 1594 (Basel UB, MS. Fr.Gr. II.1, 42, IV).

361. Scholars frequently requested portraits of one another to hang in their workrooms or other chambers; this practice reflects the importance of affective bonds as much as that of patronage relationships. On Paolo Giovio's collection, see Francis Haskell, *History and its images* (1993), 43–48.

CHAPTER THREE

1. The growing literature on "invented traditions" was sparked by Eric Hobsbawm and Terence Ranger, eds., *Invention of tradition* (1983).

2. Adam named the animals, and presumably also the plants. Solomon wrote a history of all plants, from the cedar of Lebanon to the lowly hyssop, according to I Kings 4.33.

3. On the notion of *prisca theologia*—a revelation to the gentiles that paralleled the revelation to the Hebrews—see Frances A. Yates, *Giordano Bruno* (1964).

4. Aristotle has some shadowy predecessors. Some physicians of the sixth and fifth centuries BCE discussed animals and plants in their medical and philosophical works. Menestor of Sybaris (fl. c. 450 BCE), though not a physician, left fragments indicating a concern with the nature of plants, which led Gustav Senn to nominate him as "der älteste Naturforscher und Botaniker der Griechen": Gustav Senn, *Biologische Forschungsmethode* (1933), 30. Menestor's fragments are collected in Hermann Diels, *Fragmente der Vorsokratiker* (1974–75), 1: 375–76. Aside from one passage in Iamblichus, the fragments all come from Theophrastus. They are not extensive enough to give us a clear idea of Menestor's views. The most recent survey of ancient natural history is Roger French, *Ancient natural history* (1994), which is idiosyncratic and omits the medical writers entirely.

5. G. E. R. Lloyd, *Science, folklore and ideology* (1983), 6.

6. Conrad Gessner, *Historiae animalium lib. I* (1551), sig. β2r.

7. Thomas Johnson, "To the reader," in John Gerard, *Herball* (1633), sig. ¶¶2v.

8. Stefano Perfetti, *Aristotle's zoology* (2000).

9. *Editiones principes* of the works: Aristotle, *History of animals, Parts of animals*, and *Generation of animals*, Venice 1476 (Latin, Gaza's translation), Venice 1497 (Greek) Theophrastus, *History of plants* and *Causes of plants*, Treviso 1483 (Latin, Gaza's translation), Venice 1497 (Greek, in the Aldine Aristotle). Theophrastus received little attention from Renaissance commentators: Peter Dilg, "Botanische Kommentarliteratur" (1975), 230–31. There were relatively few commentaries on Aristotle's books on animals, other than those sometimes included in the *Parva naturalia*; Charles H. Lohr, *Latin Aristotle commentaries* (1988–95), lists commentaries by Jacobus Carpentarius, Caesar Cremoninus, C. G. Fontanus, Daniel Furlanus, Christophorus Guarinonius, Conradus Horneius, Archangelus Mercenarius, Augustinus Niphus, Caesar Odo, Petrus Pomponatius, Simon Portius, Julius Caesar

Scaligerus, Jacobus Schegkius, Carolus Stain, Nicolaus Leonicus Thomaeus (not to be confused with Nicolò Leoniceno), and Franciscus Vicomercatus.

10. G. E. L. Owen, "Tithenai ta phainomena" (1961).

11. French, *Ancient natural history*, 93, 105; the latter claim is based on Hort's preface to Theophrastus, *Enquiry into plants* (1916), 1: xxiv, who calls it a "guess."

12. E.g., Werner Jaeger, as noted by Maurice Manquat, *Aristote naturaliste* (1932), 95–96.

13. French, *Ancient natural history*, 85–87; for references to the *Problemata*, see Aristotle, *Generation of animals* (1942), 2.8, 747b5, 4.7, 75b35.

14. Pliny the Elder, *Natural history*, 8.17.44.

15. Quoted by Gessner, *Historiae animalium lib. I*, sig. α5r.

16. Manquat, *Aristote naturaliste*, 95–101.

17. J. M. Bigwood, "Elephant" (1993).

18. On the garden, see Theophrastus's will in Diogenes Laertius, *Lives* (1925), 5.2.52–53.

19. Gessner, *Historiae animalium lib. I*, sig. α5r–6r; Conrad Gessner, *Historiae animalium liber II.* (1554), sig. *2r–v; Conrad Gessner, *Historia animalium liber III* (1555), sig. a4r (no mention of Alexander); Conrad Gessner, *Historiae animalium liber IIII* (1558), sig. a4v–a6r.

20. Gessner, *Historiae animalium lib. I*, sig. α5r. Hyperbole was expected in *epistolae nuncupatoriae* such as this.

21. On Albertus's natural history, see Christian Hünemörder, "Zoologie des Albertus" (1980), Jerry Stannard, "Medieval gardens" (1983), and Albertus Magnus, *On animals* (1999).

22. Anyone familiar with the Hippocratic Corpus or Galen knows that philosophical and medical ideas were hardly separated in the Hellenistic world, but philosophy, as a way of life, was quite distinct from medical teaching and practice.

23. John M. Riddle, *Dioscorides* (1985).

24. Karen M. Reeds, *Botany* (1991), 32.

25. For an overview, see Jerry Stannard, "Natural history" (1978). Among the older literature, Ernst H. F. Meyer, *Geschichte der Botanik* (1854–57), provides a detailed examination of medieval writings on plants.

26. Jerry Stannard, "Medieval herbals" (1974), 27.

27. Gerold Hayer, "Kontextüberlieferung" (1992).

28. See the fascinating discussion in Mauro Ambrosoli, *Scienziati, contadini e proprietari* (1992), chapter 1, on which my account is based.

29. Olivier de Serres, *Théâtre d'agriculture* (1600); Thomas Hill, *Gardener's labyrinth* (1987), originally published 1577.

30. Pierre Belon, *De neglecta stirpium cultura* (1589).

31. On Frederick II, see Charles Homer Haskins, *Studies in mediaeval science* (1927), and Frederick II, *Art of falconry* (1943).

32. Reeds, *Botany*, 11.

33. Claimed, usually, by those who wish to underscore the Renaissance "discovery of the world," following in the steps of Jules Michelet and Jacob Burckhardt. See Jacob Burckhardt, *Civilization of the Renaissance* (1954), 218–25.

34. Catalogued in Wilhelm Ganzenmüller, *Naturgefühl* (1914).

35. Such conventions have been noted by historians of science, e.g., I. Bernard Cohen, "Découverte et transformation" (1960); Cohen's article, however, attributes the change in the sixteenth century to the discovery of America. See below for an alternative explanation.

36. Ernst Robert Curtius, *European literature* (1953), 194–95, 197.

37. Remarked also by Dieter Hennebo, *Gärten des Mittelalters* (1962), 27.

38. Curtius, *European literature*, 184; Stannard, "Medieval gardens," 59, makes the same observation about plants.

39. Hennebo, *Gärten des Mittelalters*, 22–23; Heinrich Balss, *Albertus Magnus* (1947), 21, referring to a work cited as "J. Wilde 1936" that does not appear in Balss's bibliography. Other scholars have seen the plants in the *Capitulare de villis* as evidence not for literary borrowing but for the European climate before the "Little Ice Age" of the late Middle Ages: J. C. Russell, "Bevölkerung Europas" (1978), 31.

40. Ovid, *Metamorphoseon libri* (1975), 10.86–106.

41. Ganzenmüller, *Naturgefühl*, 27–30, 191–93, 230–33.

42. Thomas M. Greene, *Light in Troy* (1982).

43. Karl Pomeroy Harrington, *Medieval Latin* (1925), 320.

44. *Imitation of Christ [1530/1941]* (1941), 4, 154. The *Imitation* is usually attributed to Thomas à Kempis, but some modern scholars have questioned this attribution.

45. *Imitatio Christi* 2.4, quoted in Curtius, *European literature*, 320.

46. Isidore of Seville, *Etymologiae* (1911), 12.2.21. The translation from Juvenal is taken from Juvenal, *Juvenal and Persius* (1918).

47. Florence McCulloch, *Medieval bestiaries* (1960); Beryl Rowland, "The art of memory and the bestiary" (1989). The position that bestiaries were scientific works is taken by T. H. White in his introduction to *Book of beasts* (1984), and defended by Wilma George and Brunsdon Yapp, *Naming of the beasts* (1991).

48. *Book of beasts*, 49–50.

49. *Book of beasts*, 28. Isidore was not the bestiary's source (or at least not its only source), for it adds the curious detail that, if the same beaver is pursued by another hunter, "he lifts himself up and shows his members to him. And the latter, when he perceives the testicles to be missing, leaves the beaver alone."

50. Quoted in Curtius, *European literature*, 319.

51. Hans Blumenberg, *Lesbarkeit der Welt* (1983), 51.

52. Clarence J. Glacken, *Traces* (1967), 44; Brian W. Ogilvie, "Natural history" (2005). Wonder is not *necessarily* opposed to careful investigation of nature; their opposition is a historical contingency. See Lorraine Daston and Katharine Park, *Wonders* (1998).

53. Albertus Magnus, "De vegetabilibus" (1891), 6.1.1, p. 159. The notion that philosophy excluded particulars was so ingrained in medieval scholastic philosophy that Thomas Aquinas was at pains to argue that the study of God and the particular events of sacred history could count as philosophy: *Summa theologiae*, vol. 1, Ia Iae, quaestio 1, art. 2, 10–13.

54. Hence the persistence of "errors" in medieval bestiaries, such as the belief that the weasel conceives through the ear and gives birth through the mouth (or vice versa) (*Book of beasts*, 92). It would have been easy for a curious investigator

to determine that this was not the case, but the book's compiler saw no need to do so.

55. Hans Blumenberg, *Prozeß der Neugierde* (1973), esp. 103ff.; Lorraine Daston, "Curiosity" (1995), esp. 392–94.

56. Petrarch's place in the history of humanism is still being debated. Two subtle recent treatments are Carol Everhart Quillen, *Rereading the Renaissance* (1998), who identifies Petrarch as the progenitor of a humanist style of intellectual life while recognizing the many "medieval" aspects of his thought, and Ronald G. Witt, *Footsteps of the ancients* (2000), according to whom Petrarch Christianized an earlier, secular humanist movement.

57. Francesco Petrarca, "On his own ignorance" (1948), 57–59.

58. Petrarca, "On his own ignorance," 80–83.

59. Cf. Aristotle's position in *Parts of animals* 1.5: "It now remains to speak of animals and their Nature. So far as in us lies, we will not leave out any one of them, be it never so mean; for though there are animals which have no attractiveness for the senses, yet for the eye of science, for the student who is naturally of a philosophic spirit and can discern the causes of things, Nature which fashioned them provides joys which cannot be measured." Aristotle, *Parts, movement, progression* (1961), 99.

60. On this transformation in natural theology, see Ogilvie, "Natural history."

61. Francesco Petrarca, "Ascent" (1948), 36, 44.

62. Petrarca, "Ascent," 40, 44.

63. Daston and Park, *Wonders*, 75–86.

64. Hans Baron, *Crisis* (1966), 227.

65. Burckhardt, *Civilization of the Renaissance*, 222–24.

66. Conrad Celtis, "Ad Sepulum disidaemonem," in Harry C. Schnur, ed., *Lateinische Gedichte* (1978), 42.

67. Conrad Celtis, "De situ et moribus Norimbergae," in Conrad Celtis, *Opuscula* (1966).

68. Francesco Colonna, *Hypnerotomachia Poliphili* (1999), 21; cf. Francesco Colonna, *Hypnerotomachie* (1546), fol. 3v. The original Italian text was composed in the 1450s and 1460s, and first published in 1499. The sixteenth-century French translation provides a more copious account of the plants that Poliphile encounters than do the original and the modern English version.

69. Colonna, *Hypnerotomachia Poliphili*, 68.

70. Colonna, *Hypnerotomachia Poliphili*, 290–325.

71. Margherita Azzi Visentini, *L'Orto botanico di Padova* (1984), 81–83.

72. Niccolò Leoniceno, *De Plinii erroribus* (1529), 216.

73. In the wake of Charles Trinkaus, *In our image* (1970), it is difficult to maintain the older view, represented positively by Burckhardt and negatively by Ludwig von Pastor, that the Italian humanists were essentially pagans. But it is true that Christianity was much more central to Erasmus's conception of *bonae litterae* than to most of the Italian humanists' projects.

74. Cf. the emblem on this subject by Joachim Camerarius, *Symbolorum centuria altera* (1595), fol. 103, "Non tibi spiro."

75. Desiderius Erasmus, *Opera [ed. Le Clerc]* (1703), 1: 673–74.

76. Repeatedly emphasized by Cato the Elder, M. Terentius Varro, and L. Junius Columella in prose, and by Horace in verse.

77. See, for example, the younger Pliny's description of his Laurentian villa, with its garden and flowers: Pliny the Younger, *Epistulae*, 2.17. On the influence this description, see Pierre de La Ruffiniere du Prey, *Villas of Pliny* (1994).

78. See Hennebo, *Gärten des Mittelalters*, 158–78; Dieter Hennebo and Alfred Hoffmann, *Architektonischer Garten* (1963), 26–34.

79. Erasmus, *Opera [ed. Le Clerc]*, 1: 675.

80. See the brief discussion in Jonathan Israel, *Dutch Republic* (1995), and the works cited there. The schools that Erasmus attended were not operated by the Brethren, but he lived in their hostels while attending them: James McConica, *Erasmus* (1991), 6–7.

81. Subjects of two other Erasmian colloquies: the "Convivium profanum" and the "Convivium poeticum"—not to mention the "Convivium fabulosum."

82. François Rabelais, *Gargantua and Pantagruel* (1990), 58, 158.

83. Blumenberg, *Prozeß der Neugierde*, 114.

84. Julius Caesar Scaliger, *Poetices libri* (1561), 323.

85. Scaliger, *Poetices libri*, 324, 328, 333, 338, 341.

86. Gessner died of plague in 1565, before he could complete his *Historia plantarum*; material from it was first published in the late eighteenth century: Conrad Gessner, *Opera botanica* (1754–71).

87. Conrad Gessner, *De raris herbis* (1556), 43–67.

88. Gessner, *De raris herbis*, 47.

89. Gessner, *De raris herbis*, 48, 50.

90. Gessner, *De raris herbis*, 63–66.

91. Leonhart Fuchs, *De historia stirpium* (1542), fol. α2v, 3v–4v, β2v; translation by Elaine Mathers and John L. Heller in Frederick G. Meyer, Emily Emmart Trueblood, and John L. Heller, *Great Herbal of Fuchs* (1999), 1:200.

92. Thomas DaCosta Kaufmann and Virginia Roehrig Kaufmann, "Sanctification of nature" (1993).

93. Edward Lee Greene, *Landmarks* (1983); Peter Dilg, "Studia humanitatis" (1971); Karen M. Reeds, "Humanism and botany" (1976); William B. Ashworth, Jr., "Emblematic world view" (1990).

94. The literature is vast. Among moderns, the two extremes are represented by Eugenio Garin and Paul Oskar Kristeller; Garin develops the notion that humanism was a philosophical system, while Kristeller argues that humanism should be restricted to the *studia humanitatis*, the five fields of grammar, rhetoric, poetry, history, and ethics that were the basis of Renaissance educational reform. Hans Baron has argued for the importance of the narrower concept of "civic humanism" in Renaissance republican thought. See Eugenio Garin, *Italian humanism* (1965); Paul Oskar Kristeller, *Renaissance thought* (1979); Baron, *Crisis*; a useful overview is provided by Albert Rabil, Jr., "The significance of 'civic humanism'" (1988).

95. Recent overviews in Charles G. Nauert, Jr., *Humanism and Renaissance Europe* (1995); Jill Kraye, ed., *Companion to humanism* (1996); Charles B. Schmitt, *Aristotle and the Renaissance* (1983); Erika Rummel, *Humanist-scholastic debate* (1995).

96. The notion that "family resemblances" are more useful means of defining collective entities than strict definitions has been developed in several different contexts: inter alia, see David L. Hull, *Science as a process* (1988), and Florike Egmond and Peter Mason, *Mammoth and mouse* (1997), 5. Prototype theory provides a formal

account of such resemblances: George Lakoff and Mark Johnson, *Metaphors* (1980), 122–25.

97. Rummel, *Humanist-scholastic debate*, examines the relationship between humanism and scholastic philosophy through the optic of disciplinary claims and struggles over university resources, especially in northern universities where humanist studies were established in the late fifteenth and early sixteenth centuries. Her approach complements mine, which focuses more on the intellectual *habitus* that humanist studies developed.

98. A partial exception is provided by the notion of tekmeriodic proof in the Greek commentary tradition to Aristotle's logic: Donald Morrison, "Tekmeriodic proof" (1997). Another exception, the sixteenth-century logical analysis of *demonstratio quia*, may have been prompted by a desire to logically justify the kind of thing that was being taken as knowledge by sixteenth-century naturalists and other empiricists.

99. The *studia humanitatis* are defined by Kristeller, *Renaissance thought*, 22–23. Medicine and law were the chief professional faculties in Italian universities, where humanist teachers first gained a foothold and where the naturalists of the late fifteenth and early sixteenth centuries were educated (see chapter 2); faculties of theology were relatively insignificant in late medieval Italian universities. Alan B. Cobban, *Medieval universities* (1975), 70–71.

100. Paul F. Grendler, *Schooling* (1989), 162–72.

101. Grendler, *Schooling*, 164–65; Jan Pinborg, "Speculative grammar" (1982), 254–69; Brian P. Copenhaver and Charles B. Schmitt, *Renaissance philosophy* (1992), 96–101.

102. W. Keith Percival, "Renaissance grammar" (1988), 75–76; but see Copenhaver and Schmitt, *Renaissance philosophy*, 209–27, for Valla's grammatical assault on medieval metaphysics.

103. Peter Godman, *From Poliziano to Machiavelli* (1998), 38–52.

104. Example in Juan Luis Vives, "Epistola I. De ratione studii puerilis" (1782).

105. Danilo Aguzzi-Barbagli, "Humanism and poetics" (1988).

106. Aguzzi-Barbagli, "Humanism and poetics," 114–19.

107. Craig Kallendorf, *Virgil* (1999); Anthony Grafton and Lisa Jardine, *Humanism* (1986), 20–23; Grendler, *Schooling*, 244–50.

108. Grafton and Jardine, *Humanism*, and Kallendorf, *Virgil*, emphasize the relation between the humanist paraphrase-commentary method of teaching and the focus on particular elements in a text to the neglect of the work as a whole. For a contrasting look at how the most subtle Renaissance readers approached texts, see Anthony Grafton, *Commerce* (1997).

109. Fundamental studies of humanist textual scholarship are Pierre de Nolhac, *Pétrarque et l'humanisme* (1965); Rudolf Pfeiffer, *Classical scholarship 1300–1850* (1976); Anthony Grafton, *Joseph Scaliger* (1983–93), vol. 1; John F. D'Amico, *Beatus Rhenanus* (1988).

110. Jerry H. Bentley, *Renaissance Naples* (1987), 57.

111. On Perotti and the *Cornucopia*, see Anthony Grafton, *Defenders of the text* (1991), 50, 54.

112. Ronald G. Witt, "Origins of humanism" (1988); Witt, *Footsteps of the ancients*; Hanna H. Gray, "Renaissance humanism" (1963).

113. Grendler, *Schooling,* 120–21.

114. Nancy S. Struever, *Language of history* (1970). Other roots of the humanist sense of the past can be found in poetics: Greene, *Light in Troy.*

115. Lorenzo Valla, *De Constantini donatione* (1976).

116. For an overview, see Donald R. Kelley, "Humanism and history" (1988), and Eric Cochrane, *Historians* (1981).

117. Bentley, *Renaissance Naples,* 235.

118. The earliest new histories of the ancient world (as opposed to antiquarian treatises on specific aspects of it) were written in the seventeenth century: Arnaldo Momigliano, "Ancient history" (1950).

119. Desiderius Erasmus, *The Ciceronian* (1986). A measured account of Roman Ciceronianism and its cultural-political context is provided by John F. D'Amico, *Humanism in Papal Rome* (1983), 123–34. On Poliziano's response to early Ciceronians, see Greene, *Light in Troy,* 149–52, and Godman, *From Poliziano to Machiavelli,* 44–51.

120. Francesco Guicciardini, *Ricordi* (1951), no. 110; see also Felix Gilbert, *Machiavelli and Guicciardini* (1965), 247.

121. On Petrarch's struggles, see Quillen, *Rereading the Renaissance,* especially on Petrarch's reading of Augustine. Humanists' sometimes forced moral readings are examined incisively by Grafton, *Defenders of the text,* 37–38; anyone who reads Erasmus's *Enchiridion militis christianae* can see how the prince of humanists struggled to reconcile his commitment to Christian morals with his love of ancient literature.

122. My discussion in this paragraph follows Donald R. Kelley, *Foundations* (1970).

123. Quentin Skinner, *Foundations* (1978), 2: 269–70.

124. Schmitt, *Aristotle and the Renaissance.*

125. Medical humanism has been the subject of a burgeoning field of research for the last generation. For an orientation and overview, see Nancy G. Siraisi, "Some current trends" (1984); Nancy G. Siraisi, *Medieval and early Renaissance medicine* (1990); and Michael R. McVaugh and Nancy G. Siraisi, eds., *Renaissance medical learning* (1990).

126. On medical philology, in addition to the references in the previous note, see Vivian Nutton, "Humanistic surgery" (1985); "Caius and Galen" (1985); and "Prisci dissectionum professores" (1988).

127. I did not read Dilg's "Botanische Kommentarliteratur" until this book was far along in revisions; it is a useful introduction to the humanist commentaries on Pliny and Dioscorides, though the general framework of the development of botany in which it is set is no longer adequate. For the broader context of late Renaissance humanist scholarship in Italy, see (inter alia) Grafton, *Defenders of the text,* chapter 2; W. Scott Blanchard, *"O miseri philologi"* (1990); and Godman, *From Poliziano to Machiavelli.*

128. Arno Borst, *Buch der Naturgeschichte* (1995), 292, 300–317; Charles G. Nauert, Jr., "C. Plinius Secundus" (1980), esp. 304–11.

129. Borst, *Buch der Naturgeschichte,* 316–17, who however neglects the work of Beatus Rhenanus: see D'Amico, *Beatus Rhenanus.* See also Charles G. Nauert, Jr, "Humanists, scientists, and Pliny" (1979).

130. Margaret L. King, *Venetian humanism* (1986), 197–205.

131. In Giovanni Battista Egnazio, *In Dioscoridem annotamenta* (1516).

132. Reeds, "Humanism and botany," 527, citing a letter from Barbaro to Pontico Faccino, July 1484.

133. Ermolao Barbaro, *Castigationes plinianae* (1973–79), 1: 2. The contents of the work are summarized in Dilg, "Botanische Kommentarliteratur," 232–35.

134. L. D. Reynolds and N. G. Wilson, *Scribes and scholars* (1991), 208–9; E. J. Kenney, *Classical text* (1974), 4–5.

135. Kenney, *Classical text*, ch. 1, esp. 18–19.

136. John Monfasani, "The first call for press censorship" (1988).

137. Poliziano (1454–94) was from humbler social origins than the patrician Barbaro, but they were both brilliant youths with powerful patrons (in Poliziano's case, Lorenzo "il Magnifico" de' Medici), and both died within a year of one another. See Godman, *From Poliziano to Machiavelli*.

138. For those unfamiliar with classical scholarship, this might seem a self-evident procedure. However, it has methodological problems, as Kenney points out in *Classical text*. If the vulgate is founded on a bad manuscript tradition, it is possible that readings that are sensible in it are the result of earlier conjecture to correct corruption, or simply a corruption that does not result in solecism. The modern technique begins with the assumption that the entire text needs to be established before any emendations are made, rather than starting from a text of dubious quality.

139. Barbaro, *Castigationes plinianae*, 3: 875, 1: 2.

140. Barbaro, *Castigationes plinianae*, 2: 698. Borst, *Buch der Naturgeschichte*, 316, observes that Barbaro's "old" manuscripts were scarcely a century old.

141. Barbaro, *Castigationes plinianae*, 2: 810. Rackham's text in the Loeb edition reads "per parietum aspera."

142. This is also the conclusion of Barbaro's editor Giovanni Pozzi, in Barbaro, *Castigationes plinianae*, 1: lxv.

143. Barbaro, *Castigationes plinianae*, 2: 826. Modern editors have suggested "tenuia" for "bina"; cf. Pliny, *Natural history*, 20.36.93 and notes 3 and *a* in the Loeb edition.

144. Barbaro, *Castigationes plinianae*, 3: 853–54, 85; "infirmitatem virilium roborant" was in the vulgate.

145. Barbaro, *Castigationes plinianae*, 2: 739 (cf. Pliny 14.19.111), 2: 822.

146. Barbaro, *Castigationes plinianae*, 3: 872 (cf. Pliny 22.46.92), 3: 897, 2: 832–33.

147. Barbaro, *Castigationes plinianae*, 3: 853.

148. W. H. S. Jones and A. C. Andrews could not identify "culix" or "culex": see the "Index to Plants" in vol. 7 of the Loeb edition.

149. Barbaro, *Castigationes plinianae*, 3: 920–21.

150. Modern classicists still fall into the trap: "It is, moreover, often forgotten that an ancient author—and this perhaps applies especially to Pliny—may himself have made mistakes, even bad ones, that escaped the notice of is *corrector*, if he had one": W. H. S. Jones, in Pliny the Elder, *Natural history*, 7:11. See also Lynn Thorndike, *Magic and experimental science* (1923–58), 4: 601.

151. Barbaro, *Castigationes plinianae*, 1: 2, 2: 753, 3: 838.

152. Barbaro, *Castigationes plinianae*, 3: 897. Jones's text in the Loeb Pliny has "porri."

153. Barbaro, *Castigationes plinianae*, 3: 937.

154. Pliny the Elder, *Natural history*, 25.5.9, 7: 142.

155. Barbaro, *Castigationes plinianae*, 3: 857–58.

156. Contemporaries recognized this: Leonhart Fuchs, quoted by Dilg, "Botanische Kommentarliteratur," 234; and Conrad Gessner, *Bibliotheca universalis* (1545), fol. 317r. This focus on textuality, at the expense of other forms of knowledge, characterizes much humanistic scholarship. For example, Barbaro's contemporary Angelo Poliziano solved the riddle of the "Herculean knot" by reading Oribasius, after puzzling over coins, gems, and reliefs in vain: Angelo Poliziano, *Miscellaneorum centuria secunda* (1978), 53.

157. Gessner, *Bibliotheca universalis*, fol. 317r–18r.

158. For Leoniceno's biography, see Daniela Mugnai Carrara, "Profilo di Nicolò Leoniceno" (1978). Dilg, "Botanische Kommentarliteratur," 253, claims incorrectly that Leoniceno had held the medical chair in Ferrara for sixty years when *De Plinii erroribus* appeared; in fact, Leoniceno occupied the chair of moral philosophy before being promoted to that of practical medicine; he never held the chair of theoretical medicine.

159. Daniela Mugnai Carrara, "Polemica umanistico-scolastica" (1983); Daniela Mugnai Carrara, "Polemica 'de cane rabido'" (1989); William F. Edwards, "Leoniceno and method" (1976); Fridolf Kudlien, "Zwei Polemiken" (1965). Conrad Gessner called Leoniceno "oratorum omnium suae aetatis elegantissimus": Gessner, *Bibliotheca universalis*, fol. 520v.

160. For the context of this polemic in terms of Florentine intellectual politics, see Godman, *From Poliziano to Machiavelli*, 96–112, 212–34.

161. Leoniceno, *De Plinii erroribus*, sig. 3r.

162. Leoniceno, *De Plinii erroribus*, 13, 15.

163. E.g., Leoniceno, *De Plinii erroribus*, 130, on animals.

164. Though Leoniceno's judgment was long echoed by classicists and historians, Pliny's reputation as an observer has improved somewhat. His remarks on animals, for instance, indicate firsthand experience from hunting: L. Bodson, "Aspects of Pliny's zoology" (1986).

165. Leoniceno, *De Plinii erroribus*, 215, 181, 105, 109.

166. Leoniceno, *De Plinii erroribus*, 12, 116–17, 62.

167. Leoniceno, *De Plinii erroribus*, 4, 48, 89–90, 122.

168. Leoniceno, *De Plinii erroribus*, 15, 62, 109.

169. Leoniceno, *De Plinii erroribus*, 116, 216.

170. Greene, *Landmarks*, 2: 541–42.

171. Leoniceno, *De Plinii erroribus*, 65–66.

172. Paracelsus, "Herbarius" (1930), 492–94.

173. Leoniceno, *De Plinii erroribus*, 115–16, 234.

174. See, for example, the discussions at Leoniceno, *De Plinii erroribus*, 18, 62.

175. Leoniceno, *De Plinii erroribus*, 17, 239.

176. Albrecht von Haller, *Bibliotheca botanica* (1771–72), 1: 258, a judgment echoed by Daniela Mugnai Carrara, *Biblioteca di Leoniceno* (1991), 29, who notes more specifically that Leoniceno was "uno dei primi ad applicare il metodo di critica umanistica ai testi medici." The literature contains many different assessments of Leoniceno as a naturalist, depending on the author's impression of the extent of his

observations and importance of his philology. Meyer, *Geschichte der Botanik*, 4: 224–29, and Greene, *Landmarks*, 2: 528–43, are very positive. The influential treatment of Thorndike, *Magic and experimental science*, 4: 593–610, who considered Leoniceno's book a purely philological exercise, has been accepted by, *inter alia*, Reeds, "Humanism and botany," 523–24. For recent reevaluations, see Paula Findlen, *Possessing nature* (1994), 165, and Godman, *From Poliziano to Machiavelli.*

177. Here I simplify a bit: the difference between the Mediterranean flora of the ancient texts and the northern European flora explored by Brunfels, Fuchs, and Bock led to the necessity of writing new books, while the most popular sixteenth-century Italian flora, by Pietro Andrea Mattioli, was ostensibly a commentary on Dioscorides.

178. See Anna Foa, "Il nuovo e il vecchio" (1984).

179. Nancy G. Siraisi, *Avicenna* (1987), 88, remarks that Ferrara's late fifteenth-century statues are similar to early fifteenth century statues in Bologna, at a time when Avicenna formed the center of the curriculum. She adds that it is difficult to tell the degree to which these Ferrarese statues were followed in actual teaching.

180. Ferrante Borsetti Ferranti Bolani, *Historia almi Ferrariae Gymnasii* (1735), 1: 433–34, quoting a fifteenth-century set of statues: in the second year theoretical medicine course, "pro secunda lectione de mane primo legatur *Secundus Canon Avvicennae.*" This is the only work of *materia medica* mentioned in these statues.

181. See, for example, the mention of Serapion: Leoniceno, *De Plinii erroribus*, 4.

182. Leoniceno, *De Plinii erroribus*, 4, 89–90.

183. Leoniceno, *De Plinii erroribus*, 50–51. He also criticized Avicenna in his medical consilia: 89–90; cf. 216.

184. Leoniceno, *De Plinii erroribus*, 18. Elsewhere (e.g., 123–126) Leoniceno admitted that commentators, rather than Avicenna himself, may have been responsible for some errors.

185. Archivio di Stato di Ferrara, sec. XVI, busta 56, Memoriale of 1522 and 1524, quoted in Adriano Franceschini, ed., *Nuovi documenti* (1970), 18–19.

186. Discussed in Mugnai Carrara, "Polemica umanistico-scolastica."

187. Paul Oskar Kristeller, *Medieval aspects* (1992), 5–10. In philosophy, textbooks became important only in the sixteenth century: Charles B. Schmitt, "The rise of the philosophical textbook" (1988).

188. Paracelsus asserted that he had studied in Italy, but no documentary evidence proves it: Pietro Capparoni, "Ritratto di Leoniceno" (1942).

189. On collecting plants in the wild, see Valerius Cordus, *Pharmacorum conficiendorum ratio* (1548); *Antidotarium* (1561). Many apothecaries kept gardens too—Pieter Coudenbergh in Antwerp being a famous example—descended from the simple gardens of medieval monasteries.

190. Reeds, "Humanism and botany," 527, citing Philipp Melanchthon's life of Agricola. On Agricola's importance to German humanists in Melanchthon's generation, see Lisa Jardine, *Erasmus* (1993), chapter 3.

191. Leoniceno, *De Plinii erroribus*, 62, 116, 216.

192. Leoniceno, *De Plinii erroribus*, 17. See chapter 4, below, on descriptive language.

193. Euricius Cordus, *Botanologicon* (1534), 33, where Cordus refers to his "praeceptor" Leoniceno. On Cordus and the *Botanologicon*, see Greene, *Land-*

marks, 1: 360–67, and the excellent article by Dilg, "Studia humanitatis." The reference to Brunfels, published in 1532, and the dialogue's publication in 1534 fix its chronological setting; Conrad Gessner claimed that the dialogue was written in the same year it was published: Gessner, *Bibliotheca universalis*, fol. 229r.

194. Cordus, *Botanologicon*, 27.

195. Cordus, *Botanologicon*, 19, 72ff, 132. Cordus implies, with the word "mendus," that such faults are due to errors in the manuscript transmission rather than Dioscorides himself: 72.

196. Cordus, *Botanologicon*, 71–76 (with digressions), 132.

197. Such suburban gardens, located outside of the city wall and themselves fenced or walled off from the outside, were a common feature of southern German towns in the sixteenth and seventeenth century: Hennebo, *Gärten des Mittelalters*, 166–72.

198. Cordus, *Botanologicon*, 33, 72, 88, 122–23, 125.

199. Cordus, *Botanologicon*, 177.

200. Cordus, *Botanologicon*, 177.

201. Cordus, *Botanologicon*, 91. The identification was accepted by Cordus's successors, e.g., Rembert Dodoens, *Histoire des plantes* (1557), 44–45.

202. Cordus, *Botanologicon*, 109, 124, 139, 76.

203. Cordus, *Botanologicon*, 69, 133–34, 92, 160, 140, 51–52.

204. See, for example, Leoniceno, *De Plinii erroribus*, 50–51, on the medical importance of a correct knowledge of plants. Cordus, *Botanologicon*, 19, 65–66, stresses that this is a part of medicine and it behooves physicians, and not merely apothecaries, to know it.

205. Adriaan van de Spiegel, *Isagoge* (1606), 97.

206. Riddle, *Dioscorides*, ch. 3–4.

207. Cordus, *Botanologicon*, 143.

208. References in the older historiography to "false" identifications of northern with Mediterranean plants fail to consider the sometimes wide range of variations that early sixteenth-century botanists considered possible within one "species."

209. Meyer, *Geschichte der Botanik*, 4: 300, 316. The early seventeenth-century writer Adriaan van den Spiegel (see below) did not consider Brunfels worth reading, and commented that Fuchs's text was taken primarily from the ancients: Spiegel, *Isagoges*, 127.

210. Cf. Dilg, "Studia humanitatis," 78, who argued that Cordus saw the need for new experience and new discoveries. As I see it, he did not view the latter as a consequence of the former.

211. Cordus, *Botanologicon*, 178.

212. Cordus, *Botanologicon*, 102, 110, 74.

213. This contradicts the position taken by, among others, Greene, *Landmarks*, 1: 360–67. This is partly a matter of emphasis: Greene emphasizes Cordus's "progressive" attitudes, whereas I am more convinced by the total impression of conservativism.

214. Klaus A. Vogel, "America" (1995).

215. Dilg, "Botanische Kommentarliteratur," 230–31.

CHAPTER FOUR

1. André Cailleux, "Nombre d'espèces" (1953), whose figures are, however, not wholly correct.

2. Alexander von Humboldt's essay on the physiognomy of plants represents an early attempt to distinguish between the interests of descriptive and systematic botany, on the one hand, and what would become plant ecology on the other: Michael Hagner, "Physiognomik bei Humboldt" (1996), 444–50.

3. Rembert Dodoens, *Histoire des plantes* (1557), sig. *iii^{v}.

4. Carolus Clusius to Joachim Camerarius II, Sept. 4, 1582, in Friedrich Wilhelm Tobias Hunger, *Clusius* (1927–43), 2: 391.

5. Clusius to Camerarius, Dec. 1584 (Hunger, *Clusius*, 2: 402). Caspar Bauhin took time off from his other duties to write a scathing attack on the *Historia generalis plantarum* edited by Dalechamps: Caspar Bauhin, *Animadversiones* (1601).

6. Clusius to Camerarius, Sept. 4, 1576 (Hunger, *Clusius*, 2: 329).

7. Carolus Clusius, *Exoticorum libri* (1605), e.g. 62–63, 72, 88–91; Dominicus Baudius, in Clusius, *Exoticorum libri*, sig. †4r–v.

8. Marie-Noëlle Bourguet, "Collecte du monde" (1997).

9. Charles Darwin, ed., *Zoology of the Beagle* (1839–43).

10. Bourguet, "Collecte du monde," 194.

11. William T. Stearn, "Botanical exploration" (1956/57), provides an overview, though concentrating on areas outside of Europe.

12. An extreme example: Carolus Clusius took a year and a half to get from Frankfurt am Main to Montpellier, where he was to study medicine; he was observing plants along the way: Hunger, *Clusius*, 1: 18–19.

13. I. Bernard Cohen, "Découverte et transformation" (1960); see chapter 6, below, for further references.

14. Klaus A. Vogel, "America" (1995); more generally, J. H. Elliott, *Old world and the new* (1970), chapter 1, esp. 8–14.

15. Henry Lowood, "New World" (1995). See chapter 5 for a further discussion.

16. Hieronymus Bock, *Kreütterbuch* (1577), sig. c1r, fol. 18r, 196v, 71r.

17. Bock, *Kreütterbuch*, sig. b6r, c1r, fol. 226v–27r.

18. Brigitte Hoppe, *Kräuterbuch des Hieronymus Bock* (1969), esp. 44. Bock described 840 plants, but 34, according to Hoppe, are repeated.

19. Bock, *Kreütterbuch*, fol. 310r.

20. Bock, *Kreütterbuch*, fol. 32r, 163v.

21. Edward Lee Greene, *Landmarks* (1983), 1: 312.

22. Bock, *Kreütterbuch*, sig. b6r.

23. See inter alia the account in Claus Nissen, *Botanische Buchillustration* (1951), 1: 39–48.

24. Brunfels's description of his method is from his *Onomasticon medicinae*, quoted by Conrad Gessner, *Bibliotheca universalis* (1545), fol. 531r. Greene, *Landmarks*, 1: 307, notes that Fuchs tried to disguise this: but it was clear to contemporaries. Adriaan van den Spiegel did not consider Brunfels even worth reading.

25. See the detailed discussion by Hoppe, *Kräuterbuch des Hieronymus Bock*, 15ff.

26. Bock, *Kreütterbuch*, sig. b5v.

27. Leonhart Fuchs, *De historia stirpium* (1542), fol. α5v; I quote the translation by Elaine Mathers and John L. Heller in Frederick G. Meyer, Emily Emmart Trueblood, and John L. Heller, *Great Herbal of Fuchs* (1999), 1: 209.

28. Johannes Crato to Conrad Gessner, in Valerius Cordus, *Annotationes* (1561), sig. biir.

29. Crato to Gessner, in Cordus, *Annotationes*, sig. bii^{v}–biii^{r}.

30. Gessner to Heroldus, in Cordus, *Annotationes*, fol. 85r.

31. Cordus, *Annotationes*, fol. 217r ff.

32. Cordus, *Annotationes*, fol. 218r. Fol. 218r–221r contain remarks on stones and metals; fol. 221r–223v have notes on plants; fol. 223v–224r are devoted to animals.

33. Cordus, *Annotationes*, fol. 221r, 222v.

34. Cordus, *Annotationes*, fol. 221r (Muscus quidam, Fungus grandis, Coton herba [g], Herba hederaceis aut cucumeraceis foliis), 221v (Herba foliis Narcissi [g], Herba caule cubitali, Herba foliis latiusculis, Herba irinis foliis [g]), 222r (Herba cicutae foliis, Herba carduaceis foliis, Herba trientalis, Herba Ilycoctoni . . . foliis, Fruticulus tenuibus . . . virgultis [g], Herba umbellifera [g], Herba radice nucis avellanae figura [g], Fruticulus Sicilianae foliis [g]), 222v (Frutex tenuibus . . . aculeis, Herba Cicutae foliis). Those marked with a (g) were observed in gardens, others in their native locations.

35. Gessner to Heroldus, in Cordus, *Annotationes*, fol. 85r.

36. Walther Ryff praised Cordus's memory in the first edition of the latter's annotations on Dioscorides: quoted in Cordus, *Annotationes*, sig. biii^{v}.

37. Cordus, *Annotationes*, fol 223r. Other entries on this page have the same annotation.

38. Cordus, *Annotationes*, fol. 185v–90r. "De Larice," "De Oleastro Germanico," "De Tamarisco," "De Pseudocrania," "De Trifolia arbore," "De Corno foemina," "De Cotonea malo sylvestri," and "De Genista angulosa."

39. Cordus, *Annotationes*, fol. 187v (*Historia*), 223r ("Sylva").

40. Cordus, *Annotationes*, fol. 188r–89r, 86r, 186r.

41. Bock, *Kreütterbuch*, fol. 87v–88r. Bock mentions a possible identification of the "kleine Walwurtz" with Dioscorides's "Petraeum Symphiton" only to reject it immediately.

42. Cordus, *Annotationes*, fol. 221v.

43. Greene, *Landmarks*, vol. 1, chapters 7, 9.

44. Hunger, *Clusius*, 1: 135. This figure is taken by Hunger from a letter from Clusius to Johann Crato von Krafftheim.

45. Carolus Clusius, *Stirpium per Hispanias historia* (1576), sig. A2r–A3r.

46. Clusius mentioned the garden's sad fate in a letter to Camerarius, Oct. 29, 1577 (Hunger, *Clusius*, 2: 351): "One day it was completely razed, without anyone telling me, and the next day horses were let in to be exercised."

47. Some two hundred letters from Clusius to Camerarius, beginning in 1573 and going to the latter's death in 1598, are currently in the Trew-Briefsammlung of the Universitätsbibliothek Erlangen. They were transcribed and published by Hunger, *Clusius*, 2: 293–449.

48. Clusius to Camerarius, Mar. 7, 1574, Jul. 27, 1574, and Oct. 16, 1574 (Hunger, *Clusius*, 2: 297, 298, 303).

49. Clusius to Camerarius, Jun. 24, 1576 (Hunger, *Clusius*, 2: 323–24); Carolus Clusius, *Stirpium per Pannoniam historia* (1583), sig. *2v–*3r.

50. Not that Clusius considered the ancient botanists' works useless. See chapter 5, below, on the banyan tree.

51. Clusius to Camerarius, 14 Mar. 1575 (where he implies that his earlier catalogue was not complete), Aug. 24, 1575, Nov. 27, 1574, and Jul. 24, 1580 (Hunger, *Clusius*, 2: 304–5, 307–8, 310–12, 377).

52. See above; Greene, *Landmarks*, 1: 352–55.

53. Clusius to Camerarius, Aug. 24, 1575 (Hunger, *Clusius*, 2: 310).

54. Julius Caesar Scaliger, *Poetices libri* (1561), 324.

55. Clusius to Camerarius, Aug. 14, 1576 (Hunger, *Clusius*, 2: 326–27). Cf. Clusius, *Stirpium per Pannoniam historia*, 277.

56. I do not have space for a discussion of menageries as a research tool in natural history; they were insignificant in comparison with gardens, herbaria, and cabinets. Gustave Loisel's remarks on menageries are still useful, though they must be treated with caution: Gustave Loisel, *Histoire des ménageries* (1912), 2: 285–334.

57. Ernst H. F. Meyer, *Geschichte der Botanik* (1854–57), 4: 255ff.; Andrew Cunningham, "The culture of gardens" (1996). Hesso Veendorp and L. G. M. Baas Becking, *Hortus academicus* (1938), take a similar approach, though they mention a few medieval cloister gardens as predecessors.

58. Conrad Gessner, *Horti Germaniae*, (1561), fol. 239v.

59. Christopher Thacker, *History of gardens* (1979), 83–84.

60. Jerry Stannard, "Medieval gardens" (1983), 56.

61. Dieter Hennebo, *Gärten des Mittelalters* (1962), 28–36.

62. Hennebo, *Gärten des Mittelalters*, 41–47. Clemens Alexander Wimmer, *Geschichte der Gartentheorie* (1989), 23, argues however that the herbs grew around the edges of the meadow. Albertus's text is not entirely clear.

63. A selection of such miniatures is provided by Marilyn Stokstad and Jerry Stannard, *Gardens of the Middle Ages* (1983); see also Robert Defrance, ed., *Le jardin médiéval* (1990).

64. On such orti in Italy see David R. Coffin, *Gardens in Papal Rome* (1991), 3–4. For Germany, see Hennebo, *Gärten des Mittelalters*, 165, and Dieter Hennebo and Alfred Hoffmann, *Architektonischer Garten* (1963).

65. Hennebo, *Gärten des Mittelalters*, 166; Hennebo and Hoffmann, *Architektonischer Garten*, 26–34. As Claudia Swan has noted, the extent of these urban and suburban gardens has rarely been emphasized outside of garden history: Claudia Swan, "Les fleurs" (1996), 91.

66. See Matthaeus Merian, "Descriptio Basileae" (1615). A brief overview of Basel's early modern topography is given by Valentin Lötscher, "Zur Topographie," in Felix Platter, *Beschreibung und Pestbericht* (1987), 45–61.

67. Platter, *Beschreibung und Pestbericht*, 170, 172, 178, 180, 196, 238, 286, 290, 442 (nos. 527, 535, 566, 580, 683, 1008, 1274, 1291, 2239). Platter also had two fruit gardens down the street from his houses (p. 174, no. 537).

68. Euricius Cordus, *Botanologicon* (1534), 67, 89.

69. Mark Morford, "Stoic garden" (1987), 166; Hunger, *Clusius*, 1: 193, quoting Clusius's correspondence with J. van Hogheland.

70. Commercial gardening on a large scale began in the late sixteenth century in the northern Netherlands, but such commercial gardens were found earlier on the edges of cities: Jan de Vries, "Landbouw, 1490–1650" (1980), 31.

71. Hennebo, *Gärten des Mittelalters*, 175–76.

72. Desiderius Erasmus, "Convivium religiosum" (1703), col. 672, 674–76.

73. Wilfried Hansmann, *Gartenkunst* (1983), 71–73.

74. Coffin, *Gardens in Papal Rome*, 6.

75. Gessner, *Horti Germaniae*, fol. 237v–38r.

76. Gessner, *Horti Germaniae*, fol. 238r.

77. Cf. the correspondence between Carolus Clusius and Matteo Caccini, a minor Florentine noble: Leiden UB, MS. BPL 2724a, published in Carolus Clusius, *Lettere a Caccini* (1939). Caccini was primarily interested in pretty garden plants, but nonetheless was able to furnish Clusius with material he was interested in. Such examples should caution us against a strict separation of science and pleasure in the *res herbaria* of the late Renaissance.

78. Gessner, *Horti Germaniae*, fol. 238r–39v.

79. Compiled from Gessner's references (Gessner, *Horti Germaniae*, fol. 238r–39v). The one noble was the Herzog zu Bayern. "Germany" in Gessner's usage encompassed the towns along and near the Rhine from Chur to Antwerp, along with Württemberg, Bayern, and Sachsen.

80. On Coudenbergh's garden, see Leo J. Vandewiele, "Garden of Coudenberghe" (1993). Carolus Clusius had a low opinion of the Fugger gardens in the 1570s, based on what his friend Joachim Camerarius had observed there: Clusius to Camerarius, July 16, 1577 (Hunger, *Clusius*, 2: 345).

81. Gessner, *Horti Germaniae*, fol. 238r.

82. See chapter 2, above. The quasi-professional aspect is due to common training in medical schools, not to any sort of notion of botany as a professional activity—it was "professional" for only a tiny number of people who were medical professors or, possibly, garden intendants for hire such as Clusius and Jacques Plateau. On the latter see Charles de Croy to Carolus Clusius, Sept. 25, 1604 (Leiden UB, MS. VUL 101. s.v. "Principes," 1).

83. Clusius to Camerarius, Oct. 8, 1577 (Hunger, *Clusius*, 2: 350).

84. Hunger, *Clusius*, 1: 102, 112; Clusius, dedicatory letter to *Antidotarium* (1561).

85. Charles Webster, "Bauhin, Jean" (1970–80).

86. Leonhard Dold to Jacob Zwinger, Oct. 19, 1598 (Basel UB, MS. Fr.Gr. I.12, no. 130).

87. Alpino to Bauhin, May 7, 1607 (Basel UB, MS. Fr.Gr. II.1, 91).

88. Karen M. Reeds, *Botany* (1991), 89–90.

89. Veendorp and Baas Becking, *Hortus academicus*.

90. See especially Erik de Jong, *Natuur en kunst* (1993), 190ff.

91. On the Leiden garden, see the discussion below. For Paris, see Antoine Schnapper, *Le géant* (1988), 42. More generally, see Swan, "Les fleurs."

92. John Prest, *Garden of Eden* (1981); Cunningham, "The culture of gardens," 38, 41.

93. Robert Mallet, *Jardins et paradis* (1959), text facing fig. 28 (no pagination).

94. Genesis 2.8–9: "Plantaverat autem Dominus Deus paradisum voluptatis a principio, in quo posuit hominem quem formaverat. Produxitque Dominus Deus de humo omne lignum pulchrum visu, et ad vescendum suave."

95. As Cunningham, "The culture of gardens," 39, also notes.

96. John Parkinson, *Paradisus terrestris faithfully reprinted* (1904).

97. Bock, *Kreütterbuch*, sig. b1r–v.

98. Gessner, *Horti Germaniae*, fol. 239v–40r.

99. Francis Bacon, *Essays* (1973), 137.

100. Parkinson, *Paradisus terrestris faithfully reprinted*, t.p. facsimile.

101. Gessner, *Horti Germaniae*, fol. 240v.

102. Erasmus, "Convivium religiosum," col. 676.

103. Pierre Belon, *Remonstrances* (1558); Latin translation: Pierre Belon, *De neglecta stirpium cultura* (1589).

104. Kraków BJ, *Libri Picturati* A.18–30: the specific quotation is from A.19, fol. 29; references to the gardens of the "studiosi rei herbariae" (seventeen in all) occur on A.18, fol. 100; A.19, fol. 7, 29; A.23, fol. 75, 83v, 84; A.24, fol. 32, 47, 54; A.25, fol. 27; A.26, fol. 51; A.27, fol. 47, 75; A.28, fol. 29, 55; and A. 30, fol. 5, 83. References to gardens (*horti*) are much more common, but they can refer to simple flower gardens or even kitchen gardens, as the annotations on the illustration of a cucumber indicate (A.28, fol. 72). On this manuscript collection, see the discussion below.

105. P. C. Molhuysen, ed., *Bronnen* (1913–24), 1: 180* (no. 163, 1591 Aug. 12, Benoeming van Paludanus tot prefectus horti).

106. Bock, *Kreütterbuch*, fol. 328v–29r.

107. Gessner, *Horti Germaniae*, fol. 242r.

108. Cf. the description in Johannes van Meurs, *Athenae Batavae* (1625), 31–32. The gallery was added in 1599: Veendorp and Baas Becking, *Hortus academicus*, 42.

109. Camerarius, "Observationes botanicae, 1575" (Basel UB, MS. K II 9), fol. 44r.

110. Clusius to Camerarius, Sept. 7, 1580 (Hunger, *Clusius*, 2: 379).

111. Leonhard Dold to Caspar Bauhin, s.d. (ca. 1596) (Basel UB, MS. G² I 4, fol. 83–84).

112. Gessner, *Horti Germaniae*, fol. 240v.

113. Description by Outger Cluyt in a letter to Pieter Paauw, Nov. 30, 1602 (n.s.), in Molhuysen, ed., *Bronnen*, 1:436*–37*, confirmed by my visit to the Montpellier *hortus botanicus* in December 1995.

114. Nicolas-Claude Fabri de Peiresc to Clusius, Feb. 27, 1604 (Leiden UB, MS. Vulcanius 101, s.v. "Peiresc," no. 5); Peiresc sent a watercolor of the maze to Clusius. The letter, but not the illustration, is reproduced in the *Lettres de Peiresc*, ed. Philippe Tamizey de Larroque, vol. 7.

115. Emmanuel Le Roy Ladurie, *Histoire du climat* (1983), 1: 170–72, 281–87, 2: 5–17.

116. Thomas Hill, *Gardener's labyrinth* (1987), chapter 23, "What care and diligence is required of every Gardener, in the plucking up, and cleare weeding away of all unprofitable hearbs growing among the Garden plants," 77–79.

117. Schnitzer to Bauhin, Aug. 24, 1611 (Basel UB, MS. Fr.Gr. II.1, 302).

118. Clusius to curatores of Leiden University, Aug. 27/Sept. 6, 1592, in Molhuysen, ed., *Bronnen*, 1: 232*.

119. Florence Hopper, "Clusius' world" (1991), 15; see also D. Onno Wijnands, "Commercium botanicum" (1991).

120. Schnapper, *Le géant*, 42.

121. Conrad Gessner to Johannes Fabricius, Sept. 22, 1559 (Basel UB, MS. G^2 I 22, pt. 1, fol. 76).

122. Clusius, *Stirpium per Pannoniam historia*, passim (e.g., various *Auriculae ursi*, 343–50, *Cardamine alpina*, 454–55, *Aster pannonicus*, 526–34).

123. Clusius, *Stirpium per Pannoniam historia*, 193.

124. E.g., the *Iris susiana*, in Clusius, *Stirpium per Pannoniam historia*, 243–44. See Wijnands, "Commercium botanicum," 81, who writes that this plant forms no seeds and all existing specimens are probably derived through vegetative reproduction from the specimens brought to Vienna in 1573.

125. Thomas Platter to Jacob Zwinger, Nov. 4, 1596 (Basel UB, MS. Fr.Gr. I.12, no. 274).

126. See Adriaan van de Spiegel, *Isagoge* (1606), 76–77.

127. Caspar Bartholin to Caspar Bauhin, March 1, 1608 (Basel UB, MS. Fr.Gr. II.1, 50–51).

128. Gessner, *Horti Germaniae*, fol. 239v.

129. The catalogue is printed by Molhuysen, ed., *Bronnen*, 1: 317*–34*. A plan of the garden along with the catalogue is available in Hunger, *Clusius*, 1: 217–35, who copied the catalogue from Molhuysen's printed version. Leslie Tjon Sie Fat, "Clusius' garden" (1991), 6, 9, provides facsimiles of the manuscript plan and a page of the catalogue.

130. The fence in the 1610 engraving surrounds Quadrant 1, areas 2, 4, 6, and 8 in the 1594 plan. The plants in those areas are listed in Hunger, *Clusius*, 1: 219–21.

131. Hunger, *Clusius*, 1: 220–23, 234.

132. E.g., Spiegel, *Isagoge*, 115.

133. To this extent, Prest's *Garden of Eden* is on the mark.

134. On the grouping of similar plants in the *hortus*, see Fat, "Clusius' garden," 7–8. See chapter 5, below, for a discussion of classification in Renaissance natural histories.

135. Clusius, *Stirpium per Pannoniam historia*, 305.

136. Ulisse Aldrovandi to Jacob Zwinger, Oct. 4, 1596 (Basel UB, MS. Fr.Gr. I.13, no. 44).

137. De l'Obel to Bauhin, Nov. 7, 1598 (Basel UB, MS. Fr.Gr. II.1, 59); Schnitzer to Bauhin, Sept. 5, 1607 (o.s.) (Basel UB, MS. Fr.Gr. II.1, 301).

138. Clusius to Camerarius, May 27, 1581 (Hunger, *Clusius*, 2: 383). The Greek prefix *chamae-* meant "lowly, low to the ground," and was often used in a literal as well as figurative sense in Renaissance botanical nomenclature. On parallels between the two, see Allen J. Grieco, "Social politics" (1991).

139. Clusius to Camerarius, Sept. 4, 1582 (Hunger, *Clusius*, 2: 390).

140. Marie de Brimeu, princess of Chimay, called Clusius "the father of all the beautiful gardens of this country" (as mentioned above, p. 78): Hunger, *Clusius*, 1: 3 (without citation). Marie was repeating Justus Lipsius's opinion: Hopper, "Clusius' world," 17, 29 n.17.

141. Clusius, *Stirpium per Pannoniam historia*, 517.

142. Ernst Heinrich Krelage, *Bloembollenexport* (1946), 3, 7, 452–54, 465.

143. Clusius to Camerarius, Nov. 14, 1581 (Hunger, *Clusius*, 2: 386).

144. Clusius, *Stirpium per Pannoniam historia*, 286.

145. E.g., *Soldanella alpina* and various species of *Ranunculus montanus:* Clusius, *Stirpium per Pannoniam historia*, 355, 363.

146. Clusius to Camerarius, July 16, 1577 (Hunger, *Clusius*, 2: 345).

147. *Historia generalis plantarum* (1587–88), 833.

148. On *lusus naturae* generally, see Paula Findlen, "Jokes of nature" (1990).

149. Schnitzer to Bauhin, Aug. 24, 1611 (Basel UB, MS. Fr.Gr. II.1, 302).

150. *Historia generalis plantarum*, 833ff., for *Primulae*.

151. Carolus Clusius, *Rariorum plantarum historia* (1601), 272, noted that he had seen more than a thousand individuals of *Elleborine recentiorum I*.

152. Agnes Arber, *Herbals* (1986), 243.

153. An exception is provided by Clusius's inconclusive investigations into the color of tulips: Carolus Clusius, *Treatise on tulips* (1951), which is a translation of part of the *Rariorum plantarum historia*. See below, p. 205.

154. Dold to Zwinger, Jan. 25, 1598 (Basel UB, MS. Fr.Gr. I, 12, 131).

155. Spiegel, *Isagoge*, 79. The word "herbarium" referred in contemporary parlance to an herbal. However, in conformity with modern usage, I will use "herbarium" to refer to *horti sicci*. This meaning of the word "herbarium" can be dated to the late seventeenth century: Arber, *Herbals*, 142.

156. Compare the instructions in Spiegel, *Isagoge*, 80–81, with those of William Whitman Bailey, *Botanizing* (1899), 87–99, and G. Bimont, *Manuel du botaniste* (1945), 34–49.

157. Spiegel, *Isagoge*, 80–81.

158. A. Louis, *Mathieu de l'Obel* (1980), 292. Louis does not hold this position. Arber dates the first printed mention of herbaria to 1553: Arber, *Herbals*, 140.

159. Arber, *Herbals*, 139; Louis, *Mathieu de l'Obel*, 292–93; Carmelo Battiato, "Luca Ghini" (1972).

160. Arber, *Herbals*, 140.

161. Zwinger to Camerarius, Sept. 1, 1578 (MS. Erlangen Universitätsbibliothek, Trew-Sammlung, Zwinger, Theodor, 30). It is possible that Zwinger might have meant to write "exsuccandae," not "exiccandi," that is, extracting the juice. That would make his request much more comprehensible.

162. See Walther Rytz, "Herbarium Felix Platters" (1932–33).

163. Schnapper, *Le géant*, 38; Arber, *Herbals*, 142.

164. Wilfred Blunt, *Compleat naturalist* (1971), 115.

165. Meyer, *Geschichte der Botanik*, 4: 266; Jean-Baptiste Saint-Lager, "Histoire des herbiers" (1885), 1–2.

166. Saint-Lager, "Histoire des herbiers," 13–23, 24–27.

167. Arber, *Herbals*, 139.

168. Saint-Lager, "Histoire des herbiers," 29.

169. Thomas DaCosta Kaufmann and Virginia Roehrig Kaufmann, "Sanctification of nature" (1993).

170. Kaufmann and Kaufmann, "Sanctification of nature," 20–35, 38–45.

171. Kaufmann and Kaufmann, "Sanctification of nature," 14–15 (examples from Hoefnagel); 29, 35 (late fifteenth- and early sixteenth-century examples).

172. Saint-Lager opined that the origin of printing was connected with the beginnings of pressing plants in books: Saint-Lager, "Histoire des herbiers," 24, 28–29. The Kaufmanns cite the unpublished research of James Marrow in this regard: Thomas DaCosta Kaufmann, *Mastery of nature* (1993), 235 n. 56. The copy of volume two of Otto Brunfels, *Herbarum vivae eicones* (1532), in the Herzog-August-Bibliothek, Wolfenbüttel, has several dried plants pressed between the leaves, but it is impossible to tell if they are contemporary with the work or whether they were added later.

173. For an account, with bibliographic descriptions, see Nissen, *Botanische Buchillustration*.

174. Arber, *Herbals*, 55; F. David Hoeniger, "Plants and animals" (1985), 130–31.

175. Fuchs, *De historia stirpium*; facsimile with commentary in Meyer, Trueblood, and Heller, *Great Herbal of Fuchs*.

176. Saint-Lager, "Histoire des herbiers," 2.

177. Clusius, *Stirpium per Hispanias historia*, 7.

178. Clusius to Camerarius, Dec. 26, 1584 (n.s.) (Hunger, *Clusius*, 2: 403).

179. Clusius, *Exoticorum libri*, 87–88.

180. Clusius, *Exoticorum libri*, 67, 88ff.

181. Clusius, "Memorie voor die Appotteckers ende Chyrugins die den jaer 1602 op de vlote, naer Oost-Indien vaeren sullen," quoted in full by Hunger, *Clusius*, 1: 267 (original in the Algemene Rijksarchief, Den Haag).

182. Clusius, *Exoticorum libri*, 87–88; Nicolas-Claude Fabri de Peiresc to Clusius, Feb. 15, 1605 (Leiden Universiteitsbibliotheek, MS. VUL 101, s.v. "Peiresc," no. 6).

183. Saint-Lager, "Histoire des herbiers," 16–17.

184. Nicolas-Claude Fabri de Peiresc to Clusius, Feb. 25, 1604 and Feb. 15, 1606 (Leiden Univeriteitsbibliotheek, MS. VUL 101, s.v. "Peiresc," nos. 4, 8).

185. See chapter 2, above, and Paula Findlen, "Economy of scientific exchange" (1991).

186. Rytz, "Herbarium Felix Platters," 11.

187. Saint-Lager, "Histoire des herbiers," 51.

188. Hermann Friedrich Kessler, *Das älteste Herbarium Deutschlands* (1870), 11, 16.

189. Christian Callmer, "Queen Christina's herbaria" (1973), 32.

190. Saint-Lager, "Histoire des herbiers," 32–39.

191. Dodoens, *Histoire des plantes*, sig. *iiii^{v}; Caspar Bauhin, *Pinax* (1623), sig. *4v, both citing Galen.

192. Spiegel, *Isagoge*, 78.

193. Bauhin, *Pinax*, sig. *4v.

194. See the discussion on herbaria and travel above.

195. Martin Chmielecius to Jacob Zwinger, Dec. 19, 1591 (Basel UB, MS. Fr.Gr. I,12, no. 96), mentions Caspar Bauhin's identifications and synonyms for plants sent to Basel by Zwinger and his traveling companion Pascal Le Coq.

196. Hunger, *Clusius*, 1: 74.

197. Kessler, *Das älteste Herbarium Deutschlands*, 46, 50, 52, 70, 73, 74, 75, 79, 82, 83, 87 (Bl. 107, 157–60, 174, 175, 396, 428, 429, 435, 442, 448, 449, 462, 512, 515–18,

550, 567, 608). In many cases enough of the plant has survived for Kessler to propose an identification.

198. Kessler, *Das älteste Herbarium Deutschlands*, 63 (Bl. 294), for the plant Linnaeus dubbed *Scandix Pecten veneris*.

199. Rytz, "Herbarium Felix Platters," 126–90, Anhang I, "Kritische Übersicht des Herbarinhaltes."

200. Clusius, *Stirpium per Hispanias historia*, 7.

201. Clusius to Camerarius, Jan. 2, 1578, Jan. 27, 1579, and July 24, 1580 (Hunger, *Clusius*, 2: 354, 366, 377).

202. Martinus Chmielecius to Jacob Zwinger, Nov. 15, 1591 and Dec. 19, 1591 (Basel UB, MS. Fr.Gr. I, 12, 95–96).

203. Aldrovandi to Zwinger, April 5, 1593 and May 1593 (?) (Basel UB, MS. Fr. Gr. I, 12, 2–3).

204. Ulisse Aldrovandi to Cardinal Barberini, no date, quoted in translation without citation in Willy Ley, *Dawn of zoology* (1968), 158.

205. Claudia Swan, "*Ad vivum*" (1995), provides an overview and a discussion of the cognitive implications.

206. Rytz, "Herbarium Felix Platters," 74, 76–77; Walther Rytz, *Pflanzenaquarelle* (1936); Fritz Koreny, *Albrecht Dürer* (1988), 228–31, provides reproductions of two of these drawings, which were remounted in the 1930s.

207. Saint-Lager, "Histoire des herbiers," 95–96.

208. Paula Findlen, *Possessing nature* (1994), 166.

209. Gessner's drawings are available in facsimile: Conrad Gessner, *Historia plantarum* (1987). For Clusius, see P. J. P. Whitehead, G. van Vliet, and William T. Stearn, "Clusius natural history pictures" (1989).

210. Koreny, *Albrecht Dürer*, 85. "Martagon [Turk's cap lily], better than Mattioli or others have portrayed it."

211. See facsimiles in *Botany in the Low Countries* (1993).

212. *Botany in the Low Countries*, 73.

213. Facsimile in Carolus Clusius, *Fungorum historia et Codex Clusii* (1983).

214. Introduction to Clusius, *Fungorum historia et Codex Clusii.*

215. See Clusius's correspondence with his printer Jan Moretus, 1592–93: Carolus Clusius, "Huit lettres" (1895).

216. There is a deservedly growing literature on this collection. See Whitehead, Vliet, and Stearn, "Clusius natural history pictures"; H. Wille, "Albums" (1997); and especially Claudia Swan, "Lectura-imago-ostensio" (2000), with further references. A selection of 142 watercolors from the collection has been published by Claudia Swan, *Clutius botanical watercolors* (1998).

217. Swan, "Lectura-imago-ostensio," 198–203, on Clusius. Swan's attribution of the collection to Cluyt rests on direct and circumstantial evidence. The only direct evidence is the judgment of S. A. C. Dudok van Heel that the vernacular inscriptions in the *Libri Picturati* A.18–A.30 are from Dirck Cluyt's hand, as well as internal evidence that the vernacular and Latin titles are in the same "professional hand," with some exceptions. Circumstantial evidence includes a petition from Leiden medical students in 1598, calling for Cluyt's son to succeed him as prefect of the garden and mentioning the six books of illustrations, and letters from Anselm de Boodt to Clusius in 1602

in which De Boodt mentions having seen a collection of about 1050 illustrations of plants at Cluyt's widow's home.

218. Swan, "Lectura-imago-ostensio," 203 n. 30.

219. Swan has reached the same conclusion: Swan, "Lectura-imago-ostensio," 205. But it is worth underlining that her remarks address especially Clusius's mature work.

220. The folio is reproduced in Swan, *Clutius botanical watercolors*, 31.

221. Kraków BJ, *Libri Picturati* A.30, fol. 28; Swan, *Clutius botanical watercolors*, 31. One annotation uses the adverb "nunc," the other "iam"; I have translated both as "now," which seems evident from the context.

222. Molhuysen, ed., *Bronnen*, 1: *380.

223. For example, the crocus, Kraków BJ, *Libri Picturati* A.30, fol. 67 (Swan, *Clutius botanical watercolors*, 30); *Plantago Thyrsigera*, A.18, fol. 28; *Centaurium magnum*, A.18, fol. 50. Such annotations occur far too frequently to be listed in this note.

224. Kraków BJ, *Libri Picturati* A.18, fol. 75.

225. Kraków BJ, *Libri Picturati* A.18–A.30, passim; see also Swan, "Lectura-imago-ostensio," 203.

226. This is the conclusion of Swan, "Lectura-imago-ostensio."

227. On the seventeen plants cultivated by botanists, see above, p. 157 and n. 104. On the *Septifolium*, Kraków BJ, *Libri Picturati* A.18, fol. 36.

228. Kraków BJ, *Libri Picturati*, A.19, fol 29; A.20, fol. 41; A.30, fol. 81; A.28, fol. 111; A.28, fol. 30.

229. Kraków BJ, *Libri Picturati*, A.23, fol. 84; A.23, fol. 118v; A.29, fol. 28; A.22, fol. 46v (in Swan, *Clutius botanical watercolors*, 24); A.22, fol. 70; A.28, fol. 89v; A.30, fol. 24.

230. Ann Blair, "Humanism and the commonplace book" (1992).

231. Spiegel, *Isagoge*, 132–38.

232. Gessner to Bauhin, Nov. 14, 1563, in Conrad Gessner, *Vingt lettres* (1976), 28.

233. Thomas Erpenius avocated such a system for travelers: Justin Stagl, "Ars apodemica" (1992), 176.

234. Cordus, *Annotationes*, sig. biii^{v} (on his memory).

235. Spiegel, *Isagoge*, dedication, 77, 78–79.

236. See the discussion of classification below, chapter 5.

237. William T. Stearn, *Botanical Latin* (1983), 26–28. The anecdotal style that indicates a naturalist has not only observed and described a plant, but seen it in its natural habitat—and enjoyed the experience—survives in more "popular" works on natural history, including the immensely popular *Natural history of Selborne* by Gilbert White and Darwin's *Journal of researches* (both still in print).

238. I address the parallel and interwoven history of botanical description and illustration more systematically in Brian W. Ogilvie, "Image and text" (2003).

239. Dodoens, *Histoire des plantes*, provides an example of headings; marginal catch phrases are used in the *Historia generalis plantarum*.

240. Valerius Cordus, *Historia plantarum*, (1561), fol. 103v.

241. His most insightful modern commentator, Edward Lee Greene, remarked at the beginning of this century on Cordus's regular phytographic method: Greene, *Landmarks*, 1: 374–76.

242. Cordus, *Historia plantarum*, fol. 109v ff.

243. Compare the descriptions in Konrad von Weihe, ed., *Illustrierte Flora* (1972), which begin with a morphological description of stem, leaves, flower, and seed, followed by an indication of type of habitat and observed range.

244. Theophrastus, *Enquiry into plants* (1916), 1.1.9, 1.1.10.

245. This is a broad generalization. Writers like Gilbert White and John Muir arranged their observations in a temporal framework, noting down in chronological sequence what they saw, with "incidental" detail from a purely morphological standpoint.

246. Conrad Gessner, *De raris herbis* (1556), 24. "qui in hortis nostris mense Aprili & initio Maii florent, longe suavissimo odoris."

247. Clusius, *Stirpium per Hispanias historia*; Clusius, *Stirpium per Pannoniam historia*.

248. For example, Clusius, *Stirpium per Hispanias historia*, 16–17, 289, 304, 321–25.

249. Clusius to Joachim Camerarius II, Jan. 2, 1578, Jan. 27, 1579, and July 24, 1580 (Hunger, *Clusius*, 2: 354, 366, 377).

250. Clusius, *Stirpium per Hispanias historia*, 273, 289–90, 292, 12–13.

251. Clusius arrived in Vienna in the autumn of 1573 and made his first botanical expedition there in the spring of 1574. He sent the manuscript of his *Rariorum stirpium per Pannoniam historia* to Plantin in Antwerp in the spring of 1582: Clusius to Camerarius, May 22, 1582 (Hunger, *Clusius*, 2: 388).

252. Clusius, *Stirpium per Pannoniam historia*, 382–84 (quote p. 384), 483.

253. Dioscorides, *Greek herbal* (1934), 3 (see chapter 3, above).

254. Aristotle, *History of animals* (1965–91), 6.17, 571a25.

255. Clusius, *Stirpium per Pannoniam historia*, 209–11.

256. Clusius, *Stirpium per Pannoniam historia*, 183–85, 486 (quote p. 185).

257. Clusius, *Stirpium per Pannoniam historia*, 378.

258. Clusius, *Stirpium per Pannoniam historia*, 263–65.

259. Clusius, *Rariorum plantarum historia*. The title page bore the epigraph: "Plantae cuique suas vires Deus indidit, atque/Praesentem esse illum, quaelibet herba docet" ("God gave each plant its virtues, and every herb teaches that he is present"). Similar mottoes appear on other natural histories of the later sixteenth century.

260. Spiegel, *Isagoge*, 128.

261. Caspar Bauhin to Leonhard Dold, Jan. 14, 1609 (Erlangen UB, Trew-Sammlung, "Bauhinus, Caspar," no. 44.

262. Basel UB, MS. K I 12, "Praelectiones de simplicibus, in officinis usitatis."

263. Caspar Bauhin, *Prodromos* (1620), sig. 3v.

264. Molhuysen, ed., *Bronnen*, 1: 113, 154.

265. Bock, *Kreütterbuch*, fol. 265v, 266v.

266. Dodoens, *Histoire des plantes*, 443, 444.

267. Clusius, *Stirpium per Pannoniam historia*, 211–16.

268. Hunger, *Clusius*, 1: 328–31, discusses the merits of Clusius's descriptions from the standpoint of modern botany.

269. Clusius, *Stirpium per Hispanias historia*, 321.

270. Bauhin, *Prodromos*, 27. Bauhin added that he had observed this plant in the Hortus Dei near Montpellier.

271. Clusius, *Stirpium per Pannoniam historia*, 484.

272. Dioscorides, *Greek herbal*, 489 (I have checked the translation against Mattioli's version); Pietro Andrea Mattioli, *Opera omnia* (1598), gives the Latin translation of Jean Ruel.

273. Mattioli, *Opera omnia*, 788.

274. Dodoens, *Histoire des plantes*, 29.

275. Clusius, *Stirpium per Pannoniam historia*, 483–84.

276. Spiegel, *Isagoge*, 3.

277. Spiegel, *Isagoge*, 35, 76–77.

278. Theophrastus, *Enquiry into plants*, 1.1.6–1.2.6.

279. Benedict Textor, *Stirpium differentiae* (1537).

280. Textor, *Stirpium differentiae*, fol. 15r–72r.

281. Gessner, "Praefatio," in Hieronymus Bock, *De stirpium commentaria* (1552), sig. c iir.

282. The issue of accuracy in representation has been radically altered in the last century by the introduction of photographic processes: see William M. Ivins, Jr., *Prints and visual communication* (1953), esp. 143–46, 177, on photography's effects on art criticism. Historical actors, however, had a more fluid and relative notion of accuracy. Vasari, for example, could praise Giotto's realism with regard to Cimabue while still attributing a greater degree of realism to Giotto's successors.

283. The history of natural history illustration has been told repeatedly and well; I provide only a summary that brings out the elements that are most significant for understanding the science of describing. Important treatments include: Ludolf Christian Treviranus, *Anwendung des Holzschnittes* (1855); Arber, *Herbals*; Nissen, *Botanische Buchillustration*; Claus Nissen, *Illustrierte Vogelbücher* (1953); more recently, William B. Ashworth, Jr., "Persistent beast" (1985).

284. Arber, *Herbals*, 186–202; see also Harold J. Abrahams and Marion B. Savin, "The *Herbarius latinus*" (1975).

285. Treviranus, *Anwendung des Holzschnittes*, 7.

286. Arthur M. Hind, *History of woodcut* (1963), 1: 354, suggests that Erhard Reuwich, who illustrated the *Sanctae peregrinationes*, may also have been responsible for the illustrations in the *Gart*, published the previous year by the same publisher.

287. Rytz, "Herbarium Felix Platters," 74, 76–77.

288. Arber, *Herbals*, 55. But Brunfels was a lover of plants, and walked from Strasbourg to Hornbach to visit Hieronymus Bock and talk with him about botany, as Bock testifies: Bock, *Kreütterbuch*, sig. b6r.

289. Gessner, "Praefatio . . . in qua enumerantur qui de plantis quovis modo in hunc usq; diem aliquid literis prodiderunt," in Bock, *De stirpium commentaria*, sig. c iiii^{r}.

290. Koreny, *Albrecht Dürer*, 188–91; cf. the early sixteenth-century German drawing of a Turk's-cap lily on p. 193.

291. Sachiko Kusukawa, "Fuchs on pictures" (1997).

292. Fuchs, *De historia stirpium*, pref., translated by Elaine Mathers and John L. Heller, in Meyer, Trueblood, and Heller, *Great Herbal of Fuchs*, 1: 213.

293. Clusius, *Stirpium per Hispanias historia*, 289–92.

294. In this paragraph I follow Kusukawa, "Fuchs on pictures," esp. 418–23.

295. Kusukawa, "Fuchs on pictures," 426.

296. Bock, *De stirpium commentaria*, sig. d vii^{r-v}. Treviranus was unaware of the unillustrated 1539 edition and thought that the 1551 German edition was the first. His criticisms of Bock's carelessness with the illustrations (Treviranus, *Anwendung des Holzschnittes*, 15) are thus without ground.

297. Fabio Colonna, *Ecphrasis I.* 17, quoted by Treviranus, *Anwendung des Holzschnittes*, 71n.

298. Gessner in Cordus, *Annotationes*, sig. a ii^{v}, fol. 85r–v.

299. Gessner in Cordus, *Annotationes*, fol. 85v; cf. Treviranus, *Anwendung des Holzschnittes*, 18–19.

300. Gessner in Cordus, *Annotationes*, fol. 85v.

301. This point was made explicitly by eighteenth-century naturalists such as René-Antoine Ferchauld de Réaumur, *Histoire des insectes* (1737), 63–64: "Les Desseins disent bien plus vîte ce qu'ils ont à dire; ils ne peuvent pourtant pas toujours représenter tout ce qu'on voudroit qu'ils représentassent, mais ils soutiennent toujours l'imagination, & avec leur secours on lit & on entend des descriptions qu'autrement on n'entendroit ni ne liroit."

302. Clusius, *Stirpium per Hispanias historia*, 321.

303. See William M. Ivins, Jr., *How prints look* (1987), 5–22, for an overview of the woodcut process; Treviranus, *Anwendung des Holzschnittes*, remarks frequently about the quality of impressions.

304. Nineteenth-century naturalists also had to exercise firm control over their illustrators, for similar reasons—especially to make sure that the artist did not represent the peculiarities of an individual as if they were representative of the species: Lorraine Daston, "Objectivity versus truth" (1999), 22–24.

305. Treviranus, *Anwendung des Holzschnittes*, 17; two of Gessner's drawings are reproduced in Koreny, *Albrecht Dürer*, 232–35. Gessner's material for his history of plants was sold after his death by Caspar Wolf to Joachim Camerarius, who used some of the woodcuts for his *Hortus medicus* and his epitome of Mattioli (Treviranus, *Anwendung des Holzschnittes*, 19, 41; Arber, *Herbals*, 111). Gessner's drawings are now part of the Trew-Sammlung in the Universitätsbibliothek Erlangen, and a facsimile edition has been published: Gessner, *Historia plantarum*.

306. See Heinrich Zoller, "Geßner als Botaniker" (1965), 225.

307. Clusius, *Stirpium per Hispanias historia*, sig. A4v.

308. Clusius to Camerarius, Dec. 26, 1584 (n.s.) (Hunger, *Clusius*, 2: 403); Clusius to Crato, Nov. 25, 1567, in Carolus Clusius and Conrad Gessner, *Epistolae ineditae* (1830), 46. Treviranus, *Anwendung des Holzschnittes*, 35, ranks the woodcuts in this work among the most beautiful of the time.

309. Clusius, *Stirpium per Hispanias historia*, 527.

310. Bauhin to Jacob Zwinger, Oct. 28, 1595 (Basel UB, MS. Fr.Gr. I.13, no. 50). The work in question is probably Prosper Alpino, *De plantis Aegypti* (1592).

311. Bauhin, *Animadversiones*.

312. For example, Clusius's comments in Cristobal Acosta, *Aromatum liber* (1582), 5, 33; Clusius, *Stirpium per Pannoniam historia*, 56, 269.

313. Reeds, *Botany*, 145, 249 n. 43.

314. See, for example, the *Draco arbor* in Clusius, *Stirpium per Hispanias historia*, 12.

315. Treviranus, *Anwendung des Holzschnittes*, 2.

316. Agnes Arber, "Colouring of herbals" (1940); reprinted in Arber, *Herbals*, 315–18.

317. Conrad Gessner, *Historiae animalium lib. I* (1551), sig γ1v.

318. Christophe Plantin to Severinus Gobelius, Oct. 11, 1581 (letter no. 954) in Christophe Plantin, *Correspondance* (1883–1918), 6: 315–16; see n. 3 for the costs. To keep costs down, Plantin employed women—at least two of them widows—to color these illustrations: Arber, "Colouring of herbals"; reprint, Arber, *Herbals*, 317–18.

319. Plantin to Gobelius, Oct. 11, 1581, in Plantin, *Correspondance*, 6: 315–16. The book cost 113 florins colored; Carolus Clusius received some 400 florins annually a decade later as professor at the University of Leiden.

320. Arber, *Herbals*, 241–46; Nissen, *Botanische Buchillustration*, 1: 69–70.

321. Nissen, *Botanische Buchillustration*, 1: 69.

322. See Arber, *Herbals*, 243.

323. Arber, *Herbals*, 241–43; Nissen, *Botanische Buchillustration*, 2: 3 (cat. no. 21).

324. See the discussion in chapter 2.

325. Treviranus, *Anwendung des Holzschnittes*, 44ff.

326. Here I disagree with Paula Findlen's judgment that "Renaissance naturalists gave equal weight to all sensory information," unlike their seventeenth-century successors: Findlen, *Possessing nature*, 206.

327. Gessner, "Praefatio," in Bock, *De stirpium commentaria*, sig. c v^{v}.

328. Gessner, "Praefatio," in Bock, *De stirpium commentaria*, sig. c ii^{r-v}.

329. Spiegel, *Isagoge*, 97.

330. Louis, *Mathieu de l'Obel*, 279–83.

331. Textor, *Stirpium differentiae*; Stearn, *Botanical Latin*, 311–57.

332. Stearn, *Botanical Latin*, 236–37.

333. Stearn, *Botanical Latin*, 242, notes that the beginnings of modern color terminology in natural history are found in the middle of the eighteenth century.

334. Clusius, *Stirpium per Pannoniam historia*, 300, 353, 665.

335. For example, Clusius, *Stirpium per Pannoniam historia*, 68 ("acris saporis," "grato odore"), 279 ("impense amara"), 333–35 ("odore interdiu nullo, aut adeo tenui ut vix naribus percipi possit, sed sub vesperam gratissimo & suavissimo"). As the last example indicates, Clusius could be quite precise in observing when a plant did have an odor.

336. Clusius, *Stirpium per Pannoniam historia*, 146–51. An English translation of Clusius's remarks on tulips, based on the *Rariorum plantarum historia*, is provided by Clusius, *Treatise on tulips*.

337. In this regard there are illuminating parallels between natural history and other aspects of sixteenth-century intellectual life—for instance, the shift from oral to spatial ways of thinking about the world examined by Walter J. Ong, *Ramus* (1958).

338. Scott Atran, *Cognitive foundations* (1990).

CHAPTER FIVE

1. See Francisco Hernández, *Mexican treasury* (2001).

2. Caspar Bauhin, *Phytopinax* (1596), sig. α3v–α4v.

3. Brian W. Ogilvie, "Encyclopædism" (1997). Ironically, Bauhin's *Phytopinax* and *Pinax* are now used by scholars who wish to identify the plants in surviving early modern herbaria.

4. Caspar Bauhin, *Pinax* (1623), sig. *4v. On Bauhin's correspondence network, see chapter 2.

5. Justin Stagl, *History of curiosity* (1995), chapter 2.

6. Bauhin, *Pinax*, sig. *4v; cf. Caspar Bauhin, *Prodromos* (1620), sig.):(3v. Two thousand of them were still extant in the 1880s: Jean-Baptiste Saint-Lager, "Histoire des herbiers" (1885), 96.

7. Saint-Lager, "Histoire des herbiers," 95; Walther Rytz, "Herbarium Felix Platters" (1932–33).

8. Saint-Lager, "Histoire des herbiers," 95.

9. Rytz, "Herbarium Felix Platters," 6–8.

10. Hermann Friedrich Kessler, *Das älteste Herbarium Deutschlands* (1870), 37, 61.

11. William T. Stearn, *Botanical Latin* (1983), 28.

12. Bauhin, *Prodromos*, 27.

13. Carolus Clusius, *Stirpium per Pannoniam historia* (1583), 176.

14. This figure includes all the plants for which Bauhin noted a correspondent who had sent it to him or someone who had shown it to him in a herbarium: Bauhin, *Prodromos*, 1–11.

15. Bauhin, *Prodromos*, 2.

16. Bauhin, *Prodromos*, 3, 43.

17. See Paul Delaunay, *Zoologie* (1962), 217–307.Emma Spary has suggested that the historiographical focus on classification has been a significant limitation even for the history of eighteenth- and nineteenth-century natural history, in which classification was undeniably a central intellectual and practical concern: E. C. Spary, *Utopia's garden* (2000), 250–51.

18. Clusius, *Stirpium per Pannoniam historia*, 378.

19. Hieronymus Bock, *De stirpium commentaria* (1552), sig. d viii[v].

20. Theophrastus, *Enquiry into plants* (1916), 1.3.1.

21. Edward Lee Greene, *Landmarks* (1983), 1: 330–47.

22. Hieronymus Bock, *New Kreütter Buch* (1539); Rembert Dodoens, *Histoire des plantes* (1557); Carolus Clusius, *Stirpium per Hispanias historia* (1576), 1; Clusius, *Stirpium per Pannoniam historia*, 1; *Historia generalis plantarum* (1587–88).

23. Allen J. Grieco, "Social politics" (1991).

24. Clusius, *Stirpium per Pannoniam historia*, 330, 661, 343.

25. Hieronymus Bock, *Kreütterbuch* (1577), fol. 310r.

26. Agnes Arber, *Herbals* (1986), 171–73; Georges Métailié, "Histoire naturelle et humanisme" (1989).

27. *Historia generalis plantarum*, 797 (in book 7, "Plantae, qui flore placent").

28. *Historia generalis plantarum*, sig. *5r (table of contents).

29. Arber, *Herbals*, 172–73, says that the scheme was founded on "an undigested medley of ecological, medical, and morphological ideas"; only two of the groups "have any pretension to being natural."

30. *Historia generalis plantarum*, 1208, 1435.

31. *Historia generalis plantarum*, 1491.

32. *Historia generalis plantarum*, publisher's dedication, sig. *2r.

33. Arber, *Herbals*, 168, 179, 181, 169.

34. Much of the discussion that follows is based on the work of contemporary folkbiologists. For an orientation in folkbiology, see Scott Atran, *Cognitive foundations* (1990), part 1; Brent Berlin, *Ethnobiological classification* (1992); and Douglas L. Medin and Scott Atran, eds., *Folkbiology* (1999).

35. Atran, *Cognitive foundations*, 100ff.

36. For Clusius's systematic use of "species" and "genus," see Hermann Christ, "Spanische Flora des Clusius" (1912), and Hermann Christ, "Ungarisch-österreichische Flora des Clusius" (1912–13).

37. A synthesis of the literature through the 1980s is provided by Berlin, *Ethnobiological classification*.

38. Atran, *Cognitive foundations*, 260–61.

39. Jared Diamond and K. David Bishop, "Ethno-ornithology of the Ketengban" (1999).

40. Berlin, *Ethnobiological classification*, 118ff., discusses exceptions and explanations for this correspondence.

41. Ernst Mayr, *Growth of biological thought* (1982), 174–75.

42. My discussion is based on Atran, *Cognitive foundations*, chapter 2, except that I follow Berlin, *Ethnobiological classification*, chapters 2–3, in dividing Atran's "generic-specieme" taxon into generic and specific taxa.

43. Berlin, *Ethnobiological classification*, chapters 2–3, esp. pp. 60–64, 108–14.

44. Scott Atran, "Itzaj Maya taxonomy" (1999), 124, 125.

45. Berlin, *Ethnobiological classification*, 122–25, 131, 274; cf. 288–290.

46. Berlin, *Ethnobiological classification*, 272–88. Note that Berlin rejects a strictly utilitarian explanation; agricultural societies do not discriminate species more carefully than foraging societies *simply* because the differences are more useful to them. Atran, on the other hand, argues that the folkbiological regularities across many different traditional societies are the result of universal cognitive structures: Atran, *Cognitive foundations*, 1–13, 263–69; Scott Atran, "Folk biology and anthropology" (1998).

47. Scott Atran argues that Berlin's distinction between "folk species" and "folk genus" is unnecessary and misleading, preferring the term "generic species" instead, precisely because the species that are most interesting to humans in any given locality tend to be the only representatives in that locality of their scientific genus: Atran, "Itzaj Maya taxonomy." This represents a qualification of Atran's claim (Atran, *Cognitive foundations*, 27–28) that, in most localities, genera are monospecific and that the exceptions involve relatively fine discriminations between species. But I have adopted Berlin's terminology because it better corresponds to the folk terms actually used by Renaissance naturalists.

48. Scott Atran, "Species and genus concepts" (1987); Atran, *Cognitive foundations*, part 3.

49. Atran, *Cognitive foundations*, 130–38.

50. Baudouin Van den Abeele, "Albums ornithologiques de Dalechamps" (2002), 12, 20, 25.

51. Adam Zaluziansky a Zaluzian, *Methodi herbariae libri tres* (1940), p. 153/sig. Cc1v; Joachim Jungius, *Opuscula botanico-physica* (1747), 70–74. Atran, *Cognitive foundations*, 160, credits Jung with being the first to reject the distinction, but Zaluziansky wrote in 1592 that "tree" is an imprecise definition in part because trees sometimes become herbs "without any change in their nature or species."

52. Atran, *Cognitive foundations*, 135–38.

53. Atran, "Itzaj Maya taxonomy," 119. The example is borrowed from M. Jacobs: Atran, *Cognitive foundations*, 134.

54. Brigitte Hoppe, *Kräuterbuch des Hieronymus Bock* (1969), 44.

55. See also André Cailleux, "Nombre d'espèces" (1953), though Cailleux's figures contain some slight historical inaccuracies.

56. Berlin, *Ethnobiological classification*, 96–100.

57. Atran, *Cognitive foundations*, 298 n. 10.

58. For Cesalpino's biography, see Guido Moggi, "Andrea Cesalpino, botanico" (1976–78).

59. Andrea Cesalpino, *De plantis* (1583), sig. [a3]v.

60. Cesalpino, *De plantis*, sig. [a4]r, 26.

61. On Renaissance psychology as it applies to plants and animals, see Katharine Park and Eckhard Kessler, "The concept of psychology" (1988), and Katharine Park, "The organic soul" (1988).

62. Aristotle, *De anima*, 2.2. The text was a routine part of the arts curriculum in sixteenth-century universities.

63. Aristotle, *De anima*, 1.5, 411b25.

64. Cesalpino, *De plantis*, 1, 26–28. The opening sentence implies that nutrition and growth are distinct faculties, but later Cesalpino made it clear that both were aspects of the same faculty. Aristotle went further, arguing that nutrition and reproduction were caused by the same power of the soul: *De anima* 2.4, 416a19. Cesalpino allowed that some organs arose "from necessity," due to constraints imposed by the material of the plant, not its functions: *De plantis*, 23–24.

65. Cesalpino, *De plantis*, 23: Plants like bindweed "climb their neighbors as if (*quasi*) they possessed a certain sense of an adjacent body, for they creep until they find something and then fix on it."

66. Arber, *Herbals*, 143–44, 183–84, though Arber thought that Cesalpino's system was in practice too rigid to capture true natural affinities; see Guido Moggi, "Conoscenza del mondo vegetale" (1993).

67. Cesalpino, *De plantis*, 28.

68. Cesalpino, *De plantis*, 26.

69. On this point, see Atran, *Cognitive foundations*, 138–39, who argues convincingly that the "Aristotelian" species concept, according to which species are immutable, was in fact introduced by Cesalpino and has been anachronistically attributed to Aristotle by modern scholars.

70. Cesalpino, *De plantis*, 25, 27–29.

71. Cesalpino, *De plantis*, 26, 29. Cesalpino had already argued that leaves existed chiefly to provide shelter for the young shoot, the flowers, and the fruit; he held that plants nourished themselves from terrestrial humors (*De plantis*, 3–6).

72. Atran, *Cognitive foundations*, part 3, explores the ramifications of this shift.

73. I do not wish to imply that Cesalpino somehow knew that this was the way to go; if there were only a few thousand plant species in the world, a scientific taxonomy would be unnecessary for the practical ends that Cesalpino saw as the aim of taxonomy. It is also worth keeping in mind that the modern biological species concept goes beyond morphology, recognizing "crypto-species" inhabiting the same area that are morphologically identical but reproductively isolated by behavior or ecological niche.

74. Cesalpino, *De plantis*, 30.

75. On Ramus and his method, see Walter J. Ong, *Ramus* (1958); Lisa Jardine, *Francis Bacon* (1974), 41–47; Lisa Jardine, "Humanistic logic" (1988), 184–86; and Nelly Bruyère, *Méthode et dialectique* (1984). Joseph S. Freedman, "Writings of Petrus Ramus" (1993), cautions that Ramus's central European followers and commentators used his works in so many different ways that "Ramism" seems almost meaningless. Nonetheless, as Jardine notes ("Humanistic logic," 186), Ramism seemed to many sixteenth-century writers to be the method of dichotomous keying by which a subject is introduced first by defining it in the most general terms and then dividing it into more precise genera. Zaluziansky proceeds in this fashion in his *Methodus*.

76. Zaluziansky a Zaluzian, *Methodi herbariae libri tres*, pp. 3–5/sig. A4r–A5r, p. 7/sig. B1r. References to Zaluziansky include page numbers from the modern critical edition followed by signatures referring to the 1592 edition.

77. Zaluziansky a Zaluzian, *Methodi herbariae libri tres*, pp. 7–58/sig. B1r–M4r.

78. In this he is in keeping with the humanist use of "historia" discussed in chapter 3, above.

79. Zaluziansky a Zaluzian, *Methodi herbariae libri tres*, pp. 59–66/sig. M4v–N4v.

80. For an example, see Konrad von Weihe, ed., *Illustrierte Flora* (1972).

81. Zaluziansky a Zaluzian, *Methodi herbariae libri tres*, p. 4/sig. A4v–A5r.

82. Zaluziansky a Zaluzian, *Methodi herbariae libri tres*, p. 114/sig. V4v, pp. 169–74/sig. Ee2v–Ff3r.

83. Arber, *Herbals*, 144, lauds Zaluziansky for emancipating botany from medicine; cf. the discussion above in chapter 2, pp. 38–39. Peter Dilg also emphasized Zaluziansky's role in the "emancipation" of botany from medicine: Peter Dilg, "Der Beitrag Zaluzanskys" (1975). Dilg's essay does provide a useful synopsis of Zaluzansky's methodological remarks.

84. Adriaan van de Spiegel, *Isagoge* (1606), 77.

85. On seventeenth-century developments, see Mary M. Slaughter, *Universal languages* (1982); Atran, *Cognitive foundations*, part 4; and Giulio Barsanti, *Scala, mappa, albero* (1992).

86. Kurt Johannesson, *Renaissance of the Goths* (1991), chapter 5.

87. Spiegel, *Isagoge*, 131. The original, of course, is "Africa always produces something new," a Greek proverb reported by Pliny the Elder, *Histoire naturelle* (1947–85), 8.17.42.

88. I. Bernard Cohen, "Découverte et transformation" (1960); Donald F. Lach, *Century of wonder* (1977), 3:396, 427–45; Wilma George, "Sources and background" (1980); William B. Ashworth, Jr., "Emblematic world view" (1990).

89. As with Carolus Clusius, *Exoticorum libri* (1605), which included Clusius's own descriptions of exotica he had collected or borrowed, along with his translations of Pierre Belon, Garcia da Orta, Cristobal Acosta, and Nicolas Monardes; and Juan Eusebio Nieremberg, *Historia naturae, maxime peregrinae* (1635).

90. For a general discussion, see Steven Shapin, *Social history of truth* (1994).

91. On Scandinavian culture in this period, see Johannesson, *Renaissance of the Goths.*

92. Olaus Magnus wrote that his *Historia de gentibus septentrionalibus* grew out of his response to questions from fellow delegates to the Council of Trent, who asked him about his homeland during the council's many recesses: Olaus Magnus, *Historia de gentibus septentrionalibus* (1555), dedicatory letter.

93. Johannesson, *Renaissance of the Goths*, 69–70ff.

94. C. Iulius Caesar, *De Bello Gallico* (1900), 6.26–27.

95. To the extent that some refused to call the clearly cervine elk by its proper name, insisting that the true elk was goatlike; instead, they called it the "great animal" (magnum animal). The position was defended most strongly in Apollonio Menabeni, *De magno animali* (1581), which I have not been able to examine; Menabeni's arguments are summarized in Ulisse Aldrovandi, *Quadrupedum bisulcorum historia* (1621), 867 ff.

96. Pliny the Elder, *Histoire naturelle*, 8.39.

97. Pausanias, *Description of Greece* (1918–35), 5.12.1 (Elis), 12.21.3 (Boeotia).

98. Pausanias, *Description of Greece* (1965), 5: 86.

99. Albertus Magnus, *De animalibus nach der Urschrift* (1916–21), book 22, s.v. "Alches," "Equicervus," and "Rangifer"; Conrad Gessner, *Historiae animalium lib. I* (1551), 2.

100. Olaus Magnus, *Carta marina* (1964); *Historia de gentibus septentrionalibus.*

101. Olaus did correspond with Sebastian Münster, a friend and correspondent of Gessner: Sebastian Münster, *Briefe* (1964), 113.

102. Albertus Magnus, *De animalibus nach der Urschrift*, 24.19.

103. My remarks about the walrus trade slacking off are an inference from the relative ignorance of sixteenth-century naturalists and merchants about this beast (see comments below on voyages). Elk antlers and hides are mentioned by many early modern naturalists; see below.

104. Conrad Gessner, *Historiae animalium liber IIII* (1558), 249.

105. Based on its absence in later Italian natural history books and collections. It is not mentioned in Ulisse Aldrovandi, *De piscibus et de cetis* (1613), good evidence that Aldrovandi and his literary executors were not aware it had ever existed. The wall of Ferrante Imperato's museum in Naples had a walrus painted on it, out of Gessner's 1558 publication: Ferrante Imperato, *Historia naturale* (1599), unpaginated plate following t.p.

106. Gessner, *Historiae animalium liber IIII*, 250.

107. Sebastian Killermann, "Dürer's Bilder" (1912), 785.

108. Jacques Cartier, *Relations* (1986), 105.

109. Jean Alfonse, *Cosmographie* (1544), in Cartier, *Relations*, Appendice I, 221.

110. "The booke of the great and mighty Emperor of Russia . . . drawen by Richard Chancellor" (1553) in Richard Hakluyt, ed., *Principal navigations* (1903–5), 2: 224–25.

111. "The newe Navigation and discoverie of the kingdome of Moscovia, by the Northeast, in the yeere 1553," in Hakluyt, ed., *Principal navigations*, 2: 262–63.

112. "The Navigation and discoverie toward the river of Ob" (1556), in Hakluyt, ed., *Principal navigations*, 2: 328, 337–38.

113. On train oil, flax, hemp, and furs, see the Muscovy Company's letters to its agents from 1557and 1560, in Hakluyt, ed., *Principal navigations*, 2: 381–82, 387–88, 402–3. The company complained about its agents' plans to send more seal skins when the ones they had sent earlier were as yet unsold (2: 404).

114. Gerrit de Veer, *Reizen van Barents* (1917), 25.

115. Probably because Clusius did not have any walrus parts to describe. On the translation, see Veer, *Reizen van Barents*, 2: 271–72.

116. Aldrovandi, *Quadrupedum bisulcorum historia*, 869–71.

117. John Ray, *Synopsis quadrupedum* (1693), mentions that he and Willughby saw the creature in the grand duke's collection (p. 86). Charles Raven makes no mention of a visit to Florence in his account of Ray's continental tour but notes that they stopped in Pisa while coasting south from Genoa to Naples: Charles E. Raven, *John Ray* (1950), 133.

118. The museums of Ferrante Imperato in Naples and Francesco Calceolari in Verona appear to have had elk antlers but not complete animals: Imperato, *Historia naturale*, plate; Benedetto Ceruto and Andrea Chiocco, *Musaeum Calceolari* (1622), unpaginated illustration bound in front matter. Sebastian Münster's 1578 German cosmography mentions elk or "tragelaphus" antlers in an Augsburg collection, but not the rest of the beast: Sebastian Münster, *Cosmographer* (1578), Mclxxxii. They were somewhat more common in the seventeenth century. By 1683, the Leiden anatomy theater had an elk head and skin on display: *Catalogue of all the rarities* (1683) (unpaginated).

119. Not that it was impossible; cf. Clusius's description of the cassowary, or "Emeu" as he called it, in Clusius, *Exoticorum libri*, 97–99.

120. Gessner, *Historiae animalium lib. I*, sig. β2r–v.

121. Angelo Poliziano was an exception in this regard: see Anthony Grafton, *Defenders of the text* (1991), 56–57.

122. Gessner, *Historiae animalium lib. I*, 1–3, 623–24, 950–52; Gessner, *Historiae animalium liber IIII*, 245, 249–51.

123. Conrad Gessner, *Bibliotheca universalis* (1545).

124. Gessner, *Historiae animalium lib. I*, sig. e5r–γ1r.

125. In what follows, I am limiting my exegesis to the descriptions and illustrations of animals that Gessner provides; Gessner also included historical, ethical, proverbial, and other material pertaining, as he put it, more to words than things (see chapter 1, above).

126. Gessner, *Historiae animalium lib. I*, 950.

127. Gessner, *Historiae animalium lib. I*, 2.

128. Gessner, *Historiae animalium lib. I*, 1, 623–24, 950.

129. Gessner, *Historiae animalium liber IIII*, sig. a6r. On the next page Gessner exhorts the Emperor Ferdinand to have whales from the northern oceans captured and displayed in the Empire, to the emperor's great glory.

130. Gessner, *Historiae animalium liber IIII*, 245.

131. Gessner, *Historiae animalium liber IIII*, 249.

132. Gessner, *Historiae animalium liber IIII,* 249–50.

133. Imperato, *Historia naturale,* foldout plate. Considerations of scale and accuracy, as well as the lack of contemporary firsthand descriptions of the walrus, lead me to conclude that this image was painted on Imperato's ceiling. It could represent a real, stuffed walrus, but if this is so, the artist who made the engraving copied Gessner's erroneous illustration rather than the object.

134. Aldrovandi, *De piscibus et de cetis,* 670, 687.

135. Nieremberg, *Historia naturae, maxime peregrinae,* 257.

136. Imperato, *Historia naturale,* foldout plate.

137. Ray, *Synopsis quadrupedum,* 191.

138. Aldrovandi, *Quadrupedum bisulcorum historia,* 864.

139. Albertus Magnus, *De animalibus nach der Urschrift,* 24.19; Olaus Magnus, *Historia de gentibus septentrionalibus,* 757. Olaus added a few commments from humanist sources, but he did not mention Albertus, on whose text his description is primarily based.

140. Aldrovandi, *Quadrupedum bisulcorum historia,* 864 (picture on 866), 870–71.

141. *Nieuw Nederlandsch biografisch woordenboek* (1911–37), 4: 1411–12; Joannes de Laet, *Novus Orbis* (1633), 38–39.

142. John Johnston, John Ray, and Buffon all referred to Vorst's description in De Laet, though Buffon gave other descriptions as well.

143. Joannes Jonstonus, *De piscibus et cetis* (1657), 160. Johnston did, however, include Gessner's illustration, and he pointedly omitted Vorst's remark that the beast had no tail.

144. Arber, *Herbals,* 104–5; Friedrich Wilhelm Tobias Hunger, *Clusius* (1927–43), 1: 79, 88–89, 97; Garcia da Orta, *Aromatum historia* (1567).

145. Aldrovandi to Clusius, Feb. 8, 1569 (Leiden UB, MS. VUL 101, s.v. "Aldrovandi," no. 1).

146. Hunger, *Clusius,* 1: 370, 372, 373, 376, 380.

147. Hunger, *Clusius,* 1: 147, 154–55; Cristobal Acosta, *Aromatum liber* (1582).

148. Clusius in Acosta, *Aromatum liber,* 3–5.

149. Acosta, *Aromatum liber,* 7.

150. Acosta, *Aromatum liber,* 8, 9.

151. Clusius in Acosta, *Aromatum liber,* 41, 68, 50.

152. Acosta, *Aromatum liber,* 33.

153. Clusius in Acosta, *Aromatum liber,* 33.

154. For example, Willy Ley, *Dawn of zoology* (1968), 140–41, who attributes too much importance to Pigafetta's account; Fritz Koreny, *Albrecht Dürer* (1988), 100, following the account in Erwin Stresemann, "Entdeckungsgeschichte der Paradiesvögel" (1954).

155. Antonio Pigafetta, *Magellan's voyage* (1906), 105.

156. Maximilianus Transylvanus, *De Moluccis insulis* (1523 and later editions). Samuel Eliot Morison, *Southern voyages* (1974), 325, dismisses this tract as a source, but it was widely read in its time.

157. Maximilianus Transylvanus, "De Moluccis insulis" (1903), 331–32, 335.

158. Stresemann, "Entdeckungsgeschichte der Paradiesvögel," 264–65.

159. Stresemann, "Entdeckungsgeschichte der Paradiesvögel," 265–68.

160. Clusius, *Exoticorum libri*, 359.

161. Clusius, *Exoticorum libri*, 359.

162. Rotraud Bauer and Herbert Haupt, "Die Kunstkammer Rudolfs II" (1976), 10 (inventory no. 136–41).

163. Stresemann, "Entdeckungsgeschichte der Paradiesvögel," 270–73, 279ff.

164. Clusius, *Exoticorum libri*, 1.

165. Mordente was also a mathematician and inventor of a mathematical compass over which he and Giordano Bruno had a dispute: Frances A. Yates, *Giordano Bruno* (1964), 294–98. Mordente entered imperial service during or after his visit to Vienna: R. J. W. Evans, *Rudolf II* (1973), 230.

166. Clusius, *Exoticorum libri*, 1–2.

167. Clusius, *Exoticorum libri*, 2.

168. Cf. the "Anser Magellicanus," or Penguin, Clusius, *Exoticorum libri*, 101, following the text of the narrative of Mahu and De Cordes's voyage: F. C. Wieder, ed., *Reis van Mahu en De Cordes* (1923), 1: 234–36. In Clusius, *Stirpium per Pannoniam historia*, 419–20, Clusius reproduced a description and illustration of a type of geranium, which were sent by his English friend Thomas Penny.

169. Clusius, *Exoticorum libri*, 2.

170. Theophrastus, *Enquiry into plants*, 1.7.3, 4.4.4.

171. Clusius, *Exoticorum libri*, 3.

172. Clusius, *Exoticorum libri*, 85.

173. The "Oude Compagnie" was one of the so-called "forerunner companies" to the state-chartered Dutch East India Company, which was formally established in 1602. See F. S. Gaastra, "VOC in Azië" (1980), 176–78.

174. Erik de Jong, *Natuur en kunst* (1993), 216.

175. Clusius, *Exoticorum libri*, 88–92.

176. Reprinted in Hunger, *Clusius*, 1: 267. According to Hunger, the original (in Clusius's hand) is in the Algemeen Rijksarchief in the Hague. Since the VOC was formally established in 1602, for the fleet to sail that year, I presume that Clusius's instructions date from that year. They were possibly written in 1601 for one of the earlier companies which were forcibly united into the VOC.

177. The same notion was behind the more detailed instructions of Robert Boyle, *General heads* (1692), an expanded version of an article by Boyle in the *Philosophical Transactions*, no. 11 (1666), 186.

178. Literally, "one finds also in the sea various plants" ("men vindt ook in zee sommich gewas"). I presume that Clusius meant "in zee" in a broad sense, that is, in the course of the voyage.

179. Clusius, *Exoticorum libri*, 75–77.

180. Clusius, *Exoticorum libri*, sig. †7v.

181. Clusius, *Exoticorum libri*, sig. †2r–v.

182. Francis Bacon, *Advancement of learning* (1863), 184.

183. Francis Bacon, *Novum organum* (1861), 220–25; Bacon, "Parasceve" (1864), 44.

184. Bacon, *Parasceve*, 48–49; translation in Francis Bacon, "Preparative" (1864), 359.

185. Bacon, *Parasceve*, 55–56; translation in Bacon, *Preparative*, 367.

186. Francis Bacon, *Sylva sylvarum* (1862–64), 5: 147–48, nos. 960–62; Bacon, *Advancement of learning,* 233; Bacon, *Parasceve,* 56 (translation in Bacon, *Preparative,* 367).

187. Thomas Browne, *Pseudodoxia epidemica* (1981), 28–45.

188. Browne, *Pseudodoxia epidemica,* 40–42. In this respect Browne places stricter demands on natural history than Bacon, who thought that any incorrect points in his natural histories would eventually be worked out.

189. Browne, *Pseudodoxia epidemica,* 43–44.

190. Browne, *Pseudodoxia epidemica,* 105–6, 48, 330.

191. Browne, *Pseudodoxia epidemica,* 330–32 (quotation from 332).

192. Browne, *Pseudodoxia epidemica,* 257, 509, 256.

193. Ceruto and Chiocco, *Musaeum Calceolari,* 690.

194. Henshaw to Oldenburg, July 6, 1672, in Henry Oldenburg, et al., *Correspondence* (1965–86), 9:145.

195. Oldenburg, *Correspondence,* 9: 147.

196. Ray, *Synopsis quadrupedum,* sig. *A4v–5r.

197. Slaughter, *Universal languages,* 62–64; Umberto Eco, *Suche* (1994), 245, 260. Eco reproduces Ray's classification (for Wilkins's project) of the Walrus and other carnivores in the diagram on p. 249, in which the walrus and seal are amphibious, European canid animals.

198. Ray, *Synopsis quadrupedum,* sig. *A5v–6r.

199. Ray, *Synopsis quadrupedum,* sig. *A7r, pp. 191–93.

200. Buffon cited several writers, including Vorst (whom he mistook for an Englishman), but all of them firsthand observers. Georges-Louis Leclerc Buffon, comte de, *Oeuvres complètes* (s.a.), 3: 522–30.

201. Ray, *Synopsis quadrupedum,* 86–88, 227.

202. The same is true of antiquarians and historians: Arnaldo Momigliano, "Ancient history" (1955), 75–79; William McCuaig, *Carlo Sigonio* (1989), 139. Momigliano points to an increasing skepticism among antiquarians in the seventeenth century, an attitude similar to that seen among seventeenth century naturalists.

203. Münster, *Briefe,* 113. Mary B. Campbell, *Witness* (1988), traces the history of this idea through the Middle Ages and Renaissance. The idea that strange things are found far away from the center of the world goes back at least to Herodotus; Julius Caesar's commentaries and Ammianus Marcellinus's history also put the most fierce, barbarous, and weird peoples the farthest from Rome.

204. John Ray, "Appendix ad Historiam naturalium Piscium" (separately paginated), in Francis Willughby, *Historia piscium* (1686), 29–30.

205. Willughby, *Historia piscium,* 37–38.

206. The subtitle to Ray, *Synopsis quadrupedum,* claimed that the book listed "the characteristic features of the better-known animals, and complete descriptions of rarer kinds."

CHAPTER SIX

1. See Adrian Slack, *Carnivorous plants* (1979), 119–53.

2. Jeffrey K. Barnes, *Scolopendra heros* (2002).

3. The International Carnivorous Plant Society draws its membership from botanists, ecologists, and gardeners (online at http://www.carnivorousplants.org).

4. *De anima*; see the discussion above, chapter 6.

5. Charles Darwin, *Insectivorous plants*, (1989), 243 n. 10, addition by Francis Darwin (reprint p. 24).

6. Darwin, *Insectivorous plants*, (1st ed. 1875).

7. Joachim Camerarius, *Symbolorum et emblematum centuria una* (1590), emblem no. 49. Camerarius noted that the plant, which came from Peru, had been imported as an exotic and by the 1590s was grown "everywhere" in Europe.

8. Valerius Cordus, *Historia plantarum*, (1561), fol. 86. As noted in chapter 4, above, these notes were compiled in the late 1530s or early 1540s.

9. Rembert Dodoens, *Histoire des plantes* (1557), 281; the work is a translation of Rembert Dodoens, *Cruydeboeck* (1554).

10. Conrad Gessner, *De raris herbis* (1556), 37–38.

11. John Ray, *Historia plantarum* (1686–1704), 2: 1100.

12. Caspar Bauhin, *Pinax* (1623), 356.

13. Joachim Jungius, *Opuscula botanico-physica* (1747), 76, observed in the early seventeenth century that naturalists disagreed on the definition of species but generally agreed on their identity.

14. Anne Secord, "Artisan botany" (1996).

15. Jungius, *Opuscula botanico-physica*, 71–72, argued forcefully against the traditional division, on the grounds that the same plant can be a tree, a shrub, or an herb under different growing conditions. On the cognitive or perceptual grounds of this division, see the discussion of classification in chapter 5, above.

BIBLIOGRAPHY

The bibliography contains all sources that are cited in the book (except for standard reference works) and a few additional works that I consulted. Multiple publications by the same author are listed in chronological order (of the original publication, in the case of modern reprints or critical editions of Renaissance books).

ARCHIVAL SOURCES

For manuscripts, I indicate the signature or shelfmark and a title or description, taken variously from the catalogue, the manuscript, or the binding (in the case of bound volumes of correspondence).

Basel, Switzerland. Universitätsbibliothek. Handschriftenabteilung.

A N VI 16 (Caspar Bauhin's Stammbuch)

Fr. Gr. I.12 (Epistolae praesertim ad Theodorum I et ad Jacobum I Zwinger, 1565–1609)

Fr. Gr. I.13 (Epistolae praesertim ad Theodorum I et ad Iac. I Zwinger, 1563–1610)

Fr. Gr. I.22 (Epistolae Jacobi I Zwinger, 1594–1607)

Fr. Gr. II.1 (Epistolae ad Casparum Bauhinum, 1581–1626 [Apographa])

Fr. Gr. II.3 (Varia Scripta praesertim Bauhiniana)

Fr. Gr. II.4 (Epistolae praesertim ad Theodorum I et ad Jacobum I Zwinger, 156 [illeg.] 610)

Fr. Gr. II.28 (Epistolae ad Theodorum I et ad Jacobum I Zwinger, 1563–1610, II)

G II 3 (Epistolae ad J. J. Grynaeum)

G2 I 1 (Epistolae ad Casparum et ad Joh. Casparum I Bauhinum, 1581–1642)

G2 I 4 (Epistolae ad Casparum et ad Joh. Casparum I Bauhinum, 1580–1654)

G2 I 13b (Epistolae ad Casparum et ad Joh. Casparum I Bauhinum, 1593–1671)

G2 I 22 (Epistolae Variorum ad Varios, 1486–1741 [Apographa])

G2 II 37 (Various letters, apographs)

G2 II 39 (Various letters, apographs)

K I 12 (Praelectiones de Simplicibus, 17. Jh)

K II 9 (Observationes Botanicae, 1575)

K III 34 (Plantarum quae in Monte fracto prope Lucernam reperiuntur, Enumeratio et Descriptio)

K III 42 (Conrad Weigand, "Quadripartitum de Quercu," 15. Jh.)

K III 45 (Copy of Caspar Bauhin's *Catalogus Plantarum circa Basileam sponte nascentium*, 3rd ed. [Basel, 1671], with copious annotations and illustrations)

K III 51 (Plantae Horti medici Basileensis 1754. 1755.)

K IV 2 (Watercolors, mostly of grasses)

K IV 3 (Plant illustrations relating to Caspar Bauhin)

K IV 19 (Material related to plants, mostly in the hand of Caspar Bauhin)

K IV 22 (Woodcuts of plants and animals)

Autogr. Slg. Ch. Sarasin 38 (Unsigned, undated slip)

Erlangen, Germany. Universitätsbibliothek. Handschriftenabteilung.

Trew-Briefsammlung. Individual letters are filed by author. The collection, comprising approximately 20,000 letters, has been catalogued: Schmidt-Herrling, El., ed. *Die Briefsammlung des Nürnberger Arztes Christoph Jacob Trew (1695–1769) in der Universitätsbibliothek Erlangen.* Katalog der Handschriften der Universitätsbibliothek Erlangen, Neubearbeitung, 5. Erlangen: Universitätsbibliothek, 1940.

Kraków, Poland. Biblioteka Jagiellonská. Graphics department.

Libri Picturati A.18–A.30 (Watercolors of plants, sixteenth century, with annotations by Carolus Clusius and others)

Leiden, Netherlands. Universiteitsbibliotheek. Bibliotheek Dousa (Western manuscripts).

Vulcanius 101 (Letters to Carolus Clusius)

FACS UB A 123–141 (Facsimiles of letters)

BPL 948 (Plantarum seu Stirpium icones, Ordine Historiae Stirpium Remberti Dodonæi)

BPL 949 (Figuren vanden Herbarius oft Cruydt-Boeck van Rembertus Dodonaeus . . .)

BPL 2724a (Copies of letters from Carolus Clusius to Matteo Caccini)

BUR Q 19 (Epistolae Virorum Eruditorum ad J. Meursium J. Scaligeri & Aliorum Epistolae MSS)

PRINTED PRIMARY SOURCES

Acosta, Cristobal. *Aromatum & medicamentorum in Orientali India nascentium liber: Pluriumum lucis adferens iis quae a Doctore Garcia de Orta in hoc genere scripta sunt.* Translated by Carolus Clusius. Edited by Carolus Clusius. Antwerp: Ex officina Christophori Plantini, 1582.

Albertus Magnus. "De vegetabilibus." In *Opera omnia*, edited by Auguste Borgnet, vol. 10. Paris: apud Ludovicum Vives, 1891.

———. *De animalibus libri xxvi nach der Cölner Urschrift*. Edited by Hermann Stadler. 2 vols. Münster i. W.: Verlag der Aschendorffschen Verlagsbuchhandlung, 1916–21.

———. *On animals: A medieval Summa zoologica*. Translated by Kenneth Kitchell and Irven Michael Resnick. 2 vols. Baltimore: Johns Hopkins University Press, 1999.

Aldrovandi, Ulisse. *De piscibus libri v. et de cetis lib. unus*. Edited by Ioannes Cornelius Uterverius and Hieronymus Tamburinus. Bologna: Bellagambam, 1613.

———. *Quadrupedum omnium bisulcorum historia*. Edited by J. C. Uterverius, T. Dempster and H. Tamburinus. Bologna: Apud Sebastianum Bonhommium, 1621.

Alpino, Prosper. *De plantis Aegypti*. Venice: apud Franciscum de Franciscis Senensem, 1592.

Antidotarium, sive de exacta componendorum miscendorumque medicamentorum ratione libri tres. Translated by Carolus Clusius. Lyon: apud Theobaldum Paganum, 1561.

Aristotle. *Generation of animals*. Translated by A. L. Peck. Cambridge, MA: Harvard University Press, 1942.

———. *Parts of animals, Movement of animals, Progression of animals*. Translated by A. L. Peck and E. S. Forster. Revised ed. Cambridge, MA: Harvard University Press; London: William Heinemann, 1961.

———. *History of animals*. Translated by A. L. Peck and D. M. Balme. 3 vols. Cambridge, MA: Harvard University Press, 1965–91.

Bacon, Francis. *Novum organum*. In *The works of Francis Bacon*, edited by James Spedding, Robert Leslie Ellis, and Douglas Denon Heath, vol. 1, 130–539. Boston: Brown and Taggard, 1861.

———. *Sylva sylvarum; or, A natural history*. In *The works of Francis Bacon*, edited by James Spedding, Robert Leslie Ellis, and Douglas Denon Heath, vol. 4–5. Boston: Brown and Taggard; Taggard and Thompson, 1862–64.

———. *The advancement of learning*. In *The works of Francis Bacon*, edited by James Spedding, Robert Leslie Ellis, and Douglas Denon Heath, vol. 6, 77–412. Boston: Taggard and Thompson, 1863.

———. *De dignitate et augmentis scientiarum*. In *The works of Francis Bacon*, edited by James Spedding, Robert Leslie Ellis, and Douglas Denon Heath, vol. 2–3. New York: Hurd and Houghton; Boston: Taggard and Thompson, 1864.

———. *Parasceve ad historiam naturalem et experimentalem*. In *The works of Francis Bacon*, edited by James Spedding, Robert Leslie Ellis, and Douglas Denon Heath, vol. 2, 7–69. New York: Hurd and Houghton; Boston: Taggard and Thompson, 1864.

———. *Preparative towards a natural and experimental history*. In *The works of Francis Bacon*, edited by James Spedding, Robert Leslie Ellis and Douglas Denon Heath, vol. 8, 351–84. New York: Hurd and Houghton; Boston: Taggard and Thompson, 1864.

———. *Essays*. London: J. M. Dent & Sons; Rutland, VT: Charles E. Tuttle Co., 1973.

Barbaro, Ermolao. *Castigationes plinianae et in Pomponium Melam*. Edited by Giovanni Pozzi. 4 vols. Padua: in aedibus Antenoreis, 1973–79.

Bauhin, Caspar. *Phytopinax, seu Enumeratio plantarum ab herbariis nostro seculo descriptarum, cum earum differentiis*. Basel: per Sebastianum Henricpetri, 1596.

———. *Animadversiones in historiam generalem plantarum Lugduni editam*. Frankfurt am Main: excudebat Melchior Hartmann, impensis Nicolai Bassaei, 1601.

———. *Prodromos Theatri Botanici*. Frankfurt am Main: typis Pauli Iacobi, impensis Ioannis Treudelii, 1620.

———. *Catalogus plantarum circa Basileam sponte nascentium cum earundem synonomiis et locis in quibus reperiuntur*. Basel: Typis Johanni Jacobi Genathi, 1622.

———. *Pinax Theatri Botanici, sive Index in Theophrasti Dioscoridis Plinii et Botanicorum qui a seculo scripserunt opera*. Basel: sumptibus & typis Ludovici Regis, 1623.

———. *Theatri botanici sive Historiae plantarum ex veterum et recentiorum placitis propriaq. observatione concinnatae liber primus*. Edited by J. C. Bauhin. Basel: apud I. Konig, 1658.

———. *Catalogus plantarum circa Basileam sponte nascentium cum earundem Synonymiis & locis in quibus reperiuntur: In usum Scholae Medicae, quae Basileae est*. 3rd ed. Basel: Typis Johan. Rodolphi Genathii, 1671.

Bauhin, Johann. *De plantis a divis sanctisve nomen habentibus: Caput ex magno volumine de consensu & dissensu authorum circa stirpes desumptum*. Basel: Apud Conrad Waldkirck, 1591.

Bauhin, Johann, and Johann Heinrich Cherler. *Historia plantarum universalis*. Edited by Dominicus Chabraeus. 3 vols. Yverdon: Iuris vero publico fecit Fr. Lud. a Graffenried, 1650–51.

Belon, Pierre. *Remonstrances sur le defaut du labour & culture des plantes*. Paris: G. Cauellat, 1558.

———. *De neglecta stirpium cultura, atque earum cognitione libellus: Edocens qua ratione silvestres arbores cicurari & mitescere queant*. Translated by Carolus Clusius. Antwerp: Ex officina Christophori Plantini, 1589.

Besler, Basilius. *Hortus Eystettensis*. Eichstadt, 1613.

Bock, Hieronymus. *New Kreütter Buch von Underscheydt, Würckung und Namen der Kreütter so in Teütschen Landen wachsen: Auch der selbigen eygentlichem und wolgegrundtem Gebrauch in der Artznei, zu behalten und zu fürdern Leibs Gesuntheyt fast nutz und tröstlichen, vorab gemeynem Verstand*. Strasbourg: Wendel Rihel, 1539.

———. *De stirpium, maxime earum, quae in Germania nostra nascuntur, . . . Commentariorum libri tres*. Translated by David Kyber. Strasbourg: excudebat Vuendelinus Rihelius, 1552.

———. *Kreütterbuch, Darin underscheidt Namen und Wurckung der Kreütter, Staüden/Hecken unnd Beumen/sampt ihren Früchten/so inn Teutschen Landen wachsen . . .* Edited by Melchior Sebizius. Strasbourg: Josias Rihel, 1577.

The book of beasts, being a translation from a Latin bestiary of the twelfth century. Translated by T. H. White. New York: Dover Publications, 1984.

Borsetti Ferranti Bolani, Ferrante. *Historia almi Ferrariae Gymnasii*. 2 vols. Ferrara: Typis Bernardini Pomatelli, 1735.

Boyle, Robert. *General heads for the natural history of a country, great or small, drawn out for the use of travellers and navigators*. London: John Taylor and S. Holford, 1692.

Browne, Thomas. *Sir Thomas Browne's Pseudodoxia epidemica*. Edited by Robin Robbins. 2 vols. Oxford: Oxford University Press, 1981.

Brunfels, Otto. *Herbarum vivae eicones: Ad naturae imitationem, summa cum diligentia et artificio effigiatae, una cum effectibus earundem, in gratiam veteris illius & iamiam renascentis herbariae medicinae*. Strasbourg: Apud Ioannem Schottum, 1532.

Buffon, Georges-Louis Leclerc, comte de. *Oeuvres complètes*. Edited by M. Flourens. Paris: Garnier Frères, s.a.

Camerarius, Joachim. *Symbolorum et emblematum ex re herbaria desumptorum centuria una collecta a Ioachimo Camerario . . . in quib' rariores stirpium proprietates historiae ac sententiae memorabiles non paucae breviter exponuntur*. Nuremberg: Impensis J. Hofmanni et H. Camoxii, 1590.

———. *Symbolorum & emblematum ex animalibus quadrupedibus desumptorum centuria altera collecta a Ioachimo Camerario*. Nuremberg: Paulus Kaufmann, 1595.

Cartier, Jacques. *Relations*. Edited by Michel Bideaux. Montréal: Presses de l'Université de Montréal, 1986.

Castiglione, Baldesar. *The book of the courtier*. Translated by Charles S. Singleton. New York: Anchor Books, 1959.

A catalogue of all the cheifest rarities in the publick theater and Anatomie-hall of the University of Leiden. Leiden: Jacobus Voorn, 1683.

Celtis, Conrad. *Conradi Celtis Quae Vindobonae prelo subicienda curavit opuscula*. Edited by Kurt Adel. Leipzig: In aedibus B. G. Teubneri, 1966.

Ceruto, Benedetto, and Andrea Chiocco. *Musaeum Franc. Calceolari Iun. Veronensis . . . in quo multa ad naturalem moralemq; Philosophiam spectantia, non pauca ad rem Medicam pertinentia erudite proponuntur & explicantur, non sine magna rerum exoticarum supellectile, quae artifici plane manu in aes incisae, studiosis exhibentur*. Verona: Apud Angelum Tamum, 1622.

Cesalpino, Andrea. *De plantis libri XVI*. Florence: Apud Georgium Marescottum, 1583.

Clusius, Carolus. *Rariorum aliquot stirpium per Hispanias observatarum Historia, libris duobus expressa*. Antwerp: ex officina Christophori Plantini, Architypographi Regii, 1576.

———. *Rariorum aliquot stirpium, per Pannoniam, Austriam, & vicinas quasdam provincias observatarum historia, quatuor libris expressa*. Antwerp: Ex officina Christophori Plantini, 1583.

———. *Stirpium nomenclator Pannonicus*. Antwerp: ex officina Christophori Plantini, 1584.

———. *Rariorum plantarum historia*. Antwerp: ex officina Plantiniana apud Joannem Moretum, 1601.

———. *Exoticorum libri decem: quibus animalium, plantarum, aromatum, aliorumque peregrinorum Fructuum historiae describuntur*. [Leiden]: Ex Officina Plantiniana Raphelengii, 1605.

———. "Huit lettres de Charles de l'Escluse, 18 juin 1592—15 juillet 1593," ed. E. Roze. *Journal de botanique* 9 (1895): 27–32, 58–60, 99–100, 15–16.

———. *Lettere inedite di Charles de l'Escluse (Carolus Clusius) a Matteo Caccini, fioricultore fiorentino: Contributo alla storia della botanica*. Florence: L. S. Olschki, 1939.

———. *A treatise on tulips*. Translated by W. van Dijk. Haarlem: Enschede, 1951.

———. *Fungorum in Pannoniis observatum brevis historia et Codex Clusii: Mit Beiträgen von einer internationalen Autorengemeinschaft*. Edited by Stephan A. Aumüller and József Jeanplong. Budapest: Akadémiai Kiadó; Graz, Austria: Akademische Druck- und Verlagsanstalt, 1983.

Clusius, Carolus, and Conrad Gessner. *Caroli Clusii Atrebatis et Conradi Gesneri Tigurini, Epistolae ineditae*. Edited by L. C. Treviranus. Leipzig: Voss, 1830.

Colonna, Francesco. *Hypnerotomachie, ou Discours du songe de Poliphile*. Translated by Jean Martin. Paris: Pour Jacques Kerver aux deu Cochetz, 1546.

———. *Hypnerotomachia Poliphili: The strife of love in a dream*. Translated by Joscelyn Godwin. London: Thames & Hudson, 1999.

Cordus, Euricius. *Botanologicon*. Cologne: apud Ioannem Gymnicum, 1534.

Cordus, Valerius. *Pharmacorum conficiendorum ratio: vulgo vocant Dispensatorium*. Paris: apud Ioannem Roigny, 1548.

———. *Annotationes in Pedacii Dioscoridis Anazarbei de Medica materia libros V*. Edited by Conrad Gessner. Lyon: Rihel, 1561.

———. *Historia plantarum*. Edited by Conrad Gessner. In Valerius Cordus, *Annotationes in Pedacii Dioscoridis Anazarbei de Medica materia libros V*. Lyon: Rihel, 1561.

Cornut, Jacques. *Canadensium plantarum, aliarumque nondum editarum historia*. Paris: venduntur apud Simonem Le Moyne, 1635.

Darwin, Charles. *Insectivorous plants*. 2nd ed. Edited by Francis Darwin. In *The works of Charles Darwin*. New York: New York University Press, 1989. Reprint, originally published 1888.

Darwin, Charles, ed. *The zoology of the voyage of H. M. S. "Beagle," under the command of captain Fitzroy...during the years 1832 to 1836*. 3 vols. London: Smith and Elder, 1839–43.

De Toni, G. B. "Il carteggio degli italiani col botanico Carlo Clusio nella Biblioteca Leidense." *Memorie della R. Accademia di scienze, lettere ed arti di Modena* ser. 3, 10, pt. 2 (1912): 113–270.

Diels, Hermann. *Die Fragmente der Vorsokratiker: Griechisch und Deutsch*. Edited by Walther Kranz. 6th ed. 3 vols. Zürich: Weidmann, 1974–75.

Diogenes Laertius. *Lives of eminent philosophers*. Translated by R. D. Hicks. 2 vols. Cambridge, MA: Harvard University Press; London: William Heinemann, 1925.

Dioscorides. *The Greek herbal of Dioscorides*. Translated by John Goodyer. Edited by Robert T. Gunther. Oxford: Printed by John Johnson for the author at the University Press, 1934.

Dodoens, Rembert. *Cruydeboeck*. Antwerp: Jan vander Loe, 1554.

———. *Histoire des plantes, en laquelle est contenue la description entiere des herbes, c'est à dire, leures Especes, Forme, Noms, Temperament, Vertus & Operations: non seulement de celles qui croissent en ce païs, mais aussi des autres estrangeres qui viennent en usage de Medecine*. Translated by Carolus Clusius. Antwerp: de l'Imprimerie de Jean Loë, 1557.

———. *Florum, et coronariarum odoratarumque nonnullarum herbarum historia*. Antwerp: ex officina Christophori Plantini, 1568.

Egnazio, Giovanni Battista. *In Dioscoridem ab Hermolao Barbaro tralatum annotamenta*. Venice, 1516.

Erasmus, Desiderius. "Convivium religiosum." In *Opera omnia ... in decem tomos distincta*, edited by Jean Le Clerc, vol. 1. Leiden: cura & impensis Petri Vander Aa, 1703.

———. *Opera omnia ... in decem tomos distincta*. Edited by Jean Le Clerc. Leiden: cura & impensis Petri Vander Aa, 1703.

———. *The Ciceronian: A dialogue on the ideal Latin style: Dialogus Ciceronianus*. In *Collected works of Erasmus: Literary and educational writings 6, Ciceronianus/notes/indexes*. Toronto: University of Toronto Press, 1986.

———. *De copia verborum et rerum*. Edited by Betty I. Knott. In *Opera omnia Desiderii Erasmi Roterodami*, ord. 1, tom. 6. Amsterdam: North-Holland, 1988.

———. *The praise of folly and other writings*. Edited by Robert M. Adams. New York: W. W. Norton & Co., 1989.

Franceschini, Adriano, ed. *Nuovi documenti relativi ai docenti dello studio di Ferrara nel sec. XVI*. Deputazione provinciale ferrarese di storia patria, serie Monumenti, 6. Ferrara: Stabilimento Artistico Tipografico Editoriale, 1970.

Frederick II. *The art of falconry, being the De arte venandi cum avibus*. Edited by Casey A. Wood and F. Marjorie Fyfe. Stanford: Stanford University Press; London: Humphrey Milford and Oxford University Press, 1943.

Fuchs, Leonhart. *De historia stirpium commentarii insignes, maximis impensis et vigiliis elaborati, adjectis earundem vivis plusquam quingentisi imaginibus.* Basel: In Officina Isingriniana, 1542.

Gerard, John. *The herball or generall historie of plantes.* London: John Norton, 1597.

———. *The herball; or, Generall historie of plants.* Edited by Thomas Johnson. London: Printed by Adam Islip, Joice Norton, and Richard Whitakers, 1633.

Gessner, Conrad. *Bibliotheca universalis, sive Catalogus omnium scriptorum locupletissimus, in tribus linguis, Latina Graeca, & Hebraica: extantium & non extantium, veterum & recentiorum in hunc usque diem, doctorum & indoctorum, publicatorum & in Bibliothecis latentium.* Zürich: apud Christophorum Froschoverum, 1545.

———. *Historiae animalium lib. I. de quadrupedibus viviparis: Opus philosophis, medicis, grammaticis, philologis, poetis, & omnibus rerum linguarumq; variarum studiosis, utilissimum simul iucundissimumq; futurum.* Zürich: apud Christ. Froschoverum, 1551.

———. *Historiae animalium liber II. de quadrupedibus oviparis.* Zürich: excudebat C. Froschoverus, 1554.

———. *Historiae animalium liber III. qui est de avium natura.* Zürich: apud Christoph. Froschoverum, 1555.

———. *De raris et admirandis herbis, quae sive quod noctu luceant, sive alias ob causas, Lunariae nominantur, Commentariolus . . .* Zürich: Apud Andream Gesnerum F. & Iacobum Gesnerum, fratres, 1556.

———. *Historiae animalium liber IIII. qui est de piscium & aquatilium animantium natura.* Zürich: apud Christoph. Froschoverum, 1558.

———. *Horti Germaniae.* In Valerius Cordus, *Annotationes in Pedacii Dioscoridis Anazarbei de Medica materia libros V*, fol. 236–300. Lyon: Rihel, 1561.

———. *Opera botanica.* 2 vols. Nuremberg: impensis Io. Mich. Seligmanni, Typis Iosephi Fleishmanni, 1754–71.

———. *Vingt lettres à Jean Bauhin fils (1563–1565).* Translated by Augustin Sabot. Edited by Claude Longeon. Saint-Etienne: Université de Saint-Etienne, 1976.

———. *Conradi Gesneri Historia plantarum.* Edited by Heinrich Zoller, Martin Steinmann and Karl Schmid. Gesamtausgabe ed. 2 vols. Dietikon-Zürich: Urs Graf-Verlag, 1987.

Guicciardini, Francesco. *Ricordi.* Edited by Raffaele Spongano. Florence: G. C. Sansoni, 1951.

Gulyás, Pál. *A Zsámboky-könyvtár katalógusa: Bibliothecae Ioannis Sambuci catalogus librorum.* Szeged: Scriptum KFT, 1992.

Hakluyt, Richard, ed. *The principal navigations, voyages, traffiques & discoveries of the English nation.* Glasgow: J. MacLehose and sons, 1903–5.

Hernández, Francisco. *The Mexican treasury: The writings of Dr. Francisco Hernández.* Translated by Rafael Chabrán, Cynthia L. Chamberlin, and Simon Varey. Edited by Simon Varey. Stanford: Stanford University Press, 2001.

Hill, Thomas. *The gardener's labyrinth*. Edited by Richard Mabey. Oxford and New York: Oxford University Press, 1987.

Historia generalis plantarum, in libros XVIII. per certas classes artificiose digesta. 2 vols. Lyon: apud Gulielmum Rovillium, 1587–88.

The Imitation of Christ, from the first edition of an English translation made c. 1530. Translated by Richard Whitford. Edited by Edward J. Klein. New York: Harper & Brothers, 1941.

Imperato, Ferrante. *Dell'historia naturale . . . libri XXVIII, nella qvale ordinatamente si tratta della diuersa condition di miniere, e pietre; con alcune historie di piante, & animali; sin'hora non date in luce*. Naples: C. Vitale, 1599.

Isidore of Seville. *Etymologiarum sive originum libri* xx. Edited by W. M. Lindsay. 2 vols. Oxford: Oxford University Press, 1911.

Iulius Caesar, C. *Commentariorum pars prima qua continentur libri vii de Bello Gallico cum A. Hirti supplemento*. Edited by René Du Pontet. Oxford: e typographeo Clarendoniano, 1900.

Jonstonus, Joannes. *Historiae naturalis de piscibus et cetis libri V*. Amsterdam: Apud Ioannem Iacobi fil. Schipper, 1657.

Jungermann, Ludwig. *Catalogus plantarum quae circa Altorfium Noricum et vicinis quibusdam locis . . .* Altdorf, 1615.

Jungius, Joachim. *Opuscula botanico-physica*. Edited by Johann Sebastian Albrecht. Coburg: Sumtibus et typis Georgii Ottonis, 1747.

Juvenal. *Juvenal and Persius*. Translated by G. G. Ramsay. Cambridge, MA: Harvard University Press, 1918.

L'Obel, Mathieu de. *Plantarum seu stirpium historia . . . cui annexum est Adversariorum volumen*. Antwerp: ex officina Christophori Plantini, 1576.

———. *Plantarum seu stirpium icones*. Antwerp: ex officina Christophori Plantini, 1581.

L'Obel, Mathieu de, and Pierre Pena. *Stirpium adversaria nova*. London, 1570.

Laet, Joannes de. *Novus Orbis seu Descriptionis Indiae occidentalis libri xviii*. Leiden: Apud Elzevirios, 1633.

Leoniceno, Niccolò. *De Plinii et aliorum medicorum erroribus liber, cui addita sunt eiusdem autoris de herbis & fruticibus, animalibus, metallis, serpentibus, tiro seu vipera*. Basel: Henricus Petrus, 1529.

Lipsius, Justus. *Iusti Lipsi Epistolae*. Brussels: Koninklijke Academie voor Wetenschappen, Letteren en Schone Kunsten van België, 1978–.

Mattioli, Pietro Andrea. *De plantis epitome utilissima. Petri Andreae Matthioli senesis; novis iconibus et descriptionibus pluribus nunc primum diligenter aucta, a d. Ioachimo Camerario; accessit catalogus plantarum, quae in hoc compendio continentur, exactiss*. Frankfurt am Main, 1586.

———. *Opera quae extant omnia: hoc est, Commentarii in VI. libros Pedacii Dioscoridis Anazarbei de Medica materia: Adiectis in margine variis Graeci textus lectionibus, ex antiquissimis Codicis desumptis, qui Dioscoridis depravatam lectionem restituunt: Nunc a Casparo Bauhino D. Botanico at*

Anatomico Basiliensi Ordinario, post diversarum editionum collationem infinitis in locis aucti: synonymiis quoque plantarum et notis illustrati: Adiectis plantarum iconibus, supra priores editiones plus quam trecentis (quarum quamplurimae nunc primum describuntur) ad vivum delineatis. Edited by Caspar Bauhin. Frankfurt am Main: Ex officina typographica Nicolai Bassaei, 1598.

Menabeni, Apollonio. *Tractatus de magno animali.* Milan: Tinum, 1581.

Merian, Matthaeus. "Nova et genuina descriptio inclytae urbis Basileae." Basel: Merian, 1615.

Meurs, Johannes van. *Athenae Batavae; sive, De urbe Leidensi, & Academia, virisque claris; qui utramque ingenio suo, atque scriptis, illustrarunt: Libri duo.* Leiden: apud Andream Cloucquium, et Elsevirios, 1625.

Molhuysen, P. C., ed. *Bronnen tot de geschiedenis der Leidsche Universiteit.* 7 vols. The Hague: M. Nijhoff, 1913–24.

Montaigne, Michel de. *The complete essays of Montaigne.* Translated by Donald M. Frame. Stanford: Stanford University Press, 1958.

Münster, Sebastian. *Cosmographiae universalis lib. VI.* Basel: apud Henrichum Petri, 1550.

———. *Cosmographer; oder beschreibung Aller Länder herzschafftem und fürnembsten Stetten des gantzen Erdbodens . . .* Basel: in der Officin Henricpetrina, 1578.

———. *Briefe Sebastian Münsters: Lateinisch und deutsch.* Edited by Karl Heinz Burmeister. Frankfurt am Main: Insel-Verlag, 1964.

Nieremberg, Juan Eusebio. *Historia naturae, maxime peregrinae, libris xvi distincta, in quibus . . . ignota Indiarum animalia . . . describuntur.* Antwerp: Ex officina Plantiniana Balthasaris Moreti, 1635.

Olaus Magnus. *Historia de gentibus septentrionalibus, earumque diversis statibus, conditionibus, moribus, ritibus, superstitionibus, disciplinis, exercitiis, regimine, victu, bellis, structuris, instrumentis, ac mineris metallicis, & rebus mirabilibus, necnon universis pene animalibus in Septentrione degentibus, eorumque natura.* Rome: apud Ioannem Mariam, 1555.

———. *Carta marina: Karta och beskrivning över de nordiska länderna sant de underbara ting som där finnas.* Uppsala: Bokgillet, 1964.

Oldenburg, Henry, et al. *The correspondence of Henry Oldenburg.* Edited by A. Rupert Hall and Marie Boas Hall. 13 vols. Madison: University of Wisconsin Press, 1965–86.

Orta, Garcia da. *Aromatum, et simplicium aliquot medicamentorum apud Indos nascentium historia.* Translated by Carolus Clusius. Antwerp: ex off. Ch. Plantini, 1567.

Ovid. *Metamorphoseon libri i–xv.* Edited by B. A. van Proosdij. 6th ed. Leiden: E. J. Brill, 1975.

Paracelsus. "Herbarius." In *Sämtliche Werke,* edited by Bernhard Aschner, vol. 3. Jena: Gustav Fischer, 1930.

Parkinson, John. *Paradisi in sole Paradisus terrestris, by John Parkinson: Faithfully reprinted from the edition of 1629*. London: Methuen & Co., 1904.
Passe, Crispijn van de. *Hortus floridus*. Arnhem: apud Ioannem Ianssonium, 1614.
Pausanias. *Description of Greece*. 5 vols. Cambridge, MA: Harvard University Press; London: Heinemann, 1918–35.
———. *Pausanias's Description of Greece*. Translated and edited by J. G. Frazer. 6 vols. New York: Biblo and Tannen, 1965.
Petrarca, Francesco. "The ascent of Mont Ventoux." In *The Renaissance philosophy of man*, edited by Ernst Cassirer, Paul Oskar Kristeller and John Herman Randall, Jr. Chicago: University of Chicago Press, 1948.
———. "On his own ignorance and that of many others." In *The Renaissance philosophy of man*, edited by Ernst Cassirer, Paul Oskar Kristeller and John Herman Randall, Jr., 47–133. Chicago: University of Chicago Press, 1948.
Physiologus. Translated by Michael J. Curley. Austin: University of Texas Press, 1979.
Pigafetta, Antonio. *Magellan's voyage around the world*. Edited by James Alexander Robinson. 3 vols. Cleveland: The Arthur H. Clark Company, 1906.
Pighius, Stephanus Vinandus. *Annales Romanorum; qui commentarii vicem supplent in omnes veteres historiae Romanae scriptores*. Edited by Andreas Schott. 3 vols. Antwerp: ex offini Plantiniana, 1615.
Plantin, Christophe. *Correspondance de Christophe Plantin*. 9 vols. Antwerp: J. E. Buschmann, J. Denucé, [etc.], 1883–1918.
Platter, Felix. "Das Tagebuch des Felix Platters." In *Thomas und Felix Platter: Zur Sittengeschichte des XVI. Jahrhunderts*, edited by Heinrich Boos, 119–331. Leipzig: Verlag von S. Hirzel, 1878.
———. *Beschreibung der Stadt Basel, 1610, und Pestbericht, 1610/11: Synoptische Edition mit Ausschnitten aus dem Vogelschauplan von Matthäus Merian d.Ä. (1615) und dem Stadtplan von Ludwig Löffel (1862)*. Edited by Valentin Lötscher. Basel and Stuttgart: Schwabe & Co., 1987.
Platter, Thomas. *Beschreibung der Reisen durch Frankreich, Spanien, England und die Niederlande, 1595–1600*. Edited by Rut Keiser. 2 vols. Basel und Stuttgart: Schwabe & Co., 1968.
Pliny the Elder. *Natural history*. 10 vols: Loeb Classical Library.
———. *Histoire naturelle*. 37 vols. Paris: Belles Lettres, 1947–85.
Poliziano, Angelo. *Miscellaneorum centuria secunda*. Edited by Vittore Branca and Manlio Pastore Stocchi. 2nd ed. Florence: Leo S. Olschki, 1978.
Rabelais, François. *Gargantua and Pantagruel*. Translated by Burton Raffel. New York: W. W. Norton & Co., 1990.
Ray, John. *Historia plantarum; species hactenus editas aliasque insuper multas noviter inventas & descriptas complectens: In qua agitur primo de plantis in genere, earumque partibus, accidentibus & differentiis; deinde . . .* 3 vols. London: typis M. Clark, prostant apud H. Faithorne, 1686–1704.

———. *Synopsis methodica animalium quadrupedum et serpentini generis: Vulgarium notas characteristicas, rariorum descriptiones integras exhibens: cum historiis & observationibus anatomicis perquam curiosis*. London: Impensis S. Smith & B. Walford, 1693.

Réaumur, René-Antoine Ferchauld de. *Mémoires pour servir à l'histoire des insectes*. Vol. 1, pt. 1. Amsterdam: chez Pierre Mortier, 1737.

Scaliger, Julius Caesar. *In libros duos, qui inscribuntur de plantis, Aristotele autore, libri duo*. Paris: ex officina Michaelis Vascosani, 1556.

———. *Poetices libri septem*. Lyon: apud Antonium Vincentium, 1561.

———. *Commentarii, et animadversiones, in sex libros De causis plantarum Theophrasti*. Lyon: G. Rovilium, 1566.

Schnur, Harry C., ed. *Lateinische Gedichte deutscher Humanisten*. Stuttgart: Philipp Reclam Jun., 1978.

Serres, Olivier de. *Le théâtre d'agriculture et mesnage des champs*. Paris: I. Métayer, imprimeur ordinaire du roy, 1600.

Spiegel, Adriaan van de. *Isagoges in rem herbariam libri duo*. Padua: Apud Paulum Meiettum, ex typographia Laurentii Pasquati, 1606.

Textor, Benedict. *Stirpium differentiae ex Dioscoride secundum locos communes, opus ad ipsarum plantarum cognitionem admodum conducibile*. Venice: in officina divi Bernardini, 1537.

Theophrastus. *Enquiry into plants and minor works on odours and weather signs*. Translated by Arthur Hort. 2 vols. London: W. Heinemann; New York, G.P. Putnam's Sons, 1916.

Transylvanus, Maximilianus. "De Moluccis insulis." In *The Philippine Islands, 1493–1803*, edited by Emma Helen Blair and James Alexander Robertson, vol. 1, 305–37. Cleveland: The Arthur H. Clark Company, 1903.

Valla, Giorgio. *De expetendis et fugiendis rebus opus*. Venice: in aedibus Aldi Romani, impensa, ac studio Ioannis Petri Vallae filii, 1501.

Valla, Lorenzo. *De falso credita et ementita Constantini donatione*. Edited by Wolfram Setz. Weimar: Hermann Böhlaus Nachfolger, 1976.

Veer, Gerrit de. *Reizen van Willem Barents, Jacob van Heemskerck, Jan Cornelisz. Rijp en anderen naar het noorden (1594–1597)*. Edited by S. P. L'Honoré Naber. 2 vols. The Hague: Martinus Nijhoff, 1917.

Vergil, Polydore. *On discovery*. Edited and translated by Brian P. Copenhaver. Cambridge, MA: Harvard University Press, 2002.

Vesalius, Andreas. "The preface to *De fabrica corporis humani* 1543." Translated by B. Farrington. *Proceedings of the Royal Society of Medicine* 25 (1932): 1357–66.

Vives, Juan Luis. "Epistola I. De ratione studii puerilis." In *Opera omnia, distributa et ordinata in argumentorum classes praecipuas*, edited by Gregorius Majansius, vol. 1, 256–69. Valencia: in officina Benedicti Monfort, 1782.

———. "De causis corruptarum artium." In *Opera omnia, distributa et ordinata in argumentorum classes praecipuas*, edited by Gregorius Majansius, vol. 6, 8–242. Valencia: in officina Benedicti Monfort, 1785.

———. "De tradendis disciplinis; seu De institutione Christiana." In *Opera omnia, distributa et ordinata in argumentorum classes praecipuas*, edited by Gregorius Majansius, vol. 6, 243–437. Valencia: in officina Benedicti Monfort, 1785.

Wieder, F. C., ed. *De Reis van Mahu en De Cordes door de straat van Magalhães naar Zuid-Amerika en Japan, 1598–1600: Scheepsjournaal, rapporten, brieven, zeilaanwijzingen, kaarten, enz.* 3 vols. Werken uitgegeven door de Linschoten-Vereeniging, 21. The Hague: Martinus Nijhoff, 1923.

Willughby, Francis. *De historia piscium libri quatuor*. Edited by John Ray. Oxford: E Theatro Sheldoniano, 1686.

Worm, Ole. *Museum Wormianum; seu, Historia rerum rariorum, tam naturalium, quam artificialium, tam domesticarum, quam exoticarum, quae Hafniae Danorum in aedibus authoris servantur*. Amsterdam: Apud Ludovicum & Danielem Elzevirios, 1655.

Zaluziansky a Zaluzian, Adam. *Methodi herbariae libri tres*. Edited by Karel Pejml. Vol. 1 (textus). Prague: Ceské Akademie Ved a Umení, 1940.

SECONDARY SOURCES

Abrahams, Harold J., and Marion B. Savin. "The *Herbarius latinus*, Venice, 1499: The first Italian Renaissance illustrations of plants." *Episteme* 9 (1975): 225–52.

Aguzzi-Barbagli, Danilo. "Humanism and poetics." In *Renaissance humanism: Foundations, forms, and legacy*, edited by Albert Rabil, Jr., vol. 3, 85–169. Philadelphia: University of Pennsylvania Press, 1988.

Allen, D. E. *The naturalist in Britain: A social history*. 2nd ed. Princeton, NJ: Princeton University Press, 1994.

———. "Walking the swards: Medical education and the rise and spread of the botanical field class." *Archives of Natural History* 27 (2000): 335–67.

Alpers, Svetlana. *The art of describing: Dutch art in the seventeenth century*. Chicago: University of Chicago Press, 1983.

Ambrosoli, Mauro. *Scienziati, contadini e proprietari: Botanica e agricoltura nell'Europa occidentale, 1350–1850*. Torino: Giulio Einaudi editore, 1992.

Anderson, Benedict. *Imagined communities: Reflections on the origin and spread of nationalism*. London: Verso, 1983.

Anderson, Frank J. *An illustrated history of the herbals*. New York: Columbia University Press, 1977.

Arber, Agnes. "The colouring of sixteenth-century herbals." *Nature* 145 (1940): 803.

———. *Herbals: Their origin and evolution; a chapter in the history of botany, 1470–1670*. 3rd ed. Cambridge: Cambridge University Press, 1986.

Ashworth, William B., Jr. "The persistent beast: Recurring images in early zoological illustration." In *The natural sciences and the arts: Aspects of interaction from the Renaissance to the twentieth century: An international symposium*, 46–66. Uppsala: Almqvist & Wiksell, 1985.

———. "Natural history and the emblematic world view." In *Reappraisals of the Scientific Revolution*, edited by David C. Lindberg and Robert S. Westman, 303–32. Cambridge: Cambridge University Press, 1990.

———. "Emblematic natural history of the Renaissance." In *Cultures of natural history*, edited by N. Jardine, J. A. Secord and E. C. Spary. Cambridge: Cambridge University Press, 1996.

Atran, Scott. "Origin of the species and genus concepts: An anthropological perspective." *Journal of the History of Biology* 20 (1987): 195–279.

———. *Cognitive foundations of natural history: Towards an anthropology of science*. Cambridge: Cambridge University Press, 1990.

———. "Folk biology and the anthropology of science: Cognitive universals and cultural particulars" (with comments from 29 colleagues and author's response). *Behavioral and Brain Sciences* 21.4 (1998): 547–609.

———. "Itzaj Maya folkbiological taxonomy: Cognitive universals and cultural particulars." In *Folkbiology*, edited by Douglas L. Medin and Scott Atran, 119–203. Cambridge, MA: MIT Press, 1999.

Azzi Visentini, Margherita. "Il giardino di semplici di Padova: Un prodotto della cultura del Rinascimento." *Comunità* 34 (1980): 259–338.

———. *L'Orto botanico di Padova e il giardino del Rinascimento*. Milano: Edizioni il Polifilo, 1984.

Bach, Endre. *Un humaniste hungrois en France: Jean Sambucus et ses relations littéraires, 1551–1584*. Szeged: Impr. Prometheus, 1932.

Bailey, William Whitman. *Botanizing: A guide to field-collecting and herbarium work*. Providence, R.I.: Preston and Rounds Co., 1899.

Baldwin, T. W. *William Shakespere's small Latine and less Greeke*. 2 vols. Urbana: University of Illinois Press, 1944.

Balss, Heinrich. *Albertus Magnus als Biologe: Werk und Ursprung*. Stuttgart: Wissenschaftliche Verlagsgesellschaft, 1947.

Barnes, Jeffrey K. *Scolopendra heros* (web page). University of Arkansas Arthropod Museum notes, no. 13, June 21, 2002 (cited October 11, 2002). Available from http://www.uark.edu/depts/entomolo/museum/sheros.html.

Baron, Hans. *The crisis of the early Italian Renaissance*. Revised ed. Princeton, NJ: Princeton University Press, 1966.

Barrera, Antonio. "Local herbs, global medicines: Commerce, knowledge, and commodities in Spanish America." In *Merchants and marvels: Commerce, science, and art in early modern Europe*, edited by Pamela H. Smith and Paula Findlen, 163–81. New York: Routledge, 2001.

———. *Experiencing nature: The Spanish American empire and the early Scientific Revolution*. Austin: University of Texas Press, forthcoming.

Barsanti, Giulio. *La scala, la mappa, l'albero: Immagini e classificazioni della natura fra Sei e Ottocento*. Florence: Sansoni Editore, 1992.

Battiato, Carmelo. "Luca Ghini (1496–1556): Medico e botanico fondatore di orti botanici e pioniere degli erbari." *Rivista di Storia della Medicina* 16 (1972): 155–63.

Bauer, Rotraud, and Herbert Haupt. "Die Kunstkammer Kaiser Rudolfs II. in Prag: Ein Inventar aus den Jahren 1607–1611." *Jahrbuch der Kunsthistorischen Sammlungen in Wien* 72 (1976).

Baxandall, Michael. *Giotto and the orators: Humanist observers of painting in Italy and the discovery of pictorial composition, 1350–1450*. Oxford: Clarendon Press, 1971.

Bedini, Silvio A. *The pope's elephant*. 1st U.S. ed. Nashville, TN: J. S. Sanders & Co., 1998.

Bentley, Jerry H. *Politics and culture in Renaissance Naples*. Princeton, NJ: Princeton University Press, 1987.

Berger, Peter L., and Thomas Luckmann. *The social construction of reality: A treatise in the sociology of knowledge*. Garden City, N.Y.: Doubleday, 1966.

Berlin, Brent. *Ethnobiological classification: Principles of categorization of plants and animals in traditional societies*. Princeton, NJ: Princeton University Press, 1992.

Bigwood, J. M. "Aristotle and the elephant again." *American Journal of Philology* 114 (1993): 537–55.

Bimont, G. *Manuel pratique du botaniste herborisant*. Paris: Editions N. Boubée & Cie., 1945.

Black, Robert. *Humanism and education in medieval and Renaissance Italy: Tradition and innovation in Latin schools from the twelfth to the fifteenth century*. Cambridge: Cambridge University Press, 2001.

Blair, Ann. "Humanism and the commonplace book." *Journal of the History of Ideas* 53 (1992): 541–51.

Blanchard, W. Scott. "*O miseri philologi:* Codro Urceo's satire on professionalism and its context." *Journal of Medieval and Renaissance Studies* 20 (1990): 91–122.

Blumenberg, Hans. *Der Prozeß der theoretischen Neugierde*. Fränkfurt am Main: Suhrkamp, 1973.

———. *Die Lesbarkeit der Welt*. Revised ed. Frankfurt am Main: Suhrkamp, 1983.

Blunt, Wilfred. *The compleat naturalist: A life of Linnaeus*. New York: Viking Press, 1971.

Boas, Marie. *The scientific Renaissance, 1450–1630*. New York: Harper & Row, 1962.

Bodenheimer, Frederick Simon. *Materialien zur Geschichte der Entomologie bis Linné*. 2 vols. Berlin: Junk, 1928–29.

———. "Towards the history of zoology and botany in the sixteenth century." In *La science au seizième siècle: Colloque international de Royaumont, 1–4 juillet 1957*, 285–96. Paris: Hermann, 1960.

Bodson, L. "Aspects of Pliny's zoology." In *Science in the early Roman Empire: Pliny the Elder, his sources and influence*, edited by Roger French and Frank Greenaway. Totowa, N.J.: Barnes & Noble Books, 1986.

Borst, Arno. *Das Buch der Naturgeschichte: Plinius und seine Leser im Zeitalter des Pergaments*. 2nd ed. Heidelberg: Universitätsverlag C. Winter, 1995.

Botany in the Low Countries: End of the fifteenth century—ca. 1650. Antwerp: Snoeck-Ducaju & Zoon, 1993.

Bourdieu, Pierre. *Outline of a theory of practice.* Translated by Richard Nice. Cambridge: Cambridge University Press, 1977.

Bourguet, Marie-Noëlle. "La collecte du monde: Voyage et histoire naturelle, fin XVIIème siècle—début XIXème siècle." In *Le Muséum au premier siècle de son histoire,* edited by Claude Blanckaert, Claudine Cohen, Pietro Corsi, and Jean-Louis Fischer, 163–96. Paris: Éditions du Muséum national d'Histoire naturelle, 1997.

Bouwsma, William J. *The waning of the Renaissance, 1550–1640.* New Haven: Yale University Press, 2000.

Brann, Noel L. "Humanism in Germany." In *Renaissance humanism: Foundations, forms, and legacy,* edited by Albert Rabil, Jr., vol. 2. Philadelphia: University of Pennsylvania Press, 1988.

Braun, Lucien. *Conrad Gessner.* Geneva: Editions Slatkine, 1990.

Bridson, Gavin D. R. *The history of natural history: An annotated bibliography.* New York: Garland Publishing, 1994.

Bruyère, Nelly. *Méthode et dialectique dans l'œuvre de La Ramée: Renaissance et âge classique.* Paris: Librairie philosophique J. Vrin, 1984.

Burckhardt, Jacob. *The civilization of the Renaissance in Italy: An essay.* Translated by S. G. C. Middlemore. New York: The Modern Library, 1954.

Burke, Peter. *The European Renaissance: Centres and peripheries.* Oxford and Malden, MA: Blackwell Publishers, 1998.

Cailleux, André. "Progression du nombre d'espèces de plantes décrites de 1500 à nos jours." *Revue d'Histoire des Sciences* 6 (1953): 42–49.

Callmer, Christian. "Queen Christina's herbaria." In *Otium et negotium: Studies in onomatology and library science presented to Olof von Feilitzen.* Stockholm: Kungl. Boktryckeriet P. A. Norstedt & Söner, 1973.

Campbell, Mary B. *The witness and the other world: Exotic European travel writing, 400–1600.* Ithaca: Cornell University Press, 1988.

Capparoni, Pietro. "Un ritratto di Niccolò Leoniceno, maestro di Paracelso a Ferrara." *Rivista di Storia delle Scienze Mediche e Naturali* 33 (1942): 1–14.

Christ, Hermann. "Die illustrierte spanische Flora des Carl Clusius vom Jahre 1576." *Oesterreichische Botanische Zeitschrift* 62 (1912): 132-135, 189–194, 229–238, 271–275.

———. "Die ungarisch-österreichische Flora des Carl. Clusius von Jahre 1583." *Oesterreichische Botanische Zeitschrift* 62–63 (1912–13): 330–334, 426–430; 131–136, 159–167.

Cobban, Alan B. *The medieval universities: Their development and organization.* London: Methuen & Co., 1975.

Cochrane, Eric. *Florence in the forgotten centuries, 1527–1800: A history of Florence and the Florentines in the age of the grand dukes.* Chicago: University of Chicago Press, 1973.

Cochrane, Eric. *Historians and historiography in the Italian Renaissance*. Chicago: University of Chicago Press, 1981.

Coffin, David R. *Gardens and gardening in Papal Rome*. Princeton, NJ: Princeton University Press, 1991.

Cohen, I. Bernard. "La découverte du nouveau monde et la transformation de l'idée de la nature." In *La science au seizième siècle: Colloque international de Royaumont, 1–4 juillet 1957*, 189–210. Paris: Hermann, 1960.

Cook, Harold J. "Physicians and natural history." In *Cultures of natural history*, edited by N. Jardine, J. A. Secord and E. C. Spary, 91–105. Cambridge: Cambridge University Press, 1996.

Cooper, Alix. "The museum and the book: The *Metallotheca* and the history of an encyclopaedic natural history in early modern Italy." *Journal of the History of Collections* 7.1 (1995): 1–23.

———. "Inventing the indigenous: Local knowledge and natural history in the early modern German territories." Ph.D. dissertation, Harvard University, 1998.

Copenhaver, Brian P., and Charles B. Schmitt. *Renaissance philosophy*. Oxford: Oxford University Press, 1992.

Corbett, Margery. "The emblematic title-page to *Stirpium adversaria nova* by Petro Pena and Mathias de L'Obel (1570)." *Archives of Natural History* 10 (1981): 111–17.

Crawford, M. H., ed. *Antonio Agustín between Renaissance and Counter-Reform*. London: Warburg Institute, University of London, 1993.

Cunningham, Andrew. "The culture of gardens." In *Cultures of natural history*, edited by N. Jardine, J. A. Secord and E. C. Spary, 38–56. Cambridge: Cambridge University Press, 1996.

Curtius, Ernst Robert. *European literature and the Latin Middle Ages*. Translated by Willard R. Trask. Princeton, NJ: Princeton University Press, 1953.

D'Amico, John F. *Renaissance humanism in papal Rome: Humanists and churchmen on the eve of the Reformation*. Baltimore: Johns Hopkins University Press, 1983.

———. *Theory and practice in Renaissance textual criticism: Beatus Rhenanus between conjecture and history*. Berkeley and Los Angeles: University of California Press, 1988.

Dannenfeldt, Karl H. *Leonhard Rauwolf: Sixteenth-century physician, botanist, and traveler*. Cambridge, MA: Harvard University Press, 1968.

Danto, Arthur C. *Narration and knowledge*. New York: Columbia University Press, 1985.

Daston, Lorraine. "Baconian facts, academic civility, and the prehistory of objectivity." *Annals of Scholarship* 8 (1991): 337–63.

———. "The ideal and reality of the Republic of Letters in the Enlightenment." *Science in Context* 4 (1991): 367–86.

———. "Marvelous facts and miraculous evidence in early modern Europe." *Critical Inquiry* 18 (1991): 93–124.

———. "The several contexts of the scientific revolution." Review of *The Scientific Revolution in national context*, edited by Roy Porter and Mikulas Teich. *Minerva* 32 (1994): 108–14.

———. "Curiosity in early modern science." *Word & image* 11 (1995): 391–404.

———. "The moralized objectivities of science." In *Wahrheit und Geschichte: Ein Kolloquium zu Ehren des 60. Geburtstages von Lorenz Krüger*, edited by Wolfgang Carl and Lorraine Daston, 78–100. Göttingen: Vandenhoeck & Ruprecht, 1999.

———. "Objectivity versus truth." In *Wissenschaft als kulturelle Praxis, 1750–1900*, edited by Hans Erich Bödeker, Peter Hanns Reill and Jürgen Schlumbohm, 17–32. Göttingen: Vandenhoeck & Ruprecht, 1999.

Daston, Lorraine, and Peter Galison. "The image of objectivity." *Representations* 40 (1992): 81–128.

Daston, Lorraine, and Katharine Park. *Wonders and the order of nature, 1150–1750*. New York: Zone Books, 1998.

Dear, Peter. *Mersenne and the learning of the schools*. Ithaca: Cornell University Press, 1988.

———. "Miracles, experiments, and the ordinary course of nature." *Isis* 81 (1990): 663–83.

———. "Narratives, anecdotes, and experiments: Turning experience into science in the seventeenth century." In *The literary structure of scientific argument: Historical studies*, edited by Peter Dear, 135–63. Philadelphia: University of Pennsylvania Press, 1991.

———. *Discipline & experience: The mathematical way in the scientific revolution*. Chicago: University of Chicago Press, 1995.

Debus, Allen G. *The chemical philosophy: Paracelsian science and medicine in the sixteenth and seventeenth centuries*. 2 vols. New York: Science History Publications, 1977.

Defrance, Robert, ed. *Le jardin médiéval*. Cahiers de l'Abbaye de Saint-Arnoult. Warluis: Editions A.D.A.M.A., 1990.

Delaunay, Paul. *La zoologie au seizième siècle*. Paris: Hermann, 1962.

Denecke, Dietrich. "Straßen, Reiserouten und Routenbücher (Itinerare) im späten Mittelalter und in der frühen Neuzeit." In *Reisen und Reiseliteratur im Mittelalter und in der Frühen Neuzeit: Vorträge eines interdiziplinären Symposiums vom 3.-8. Juni 1991 an der Justus-Liebig-Universität Gießen*, edited by Xenja von Ertzdorff and Dieter Neukirch, 227–53. Amsterdam and Atlanta: Rodopi, 1992.

Diamond, Jared, and K. David Bishop. "Ethno-ornithology of the Ketengban people, Indonesian New Guinea." In *Folkbiology*, edited by Douglas L. Medin and Scott Atran, 17–45. Cambridge, MA: MIT Press, 1999.

Dibon, Paul. "Communication in the Respublica literaria of the seventeenth century." *Res publica litterarum* 1 (1978): 43–55.

Dilg, Peter. *Das Botanologicon des Euricius Cordus: Ein Beitrag zur botanischen Literatur der Renaissance*. Marburg: Erich Mauersberger, 1969.

———. "Studia humanitatis et res herbaria: Euricus Cordus als Humanist und Botaniker." *Rete* 1 (1971): 71–85.

———. "Der Beitrag Adam Zaluzanskys zur Ausbildung einer wissenschaftlichen Botanik." In *Proceedings of the fourteenth International Congress on the History of Sciences (Tokyo and Kyoto, 1974)*, vol. 4, 11–14. Tokyo: Science Council of Japan, 1975.

———. "Die botanische Kommentarliteratur Italiens um 1500 und ihr Einfluß auf Deutschland." In *Der Kommentar in der Renaissance*, edited by August Buck and Otto Herding, 225–52. Bonn-Bad Godesberg: Deutsche Forschungsgemeinschaft; Boppard: Harald Boldt Verlag, 1975.

Durling, Richard J. "Conrad Gesner's *Liber amicorum*." *Gesnerus* 22 (1965): 134–59.

Eco, Umberto. *Die Suche nach der vollkommenen Sprache*. Translated by Burkhart Kroeber. Munich: C. H. Beck, 1994.

Edwards, William F. "Niccolò Leoniceno and the origins of humanist discussion of method." In *Philosophy and humanism: Renaissance essays in honor of Paul Oskar Kristeller*, edited by Edward P. Mahoney, 283–305. New York: Columbia University Press, 1976.

Egmond, Florike, and Peter Mason. *The mammoth and the mouse: Microhistory and morphology*. Baltimore: Johns Hopkins University Press, 1997.

Eisenstein, Elizabeth L. *The printing press as an agent of change: Communications and cultural transformations in early-modern Europe*. Cambridge: Cambridge University Press, 1979.

Elliott, J. H. *The old world and the new, 1492–1650*. Cambridge: Cambridge University Press, 1970.

Engel, Hendrik. *Hendrik Engel's Alphabetical list of Dutch zoological cabinets and menageries*. Edited by Pieter Smit, A. P. M. Sanders and J. P. F. van der Veer. Amsterdam: Editions Rodopi, 1986.

Evans, R. J. W. *Rudolf II and his world: A study in intellectual history, 1576–1612*. Oxford: Clarendon Press, 1973.

Fat, Leslie Tjon Sie. "Clusius' garden: A reconstruction." In *The authentic garden: A symposium on gardens*, edited by Leslie Tjon Sie Fat and Erik de Jong, 3–12. Leiden: Clusius Foundation, 1991.

Ferguson, Wallace K. *The Renaissance in historical thought: Five centuries of interpretation*. Boston: Houghton Mifflin Company, 1948.

Findlen, Paula. "Jokes of nature and jokes of knowledge: The playfulness of scientific discourse in early modern Europe." *Renaissance Quarterly* 43 (1990): 292–331.

———. "The economy of scientific exchange in early modern Italy." In *Patronage and institutions: Science, technology, and medicine at the European court, 1500–1750*, edited by Bruce T. Moran, 5–24. Rochester, NY: Boydell Press, 1991.

———. *Possessing nature: Museums, collecting, and scientific culture in early modern Italy*. Berkeley and Los Angeles: University of California Press, 1994.

———. "The formation of a scientific community: Natural history in sixteenth-century Italy." In *Natural particulars: Nature and the disciplines in Renaissance Europe*, edited by Anthony Grafton and Nancy Siraisi, 369–400. Cambridge, MA: MIT Press, 1999.

Fischer, H., ed. *Conrad Gessner 1516–1565: Universalgelehrter, Naturforscher, Arzt*. Zürich: Orell Füssli Verlag, 1967.

Foa, Anna. "Il nuovo e il vecchio: L'insorgere della sifilide (1494–1530)." *Quaderni Storici* 19, no. 55, (1984): 11–34.

Foucault, Michel. *Les mots et les choses: Une archéologie des sciences humaines*. Paris: Gallimard, 1966.

Freedberg, David. *The eye of the lynx: Galileo, his friends, and the beginnings of modern natural history*. Chicago: University of Chicago Press, 2002.

Freedman, Joseph S. "The diffusion of the writings of Petrus Ramus in central Europe, c. 1570–c. 1630." *Renaissance Quarterly* 46 (1993): 98–152.

French, Roger. *Ancient natural history: Histories of nature*. London and New York: Routledge, 1994.

Fubini, Riccardo. "Umanesimo ed enciclopedismo: A propositi di contributi recenti su Giorgio Valla." *Il pensiero politico* 16.2 (1983): 251–69.

Fumaroli, Marc. *L'âge de l'éloquence: Rhétorique et "Res literaria" de la Renaissance au seuil de l'époque classique*. Geneva: Droz, 1980.

Gaastra, F. S. "De VOC in Azië tot 1680." In *Algemene geschiedenis der Nederlanden*, vol. 7, 174–219. Haarlem: Fibula-Van Dishoeck, 1980.

Ganzenmüller, Wilhelm. *Das Naturgefühl im Mittelalter*. Leipzig and Berlin: B. G. Teubner, 1914.

Garber, Daniel. "Experiment, community, and the constitution of nature in the seventeenth century." *Perspectives on Science* 3 (1995): 173–205.

Garin, Eugenio. *Italian humanism: Philosophy and civic life in the Renaissance*. New York: Harper and Row, 1965.

Geertz, Clifford. *The interpretation of culture: Selected essays*. New York: Basic Books, 1973.

George, Wilma. "Sources and background to discoveries of new animals in the sixteenth and seventeenth centuries." *History of Science* 18 (1980): 79–104.

George, Wilma, and Brunsdon Yapp. *The naming of the beasts: Natural history in the medieval bestiary*. London: Duckworth, 1991.

Gerbi, Antonello. *Nature in the New World: From Christopher Columbus to Gonzalo Fernández de Oviedo*. Translated by Jeremy Moyle. Pittsburgh: University of Pittsburgh Press, 1985.

Gerlo, Alois. *Inventaire de la correspondance de Juste Lipse, 1564–1606*. Antwerp: Editions Scientifiques Erasme, 1968.

Giddens, Anthony. *Central problems in social theory: Action, structure and contradiction in social analysis*. Berkeley and Los Angeles: University of California Press, 1979.

Gilbert, Felix. *Machiavelli and Guicciardini: Politics and history in sixteenth-century Florence*. Princeton, NJ: Princeton University Press, 1965.

Glacken, Clarence J. *Traces on the Rhodian shore: Nature and culture in Western thought from ancient times to the end of the eighteenth century*. Berkeley: University of California Press, 1967.

Glamann, Kristof. "Der europäische Handel, 1500–1750." In *Europäische Wirtschaftsgeschichte*, edited by Carlo Cipolla and K. Borchardt, vol. 2. Stuttgart and New York: Gustav Fischer, 1979.

Godman, Peter. *From Poliziano to Machiavelli: Florentine humanism in the High Renaissance*. Princeton, NJ: Princeton University Press, 1998.

———. *The silent masters: Latin literature and its censors in the High Middle Ages*. Princeton, NJ: Princeton University Press, 2000.

Goldthwaite, Richard A. *Wealth and the demand for art in Italy, 1300–1600*. Baltimore: Johns Hopkins University Press, 1993.

Goodman, Dena. *The Republic of Letters: A cultural history of the French Enlightenment*. Ithaca: Cornell University Press, 1994.

Gordon, Daniel. *Citizens without sovereignty: Equality and sociability in French thought, 1670–1789*. Princeton, NJ: Princeton University Press, 1994.

Grafton, Anthony. *Joseph Scaliger: A study in the history of classical scholarship*. 2 vols. Oxford: Clarendon Press, 1983–93.

———. "Close encounters of the learned kind: Joseph Scaliger's table talk." *American Scholar* 57 (1988): 581–88.

———. *Defenders of the text: The traditions of scholarship in an age of science, 1450–1800*. Cambridge, MA: Harvard University Press, 1991.

———. *Commerce with the classics: Ancient books and Renaissance readers*. Ann Arbor: University of Michigan Press, 1997.

Grafton, Anthony, and Lisa Jardine. *From humanism to the humanities: Education and the liberal arts in fifteenth- and sixteenth-century Europe*. Cambridge, MA: Harvard University Press, 1986.

Grafton, Anthony, April Shelford, and Nancy Siraisi. *New worlds, ancient texts: The power of tradition and the shock of discovery*. Cambridge, MA: Harvard University Press, 1992.

Grafton, Anthony, and Nancy Siraisi, eds. *Natural particulars: Nature and the disciplines in Renaissance Europe*. Cambridge, MA: MIT Press, 1999.

Gray, Hanna H. "Renaissance humanism: The pursuit of eloquence." *Journal of the History of Ideas* 24 (1963): 497–514.

Greene, Edward Lee. *Landmarks of botanical history*. Edited by Frank N. Egerton. 2 vols. Stanford: Stanford University Press, 1983.

Greene, Thomas M. *The light in Troy: Imitation and discovery in Renaissance poetry*. New Haven: Yale University Press, 1982.

Grendler, Paul F. *Schooling in Renaissance Italy: Literacy and learning, 1300–1600.* Baltimore: Johns Hopkins University Press, 1989.

Grendler, Paul F., ed. "Education in the Renaissance and Reformation." *Renaissance Quarterly* 43 (1990): 774–824.

Grieco, Allen J. "The social politics of pre-Linnean botanical classification." *I Tatti Studies* 4 (1991): 131–49.

Habermas, Jürgen. *Strukturwandel der Öffentlichkeit: Untersuchungen zu einer Kategorie der bürgerlichen Gesellschaft.* Frankfurt am Main: Suhrkamp, 1990.

Hagner, Michael. "Zur Physiognomik bei Alexander von Humboldt." In *Geschichten der Physiognomik: Text, Bild, Wissen,* edited by Rüdiger Campe and Manfred Schneider, 431–52. Freiburg: Rombach, 1996.

Hale, John. *The civilization of Europe in the Renaissance.* New York: Simon & Schuster, 1995.

Hall, Thomas S. *History of general physiology, 600 B.C. to A.D. 1900.* 2 vols. Chicago: University of Chicago Press, 1969.

Haller, Albrecht von. *Bibliotheca botanica, qua scripta ad rem herbariam facientia a rerum initiis recensentur.* 2 vols. Zürich: Orell, Gessner, Fuessli, et Socc., 1771–72.

Hansmann, Wilfried. *Gartenkunst der Renaissance und des Barock.* Cologne: DuMont Buchverlag, 1983.

Harms, Wolfgang. "On natural history and emblematics in the sixteenth century." In *The natural sciences and the arts: Aspects of interaction from the Renaissance to the twentieth century: An international symposium.* Uppsala: Almqvist & Wiksell, 1985.

Harrington, Karl Pomeroy. *Medieval Latin.* Boston: Allyn and Bacon, 1925.

Haskell, Francis. *History and its images: Art and the interpretation of the past.* New Haven: Yale University Press, 1993.

Haskins, Charles Homer. *Studies in the history of mediaeval science.* 2nd ed. Cambridge, MA: Harvard University Press, 1927.

Hayer, Gerold. "Zu Kontextüberlieferung und Gebrauchsfunktion von Konrads von Megenberg "Buch der Natur"." In *Latein und Volkssprache im deutschen Mittelalter, 1100–1500: Regensburger Colloquium 1988,* edited by Nikolaus Henkel and Nigel F. Palmer, 62–73. Tübingen: Max Niemeyer Verlag, 1992.

Hennebo, Dieter. *Gärten des Mittelalters.* Hamburg: Broschek Verlag, 1962.

Hennebo, Dieter, and Alfred Hoffmann. *Der architektonische Garten: Renaissance und Barock.* Hamburg: Broschek Verlag, 1963.

Hind, Arthur M. *An introduction to a history of woodcut.* 2 vols. New York: Dover Books, 1963.

Hobsbawm, Eric, and Terence Ranger, eds. *The invention of tradition.* Cambridge and New York: Cambridge University Press, 1983.

Hodgson, Marshall G. S. *The venture of Islam: Conscience and history in a world civilization.* 3 vols. Chicago: University of Chicago Press, 1974.

Hoeniger, F. David. "How plants and animals were studied in the mid-sixteenth century." In *Science and the arts in the Renaissance*, edited by John W. Shirley and F. David Hoeniger. Washington, D.C.: Folger Shakespeare Library; London and Toronto: Associated University Presses, 1985.

Hoppe, Brigitte. *Das Kräuterbuch des Hieronymus Bock, wissenschaftshistorische Untersuchung: Mit einem Verzeichnis sämtlicher Pflanzen des Werkes, der literarischen Quellen der Heilanzeigen und der Anwendungen der Pflanzen.* Stuttgart: Hiersemann, 1969.

Hopper, Florence. "Clusius' world: The meeting of science and art." In *The authentic garden: A symposium on gardens*, edited by Leslie Tjon Sie Fat and Erik de Jong, 13–36. Leiden: Clusius Foundation, 1991.

Hull, David L. *Science as a process: An evolutionary account of the social and conceptual development of science.* Chicago: University of Chicago Press, 1988.

Hünemörder, Christian. "Die Zoologie des Albertus Magnus." In *Albertus Magnus, Doctor Universalis, 1280/1980*, edited by Gerbert Meyer and Albert Zimmermann, 235–48. Mainz: Matthias-Grünewald-Verlag, 1980.

Hunger, Friedrich Wilhelm Tobias. *Charles de l'Escluse (Carolus Clusius) Nederlandsch kruidkundige, 1526–1609.* 2 vols. The Hague: M. Nijhoff, 1927–43.

Huppert, George. *Public schools in Renaissance France.* Urbana and Chicago: University of Illinois Press, 1984.

Hyde, Elizabeth. "Cultivated power: Flowers, culture, and politics in early modern France." Ph.D. dissertation, Harvard University, 1998.

Impey, Oliver, and Arthur MacGregor, eds. *The origins of museums: The cabinet of curiosities in sixteenth- and seventeenth-century Europe.* Oxford: Clarendon Press, 1985.

Israel, Jonathan. *The Dutch Republic: Its rise, greatness, and fall.* Oxford: Clarendon Press, 1995.

Istvánffi, Gyula. *A Clusius-codex mykologiaia méltatása, adatokkal Clusius életrajzához: Études et commentaires sur le Code de l'Escluse, augmentés de quelques notices biographiques.* Budapest: chez l'auteur, 1900.

Ivins, William M., Jr. *Prints and visual communication.* Cambridge, MA: M.I.T. Press, 1953.

———. *How prints look: Photographs with commentary.* Edited by Marjorie B. Cohn. Revised ed. Boston: Beacon Press, 1987.

Jahn, Ilse. *Grundzüge der Biologiegeschichte.* Jena: Gustav Fischer Verlag, 1990.

———. "La descrizione nella storia naturale: Dall'empiria a regole e leggi." *Intersezioni* 11.1 (1991): 55–73.

Jansen, Dirk Jacob. "The instruments of patronage: Jacopo Strada at the court of Maximilian II: A case-study." In *Kaiser Maximilian II.: Kultur und Politik im 16. Jahrhundert*, edited by Friedrich Edelmayer and Alfred Kohler, 182–202. Vienna: Verlag für Geschichte und Politik; Munich: R. Oldenbourg Verlag, 1992.

Jardine, Lisa. *Francis Bacon: Discovery and the art of discourse.* Cambridge: Cambridge University Press, 1974.

———. "Humanistic logic." In *The Cambridge history of Renaissance philosophy,* edited by Charles B. Schmitt, Quentin Skinner, Eckhard Kessler, and Jill Kraye, 173–98. Cambridge: Cambridge University Press, 1988.

———. *Erasmus, man of letters: The construction of charisma in print.* Princeton, NJ: Princeton University Press, 1993.

Jardine, N., J. A. Secord, and E. C. Spary, eds. *Cultures of natural history.* Cambridge: Cambridge University Press, 1996.

Johannesson, Kurt. *The Renaissance of the Goths in sixteenth-century Sweden: Johannes and Olaus Magnus as politicians and historians.* Translated by James Larson. Berkeley and Los Angeles: University of California Press, 1991.

Jong, Erik de. *Natuur en kunst: Nederlandse tuin- en landschapsarchitectuur, 1650–1740.* Amsterdam: Uitgeverij Thoth, 1993.

Kallendorf, Craig. *Virgil and the myth of Venice: Books and readers in the Italian Renaissance.* New York: Oxford University Press, 1999.

Kaufmann, Thomas DaCosta. *The mastery of nature: Aspects of art, science, and humanism in the Renaissance.* Princeton, NJ: Princeton University Press, 1993.

Kaufmann, Thomas DaCosta, and Virginia Roehrig Kaufmann. "The santification of nature: Observations on the origins of trompe l'oeil in Netherlandish book painting of the fifteenth and sixteenth centuries." In *The mastery of nature: Aspects of art, science, and humanism in the Renaissance,* 11–48. Princeton, NJ: Princeton University Press, 1993.

Kelley, Donald R. *Foundations of modern historical scholarship: Language, law, and history in the French Renaissance.* New York: Columbia University Press, 1970.

———. "Humanism and history." In *Renaissance humanism: Foundations, forms, and legacy,* edited by Albert Rabil, Jr., vol. 3, 236–70. Philadelphia: University of Pennsylvania Press, 1988.

Kenney, E. J. *The classical text: Aspects of editing in the age of the printed book.* Berkeley and Los Angeles: University of California Press, 1974.

Kenseth, Joy, ed. *The age of the marvelous.* Hanover, N.H.: Hood Museum of Art, Dartmouth College, 1991.

Kessler, Hermann Friedrich. *Das älteste und erste Herbarium Deutschlands, von Dr. Caspar Ratzenberger angelegt, gegenwärtig noch im Königlichen Museum zu Cassel befindlich.* Kassel: Verlag von August Freyschmidt, 1870.

Killermann, Sebastian. "A. Dürer's Bilder vom Walroß, Wisent und Elentier." *Naturwissenschaftliche Wochenschrift* 27 (1912): 785–89.

King, Margaret L. *Venetian humanism in an age of patrician dominance.* Princeton, NJ: Princeton University Press, 1986.

Koreny, Fritz. *Albrecht Dürer and the animal and plant studies of the Renaissance.* Translated by Pamela Marwood and Yehuda Shapiro. Boston: Little, Brown and Company, 1988.

Koselleck, Reinhart. *Kritik und Krise: Eine Studie zur Pathogenese der bürgerlichen Welt*. Frankfurt am Main: Suhrkamp, 1973.

Kraye, Jill, ed. *The Cambridge companion to Renaissance humanism*. Cambridge companions to literature. Cambridge: Cambridge University Press, 1996.

Krelage, Ernst Heinrich. *Drie eeuwen bloembollenexport: De geschiedenis van den bloembollenhandel en der hollandsche bloembollen tot 1938*. The Hague: Rijksuitgeverij, Dienst van de Nederlandsche staatscourant, 1946.

Kristeller, Paul Oskar. *Renaissance thought and its sources*. Edited by Michael Mooney. New York: Columbia University Press, 1979.

———. *Medieval aspects of Renaissance learning*. Edited by Edward P. Mahoney. New York: Columbia University Press, 1992.

Kudlien, Fridolf. "Zwei medizinisch-philologische Polemiken am Ende des 15. Jahrhunderts: Marzio gegen Merula und Leoniceno gegen einen Anonymus." *Gesnerus* 22 (1965): 85–92.

Kuhn, Thomas S. *The structure of scientific revolutions*. 2nd ed. Chicago: University of Chicago Press, 1970.

Kusukawa, Sachiko. "Leonhart Fuchs on the importance of pictures." *Journal of the History of Ideas* 58 (1997): 403–27.

La Ruffiniere du Prey, Pierre de. *The villas of Pliny from antiquity to posterity*. Chicago: University of Chicago Press, 1994.

Lach, Donald F. *Asia in the making of Europe, vol. 2, A century of wonder*. Chicago: University of Chicago Press, 1977.

Lakoff, George, and Mark Johnson. *Metaphors we live by*. Chicago: University of Chicago Press, 1980.

Landolt, Elisabeth. "Materialien zu Felix Platter als Sammler und Kunstfreund." *Basler Zeitschrift für Geschichte und Altertumskunde* 72 (1972): 245–306.

Le Roy Ladurie, Emmanuel. *Histoire du climat depuis l'an mil*. 2 vols. Paris: Flammarion, 1983.

Lefebvre, Joël. "Le latin et l'allemand dans la correspondance humaniste." In *Acta Conventus Neo-Latini Turonensis (6–10 Septembre 1976)*, edited by Jean-Claude Margolin, 1: 501–11. Paris: Vrin, 1980.

Legré, Ludovic. *La botanique en Provence au XVIe siècle: Les deux Bauhin, Jean Henri Cherler, et Valerand Dourez*. Marseille, 1904.

Lepenies, Wolf. *Das Ende der Naturgeschichte: Wandel kultureller Selbstverständlichkeiten in den Wissenschaften des 18. und 19. Jahrhunderts*. Munich: Hanser Verlag, 1976.

Leu, Urs B. *Conrad Gesner als Theologe: Ein Beitrag zur Zürcher Geistesgeschichte des 16. Jahrhunderts*. Berne: Peter Lang, 1990.

Levine, Joseph M. "Natural history and the history of the scientific revolution." *Clio* 13 (1983): 57–73.

Ley, Willy. *Dawn of zoology*. Englewood Cliffs, N.J.: Prentice-Hall, 1968.

Liebenau, Theodor von. "Felix Plater von Basel und Rennward Cysat von Luzern." *Basler Jahrbuch* (1900): 85–109.

Lloyd, G. E. R. *Science, folklore and ideology: Studies in the life sciences in ancient Greece.* Cambridge: Cambridge University Press, 1983.

Lohr, Charles H. *Latin Aristotle commentaries.* 3 vols. Florence: L. S. Olschki, 1988–95.

Loisel, Gustave. *Histoire des ménageries de l'antiquité à nos jours.* 3 vols. Paris: Octave Doin et Fils; Henri Laurens, 1912.

López Piñero, José Maria. "The Pomar Codex (ca. 1590): Plants and animals of the Old World and from the Hernandez expedition to America." *Nuncius* 7.1 (1992): 35–52.

———. "Las 'nuevas medicinas' americanas en la obra (1565–1574) de Nicolás Monardes." *Asclepio* 42.1 (1990): 3–67.

Louis, A. *Mathieu de l'Obel, 1538–1616: Episode de l'histoire de la botanique.* Ghent and Louvain: Story-Scientia, 1980.

Lovejoy, Arthur O. *The great chain of being: A study of the history of an idea.* Cambridge, MA: Harvard University Press, 1964 [1936].

Lowood, Henry. "The New World and the European catalog of nature." In *America in European consciousness, 1493–1750,* edited by Karen Ordahl Kupperman, 295–323. Chapel Hill, NC: University of North Carolina Press, for the Institute of Early American History and Culture, Williamsburg, VA, 1995.

Lütjeharms, Wilhelm Jan. *Zur Geschichte der Mykologie: Das 18. Jahrhundert.* Gouda: Koch & Knuttel, 1936.

Maccagni, Carlo. "Le raccolte e i musei di storia naturale e gli orti botanici come istituzioni alternative e complementari rispetto alla cultura delle Università e delle Accademie." In *Università, accademie e società scientifiche in Italia e in Germania dal cinquecento al settecento,* edited by Laetitia Boehn and Ezio Raimondi, 283–310. Bologna: Il Mulino, 1981.

MacDougall, Elisabeth B. "A paradise of plants: Exotica, rarities, and botanical fancies." In *The age of the marvelous,* edited by Joy Kenseth, 145–57. Hanover, NH: Hood Museum of Art, Dartmouth College, 1991.

Maclean, Ian. "Foucault's Renaissance episteme reassessed: An Aristotelian counterblast." *Journal of the History of Ideas* 59.1 (1998): 149–66.

Mallet, Robert. *Jardins et paradis.* Paris: Gallimard, 1959.

Manquat, Maurice. *Aristote naturaliste.* Paris: J. Vrin, 1932.

Mayr, Ernst. *The growth of biological thought: Diversity, evolution, and inheritance.* Cambridge, MA: Harvard University Press, 1982.

McClure, George W. *Sorrow and consolation in Italian humanism.* Princeton, NJ: Princeton University Press, 1991.

McConica, James. *Erasmus.* Oxford: Oxford University Press, 1991.

McCuaig, William. *Carlo Sigonio: The changing world of the late Renaissance.* Princeton, NJ: Princeton University Press, 1989.

McCulloch, Florence. *Medieval Latin and French bestiaries.* Chapel Hill: University of North Carolina Press, 1960.

McVaugh, Michael R., and Nancy G. Siraisi, eds. *Renaissance medical learning: Evolution of a tradition*. Osiris, second series, 6. Canton, Mass.: History of Science Society, 1990.

Medin, Douglas L., and Scott Atran, eds. *Folkbiology*. Cambridge, MA: MIT Press, 1999.

Métailié, Georges. "Histoire naturelle et humanisme en Chine et en Europe au XVIe siècle: Li Shizhen et Jacques Dalechamp." *Revue d'Histoire des Sciences* 42 (1989): 353–74.

Metzger, Hélène. "L'historien des sciences doit-il se faire contemporain des savants dont il parle?" *Archeion* 15..1 (1933): 34–46.

Meyer, Ernst H. F. *Geschichte der Botanik*. 4 vols. Königsberg: Verlag der Gebrüder Bornträger, 1854–57.

Meyer, Frederick G., Emily Emmart Trueblood, and John L. Heller. *The Great Herbal of Leonhart Fuchs: De historia stirpium commentarii insignes, 1542 (Notable commentaries on the history of plants)*. 2 vols. Stanford: Stanford University Press, 1999.

Moggi, Guido. "Andrea Cesalpino, botanico." *Atti e Memorie della Reale Accademia Petrarca di Lettere, Arti e Scienze* 42 (1976–78): 235–49.

———. "La conoscenza del mondo vegetale prima e dopo Andrea Cesalpino." In *Le monde végétal: Savoirs et usages sociaux*, edited by Allen J. Grieco, Odile Redon, and Lucia Tongiorgi Tomasi, 123–40. Saint-Denis: Presses Universitaires de Vincennes, 1993.

Momigliano, Arnaldo. "Ancient history and the antiquarian." *Journal of the Warburg and Courtauld Institutes* 13 (1950): 285–315.

———. "Ancient history and the antiquarian." In *[Primo] Contributo alla storia degli studi classici*, 67–106. Rome: Edizioni di storia e letteratura, 1955.

Monfasani, John. "The first call for press censorship: Niccolò Perotti, Giovanni Andrea Bussi, Antonio Moreto, and the editing of Pliny's *Natural history*." *Renaissance Quarterly* 41 (1988): 1–31.

———. "Humanism and rhetoric." In *Renaissance humanism: Foundations, forms, and legacy*, edited by Albert Rabil, Jr., vol. 3, 171–235. Philadelphia: University of Pennsylvania Press, 1988.

Moran, Bruce T. "Princes, machines and the evaluation of precision in the sixteenth century." *Sudhoffs Archiv* 61 (1977): 209–28.

Morford, Mark. "The Stoic garden." *Journal of Garden History* 7 (1987): 151–75.

Morison, Samuel Eliot. *The European discovery of America: The southern voyages, A.D. 1492–1616*. Oxford: Oxford University Press, 1974.

Morrison, Donald. "Philoponus and Simplicius on tekmeriodic proof." In *Method and order in Renaissance philosophy of nature: The Aristotle commentary tradition*, edited by Daniel A. DiLiscia, Eckhard Kessler, and Charlotte Methuen, 1–22. Aldershot: Ashgate, 1997.

Mugnai Carrara, Daniela. "Profilo di Nicolò Leoniceno." *Interpres* 2 (1978): 169–212.

———. "Una polemica umanistico-scolastica circa l'interpretazione delle tre dottrine ordinate di Galeno." *Annali dell'Istituto e Museo di Storia della Scienza di Firenze* 8.1 (1983): 31–57.

———. "La polemica 'de cane rabido' di Nicolò Leoniceno, Nicolò Zocca e Scipione Carteromaco: Un episodio di filologia medico-umanistica." *Interpres* 9 (1989): 196–236.

———. *La biblioteca di Nicolò Leoniceno: Tra Aristotele e Galeno: Cultura e libri di un medico umanista.* Florence: Olschki, 1991.

Mühlberger, Kurt. "Bildung und Wissenschaft: Kaiser Maximilian II. und die Universität Wien." In *Kaiser Maximilian II.: Kultur und Politik im 16. Jahrhundert*, edited by Friedrich Edelmayer and Alfred Kohler, 203–30. Vienna: Verlag für Geschichte und Politik; Munich: R. Oldenbourg Verlag, 1992.

Nauert, Charles G., Jr. "Humanists, scientists, and Pliny: Changing approaches to a classical author." *American Historical Review* 84 (1979): 72–85.

———. "C. Plinius Secundus (Naturalis historia)." In *Catalogus translationum et commentariorum*, edited by F. E. Cranz, vol. 4, 297–422. Washington: Catholic University of America Press, 1980.

———. *Humanism and the culture of Renaissance Europe.* Cambridge: Cambridge University Press, 1995.

Nieuw Nederlandsch biografisch woordenboek. 10 vols. Leiden: A. W. Sijthoff's Uitgeversmaatschappij, 1911–37.

Nissen, Claus. *Die botanische Buchillustration: Ihre Geschichte und Bibliographie.* 2 vols. Stuttgart: Hiersemann, 1951.

———. *Die illustrierten Vogelbücher: Ihre Geschichte und Bibliographie.* Stuttgart: Hiersemann Verlag, 1953.

Nolhac, Pierre de. *Pétrarque et l'humanisme.* Nouvelle ed. 2 vols. Paris: Librairie Honoré Champion, 1965.

Novick, Peter. *That noble dream: The "objectivity question" and the American historical profession.* Cambridge: Cambridge University Press, 1988.

Nutton, Vivian. "Humanistic surgery." In *The medical Renaissance of the sixteenth century*, edited by Andrew Wear, Roger K. French and I. M. Lonie, 75–99. Cambridge: Cambridge University Press, 1985.

———. "John Caius and the Eton Galen: Medical philology in the Renaissance." *Medizinhistorisches Journal* 20 (1985): 227–52.

———. "'Prisci dissectionum professores': Greek texts and Renaissance anatomists." In *The uses of Greek and Latin: Historical essays*, edited by A. C. Dionisotti, Anthony Grafton, and Jill Kraye. London: Warburg Institute, 1988.

Ogilvie, Brian W. "Encyclopædism in Renaissance botany: From *historia* to *pinax.*" In *Pre-modern encyclopædic texts: Proceedings of the Second COMERS Congress, Groningen, 1–4 July 1996*, edited by Peter Binkley, 89–99. Leiden: Brill, 1997.

———. "Image and text in natural history, 1500–1700." In *The power of images in early modern science*, edited by Wolfgang Lefèvre, Jürgen Renn and Urs Schöpflin, 141–66. Basel: Birkhäuser Verlag, 2003.

———. "The many books of nature: Renaissance naturalists and information overload." *Journal of the History of Ideas* 64.1 (2003): 29–40.

———. "Natural history, ethics, and physico-theology." In *Historia: Empiricism and erudition in early modern Europe*, edited by Gianna Pomata and Nancy Siraisi. Cambridge, MA: MIT Press, 2005.

Olmi, Giuseppe. *Ulisse Aldrovandi: Scienza e natura nel secondo cinquecento.* Trent: Libera Università degli studi di Trento, Gruppi di teoria e storia sociale, 1978.

———. "'Molti amici in varij luoghi': Studio della natura e rapporti epistolari nel secolo XVI." *Nuncius* 6. 1 (1991): 3–31.

———. *L'inventario del mondo: Catalogazione della natura e luoghi del sapere nella prima età moderna*. Bologna: Il Mulino, 1992.

Ong, Walter J. *Ramus, method, and the decay of dialogue: From the art of discourse to the art of reason.* Cambridge, MA: Harvard University Press, 1958.

Owen, G. E. L. "Tithenai ta phainomena." In *Aristote et les problèmes de méthode: Communications présentées au Symposium aristotelicum tenu à Louvain du 24 août au 1er septembre 1960*, 83–103. Louvain: Publications universitaires; Paris: Béatrice-Nauwelaerts, 1961.

Panofsky, Erwin. *Renaissance and renascences in Western art.* New York: Icon Editions, 1972.

Pardo Tomás, José. *Ciencia y censura: La Inquisición española y los libros científicos en los siglos XVI y XVII.* Madrid: Consejo Superior de Investigaciones Científicas, 1991.

Pardo Tomás, José, and María Luz López Terrada. *Las primeras noticias sobre plantas americanas en las relaciones de viajes y cronicas de Indias (1493–1553).* Valencia: Instituto de Estudios Documentales e Históricos sobre la Ciencia, Universidad de Valencia, 1993.

Park, Katharine. "The organic soul." In *The Cambridge history of Renaissance philosophy*, edited by Charles B. Schmitt, Quentin Skinner, Eckhard Kessler, and Jill Kraye, 464–84. Cambridge: Cambridge University Press, 1988.

———. "The criminal and the saintly body: Autopsy and dissection in Renaissance Italy." *Renaissance Quarterly* 47.1 (1994): 1–33.

Park, Katharine, and Eckhard Kessler. "The concept of psychology." In *The Cambridge history of Renaissance philosophy*, edited by Charles B. Schmitt, Quentin Skinner, Eckhard Kessler, and Jill Kraye, 455–63. Cambridge: Cambridge University Press, 1988.

Percival, W. Keith. "Renaissance grammar." In *Renaissance humanism: Foundations, forms, and legacy*, edited by Albert Rabil, Jr., vol. 3, 67–83. Philadelphia: University of Pennsylvania Press, 1988.

Perfetti, Stefano. *Aristotle's zoology and its Renaissance commentators*. Leuven: Leuven University Press, 2000.

Pfeiffer, Rudolf. *History of classical scholarship from 1300 to 1850*. Oxford: Clarendon Press, 1976.

Pils, Susanne Claudine. "Die Stadt als Lebensraum: Wien im Spiegel der Oberkammeramtsrechnungen, 1556–1576." *Jahrbuch des Vereins für Geschichte der Stadt Wien* 49 (1993): 119–72.

Pinborg, Jan. "Speculative grammar." In *The Cambridge history of later medieval philosophy*, edited by Norman Kretzmann, Anthony Kenny, Jan Pinborg and Eleonore Stump. Cambridge: Cambridge University Press, 1982.

Pinon, Laurent. *Livres de zoologie à la Renaissance: Une anthologie, 1450–1700*. Paris: Klincksieck, 1995.

Polanyi, Michael. *Personal knowledge: Towards a post-critical philosophy*. Corrected ed. Chicago: University of Chicago Press, 1962.

Pomian, Krzysztof. *Collectionneurs, amateurs et curieux: Paris, Venise, XVIe–XVIIIe siècle*. Paris: Gallimard, 1987.

Poovey, Mary. *A history of the modern fact: Problems of knowledge in the sciences of wealth and society*. Chicago: University of Chicago Press, 1998.

Porter, Roy, and Mikulás Teich, eds. *The Scientific Revolution in national context*. Cambridge: Cambridge University Press, 1992.

Prest, John. *The garden of Eden: The botanic garden and the re-creation of paradise*. New Haven: Yale University Press, 1981.

Quillen, Carol Everhart. *Rereading the Renaissance: Petrarch, Augustine, and the language of humanism*. Ann Arbor: University of Michigan Press, 1998.

Rabil, Albert, Jr. "The significance of 'civic humanism' in the interpretation of the Italian Renaissance." In *Renaissance humanism: Foundations, forms, and legacy*, edited by Albert Rabil, Jr., vol. 1, 141–74. Philadelphia: University of Pennsylvania Press, 1988.

Rabil, Albert, Jr., ed. *Renaissance humanism: Foundations, forms, and legacy*. 3 vols. Philadelphia: University of Pennsylvania Press, 1988.

Raven, Charles E. *English naturalists from Neckam to Ray: A study of the making of the modern world*. Cambridge: Cambridge University Press, 1947.

———. *John Ray, naturalist: His life and works*. Cambridge: Cambridge University Press, 1950.

Reeds, Karen M. "Renaissance humanism and botany." *Annals of Science* 33 (1976): 519–42.

———. *Botany in medieval and Renaissance universities*. New York: Garland Publishing, Inc., 1991.

Reynolds, L. D., ed. *Texts and transmission: A survey of the Latin classics*. Oxford: Clarendon Press, 1983.

Reynolds, L. D., and N. G. Wilson. *Scribes and scholars: A guide to the transmission of Greek and Latin literature*. 3rd ed. Oxford: Oxford University Press, 1991.

Riddle, John M. *Dioscorides on pharmacy and medicine.* Austin: University of Texas Press, 1985.

Riedl-Dorn, Christa. *Wissenschaft und Fabelwesen: Ein kritischer Versuch über Conrad Gesner und Ulisse Aldrovandi.* Wien: Böhlau, 1989.

Rouse, R. H. "The transmission of the texts." In *The legacy of Rome: A new appraisal,* edited by Richard Jenkyns. Oxford: Oxford University Press, 1992.

Rowland, Beryl. "The art of memory and the bestiary." In *Beasts and birds of the Middle Ages: The bestiary and its legacy,* edited by Willene B. Clark and Meradith T. McMunn, 12–25. Philadelphia: University of Pennsylvania Press, 1989.

Roze, E. *Charles de l'Ecluse d'Arras: Le propagateur de la pomme de terre au XVIe siècle: Sa biographie et sa correspondance.* Paris: J. Rothschild, 1899.

Rudwick, Martin J. S. *The meaning of fossils: Episodes in the history of palaeontology.* 2nd ed. Chicago: University of Chicago Press, 1985.

Rummel, Erika. *The humanist-scholastic debate in the Renaissance and Reformation.* Cambridge: Harvard University Press, 1995.

Russell, J. C. "Die Bevölkerung Europas 500–1500." In *Europäische Wirtschaftsgeschichte,* edited by Carlo M. Cipolla and K. Borchardt, vol. 1. Stuttgart and New York: Gustav Fischer Verlag, 1978.

Rytz, Walther. "Das Herbarium Felix Platters: Ein Beitrag zur Geschichte der Botanik des XVI. Jahrhunderts." *Verhandlungen der Naturforschenden Gesellschaft in Basel* 44.1 (1932–33): 1–222.

———. *Pflanzenaquarelle des Hans Weiditz aus dem Jahre 1529.* Bern: P. Haupt, 1936.

Saint-Lager, Jean-Baptiste. "Histoire des herbiers." *Annales de la Société Botanique de Lyon. Notes et memoires* 13 (1885): 1–120.

Schierbeek, A. *Jan Swammerdam (12 February 1637–17 February 1680): His life and works.* Amsterdam: Swets & Zeitlinger, 1967.

Schmitt, Charles B. *Aristotle and the Renaissance.* Cambridge, MA: Harvard University Press, 1983.

———. "The rise of the philosophical textbook." In *The Cambridge history of Renaissance philosophy,* edited by Charles B. Schmitt, Quentin Skinner, Eckhard Kessler and Jill Kraye, 792–804. Cambridge: Cambridge University Press, 1988.

Schmitt, Charles B., Quentin Skinner, Eckhard Kessler, and Jill Kraye, eds. *The Cambridge history of Renaissance philosophy.* Cambridge: Cambridge University Press, 1988.

Schmitz, R. "Apotheke." In *Lexikon des Mittelalters,* vol. 1. Munich and Zürich: Artemis Verlag, 1980.

Schnapper, Antoine. *Le géant, la licorne et la tulipe: Collections et collectionneurs dans la France du XVIIe siècle.* Paris: Flammarion, 1988.

Schneppen, Heinz. *Niederländische Universitäten und deutsches Geistesleben: Von der Gründung der Universität Leiden bis ins späte 18. Jahrhundert.* Münster: Aschendorffsche Verlagsbuchhandlung, 1960.

Schultz, Bernard. *Art and anatomy in Renaissance Italy.* Ann Arbor: UMI Research Press, 1985.

Secord, Anne. "Artisan botany." In *Cultures of natural history*, edited by Nicholas Jardine, James A. Secord and Emma C. Spary, 378–93. Cambridge: Cambridge University Press, 1996.

Senn, Gustav. *Die Entwicklung der biologischen Forschungsmethode in der Antik und ihre grundsätzliche Förderung durch Theophrast von Eresos.* Aarau: H. R. Sauerländer & Co., 1933.

Sewell, William H., Jr. "A theory of structure: Duality, agency, and transformation." *American Journal of Sociology* 98.1 (1992): 1–29.

Seznec, Jean. *The survival of the pagan gods: The mythological tradition and its place in Renaissance humanism and art.* Princeton, NJ: Princeton University Press, 1972.

Shapin, Steven. *A social history of truth: Civility and science in seventeenth-century England.* Chicago: University of Chicago Press, 1994.

———. *The scientific revolution.* Chicago: University of Chicago Press, 1996.

Shapin, Steven, and Simon Schaffer. *Leviathan and the air-pump: Hobbes, Boyle, and the experimental life.* Princeton, NJ: Princeton University Press, 1985.

Shapiro, Barbara. "History and natural history in sixteenth- and seventeenth-century England: An essay on the relationship between humanism and science." In *English scientific virtuosi in the sixteenth and seventeenth centuries: Papers read at a Clark Library Seminar, 5 February 1977*, 1–55. Los Angeles: William Andrews Clark Memorial Library, University of California, 1979.

———. *A culture of fact: England, 1550–1720.* Ithaca: Cornell University Press, 2000.

Siraisi, Nancy G. "Some current trends in the study of Renaissance medicine." *Renaissance Quarterly* 37 (1984): 585–600.

———. *Avicenna in Renaissance Italy: The Canon and medical teaching in Italian universities after 1500.* Princeton, NJ: Princeton University Press, 1987.

———. *Medieval and early Renaissance medicine: An introduction to knowledge and practice.* Chicago: University of Chicago Press, 1990.

Skinner, Quentin. *The foundations of modern political thought.* 2 vols. Cambridge: Cambridge University Press, 1978.

Slack, Adrian. *Carnivorous plants.* Cambridge, MA: MIT Press, 1979.

Slaughter, Mary M. *Universal languages and scientific taxonomy in the seventeenth century.* Cambridge: Cambridge University Press, 1982.

Smit, Pieter. "Carolus Clusius and the beginning of botany in Leiden University." *Janus* 60 (1973): 87–92.

Smith, Pamela H., and Paula Findlen, eds. *Merchants and marvels: Commerce, science, and art in early modern Europe.* New York: Routledge, 2001.

Spary, E. C. *Utopia's garden: French natural history from Old Regime to Revolution*. Chicago: University of Chicago Press, 2000.

Stagl, Justin. "Ars apodemica: Bildungsreise und Reisemethodik von 1560 bis 1600." In *Reisen und Reiseliteratur im Mittelalter und in der Frühen Neuzeit: Vorträge eines interdiziplinären Symposiums vom 3.-8. Juni 1991 an der Justus-Liebig-Universität Gießen*, edited by Xenja von Ertzdorff and Dieter Neukirch, 141–89. Amsterdam and Atlanta: Rodopi, 1992.

———. *A history of curiosity: The theory of travel, 1550–1800*. Chur: Harwood Academic Publishers, 1995.

Stannard, Jerry. "Medieval herbals and their development." *Clio medica* 9 (1974): 23–33.

———. "Natural history." In *Science in the Middle Ages*, edited by David C. Lindberg, 429–60. Chicago: University of Chicago Press, 1978.

———. "Medieval gardens and their plants." In *Gardens of the Middle Ages*, edited by Marilyn Stokstad and Jerry Stannard, 37–69. Lawrence, KS: Spencer Museum of Art, University of Kansas, 1983.

Stearn, William T. "Botanical exploration to the time of Linnaeus." *Proceedings of the Linnaean Society of London* 169 (1956/57): 173–96.

———. *Botanical Latin: History, grammar, syntax, terminology and vocabulary*. 3rd ed. Newton Abbot, London, and North Pomfret, VT: David & Charles, 1983.

Stewart, David, and Algis Mickunas. *Exploring phenomenology: A guide to the field and its literature*. 2nd ed. Athens, OH: Ohio University Press, 1990.

Stokstad, Marilyn, and Jerry Stannard. *Gardens of the Middle Ages*. Lawrence, KS: Spencer Museum of Art, University of Kansas, 1983.

Stresemann, Erwin. "Die Entdeckungsgeschichte der Paradiesvögel." *Journal für Ornithologie* 95, no. Heft 3–4 (1954): 263–91.

Struever, Nancy S. *The language of history in the Renaissance: Rhetoric and historical consciousness in Florentine humanism*. Princeton, NJ: Princeton University Press, 1970.

Suutala, Maria. *Tier und Mensch im Denken der deutschen Renaissance*. Helsinki: Societas Historica Finlandiae, 1990.

Swan, Claudia. "*Ad vivum, naer het leven*, from the life: Defining a mode of representation." *Word & Image* 11.4 (1995): 353–72.

———. "Les fleurs comme *curiosa*." In *L'empire de flore: Histoire et représentation des fleurs en Europe du XVIe au XIXe siècle*, edited by Sabine van Sprang, 86–99. Brussels: La Renaissance du livre, 1996.

———. "Jacques de Gheyn II and the representation of the natural world in the Netherlands, ca. 1600." Ph.D. dissertation, Columbia University, 1997.

———. *The Clutius botanical watercolors: Plants and flowers of the Renaissance*. New York: Harry N. Abrams, Inc., 1998.

———. "Lectura-imago-ostensio: The role of the *Libri picturati* A.18–A.30 in medical instruction at the Leiden University." In *Natura-cultura: L'interpretazione del mondo fisico nei testi e nelle immagine; Atti del Convegno Internazionale di Studi Mantova, 5–8 ottobre 1996*, edited by Giuseppe Olmi, Lucia Tongiorgi Tomasi, and Attilio Zanca, 189–214. Florence: Leo S. Olschki, 2000.

Swerdlow, Noel. "Science and humanism in the Renaissance: Regiomontanus's oration on the dignity and utility of the mathematical sciences." In *World changes: Thomas Kuhn and the nature of science*, edited by Paul Horwich, 131–68. Cambridge, MA: MIT Press, 1993.

Thacker, Christopher. *The history of gardens*. Berkeley and Los Angeles: University of California Press, 1979.

Theunisz, Johannes. *Carolus Clusius: Het merkwaardige leven van een pionier der wetenschap*. Amsterdam: P. N. van Kampen & Zoon, 1939.

Thomas, Christiane. "Wien als Residenz unter Kaiser Ferdinand I." *Jahrbuch des Vereins für Geschichte der Stadt Wien* 49 (1993): 101–17.

Thorndike, Lynn. *A history of magic and experimental science*. 8 vols. New York: Columbia University Press, 1923–58.

Tillyard, E. M. W. *The Elizabethan world picture*. 1st American ed. New York: Macmillan, 1943.

Treviranus, Ludolf Christian. *Die Anwendung des Holzschnittes zur bildlichen Darstellung von Pflanzen, nach Entstehung, Blüthe, Verfall und Restauration*. Leipzig, 1855.

Tribby, Jay. "Body/building: Living the museum life in early modern Europe." *Rhetorica* 10 (1992): 139–63.

Trinkaus, Charles. *In our image and likeness: Humanity and divinity in Italian humanist thought*. 2 vols. Chicago: University of Chicago Press, 1970.

Ullman, B. L. *The origin and development of humanistic script*. Rome: Edizioni di Storia e letteratura, 1960.

Van den Abeele, Baudouin. "Les albums ornithologiques de Jacques Dalechamps, médecin et naturaliste à Lyon (1513–1588)." *Archives Internationales d'Histoire des Sciences* 52 (2002): 3–45.

Vandewiele, Leo J. "The garden of Pieter van Coudenberghe." In *Botany in the Low Countries (end of the 15th century-ca. 1650): Plantin-Moretus Museum exhibition*, 23–31. Antwerp: The Plantin-Moretus Museum and the Stedelijk Prentenkabinet, 1993.

Veendorp, Hesso, and L. G. M. Baas Becking. *1587–1937 Hortus Academicus Lugduno Batavus: The development of the gardens of Leyden University*. Haarlem: Typographia Enschedaiana, 1938.

Voet, Leon. *The Golden Compasses: A history and evaluation of the printing and publishing activities of the Officina Plantiniana at Antwerp*. 2 vols. Amsterdam: Vangendt & Co., 1969–72.

Vogel, Klaus A. "'America': Begriff, geographische Konzeption und frühe Entdeckungsgeschichte in der Perspektive der deutschen Humanisten." In *Von der Weltkarte zum Kuriositätenkabinett: Amerika im deutschen Humanismus und Barock*, edited by Karl Kohut, 11–43. Frankfurt am Main: Vervuert Verlag, 1995.

Vries, Jan de. "Landbouw in de Noordelijke Nederlanden, 1490–1650." In *Algemene geschiedenis der Nederlanden*, vol. 7. Haarlem: Fibula-Van Dishoeck, 1980.

Waquet, Françoise. *Le modèle français et l'Italie savante: Conscience de soi et perception de l'autre dans la république des lettres, 1660–1750*. Rome: Ecole Française de Rome, 1989.

———. "Qu'est-ce que la République des Lettres? Essai de sémantique historique." *Bibliothèque de l'Ecole des Chartes* 1 (1989): 473–502.

———. *Le latin, ou l'empire d'un signe, XVIe-XXe siècle*. Paris: Albin Michel, 1998.

Webster, Charles. "Bauhin, Jean." In *Dictionary of scientific biography*, vol. 1, 525–27. New York: Charles Scribner's Sons, 1970–1980.

Weihe, Konrad von, ed. *Illustrierte Flora: Deutschland und angrenzende Gebiete: Gefäßkryptogamen und Blütenpflanzen*. Berlin and Hamburg: Verlag Paul Parey, 1972.

Weiß, Karl. *Geschichte der Stadt Wien*. 2nd ed. Vienna: Verlag von Rudolf Lechner, 1883.

Wellisch, Hans H. *Conrad Gessner: A bio-bibliography*. 2nd ed. Zug: IDC, 1984.

Whitehead, P. J. P., G. van Vliet, and William T. Stearn. "The Clusius and other natural history pictures in the Jagiellon Library, Kraków." *Archives of Natural History* 16 (1989): 15–32.

Wijnands, D. Onno. "Commercium botanicum: The diffusion of plants in the sixteenth century." In *The authentic garden: A symposium on gardens*, edited by Leslie Tjon Sie Fat and Erik de Jong, 75–84. Leiden: Clusius Foundation, 1991.

Wille, H. "The albums of Karel van Sint Omaars, 1533–1569 (*Libri picturati A.16-A.31*)." *Archives of Natural History* 24 (1997): 423–37.

Wimmer, Clemens Alexander. *Geschichte der Gartentheorie*. Darmstadt: Wissenschaftliche Buchgesellschaft, 1989.

Witt, Ronald G. "Medieval Italian culture and the origins of humanism as a stylistic ideal." In *Renaissance humanism: Foundations, forms, and legacy*, edited by Albert Rabil, Jr., vol. 1, 29–70. Philadelphia: University of Pennsylvania Press, 1988.

———. *"In the footsteps of the ancients": The origins of humanism from Lovato to Bruni*. Leiden: Brill, 2000.

Yates, Frances A. *Giordano Bruno and the Hermetic tradition*. Chicago: University of Chicago Press, 1964.

Zoller, Heinrich. "Konrad Geßner als Botaniker." *Gesnerus* 22 (1965): 216–27.

INDEX